Integral Methods in Science and Engineering

Analytic and Numerical Techniques

C. Constanda
M. Ahues
A. Largillier
Editors

Springer Science+Business Media, LLC

C. Constanda
University of Tulsa
Department of Mathematical
  and Computer Sciences
Tulsa, OK 74104
USA

M. Ahues
Université de St. Étienne
Équipe d'Analyse Numérique
42100 St. Étienne
France

A. Largillier
Université de St. Étienne
Équipe d'Analyse Numérique
42100 St. Étienne
France

Library of Congress Cataloging-in-Publication Data

Integral methods in science and engineering : analytic and numerical techniques / C.
   Constanda, Mario Ahues, Alain Largillier, editors.
      p. cm.
   Includes bibliographical references and index.
   ISBN 978-1-4612-6479-8   ISBN 978-0-8176-8184-5 (eBook)
   DOI 10.1007/978-0-8176-8184-5
      1. Integral equations–Numerical solutions–Congresses. 2. Mathematical
   analysis–Congresses. I. Constanda, C. (Christian) II. Ahues, Mario. III. Largillier,
   Alain. IV. International Conference on Integral Methods in Science and Engineering (7th
   2002 : Saint Étienne, France)

   QA431.I14 2004
   515–dc22   2003063692

AMS Subject Classifications: 45-06, 65-06, 74-06, 76-06

ISBN 978-1-4612-6479-8   Printed on acid-free paper.

© 2004 Springer Science+Business Media New York
Originally published by Birkhäuser Boston in 2004
Softcover reprint of the hardcover 1st edition 2004

9 8 7 6 5 4 3 2 1   SPIN 10925157

www.birkhäuser-science.com

Contents

26 On Stabilization of Solutions of Elliptic Equations Containing Bessel Operators
Andrey B. Muravnik   **157**

27 New Zonal, Spectral Solutions for Compressible Navier–Stokes Partial Differential Equations
Adriana Nastase   **163**

28 Crack Problems and Boundary Variational Inequalities
David Natroshvili   **169**

29 Spline Approximations for Weakly Singular Volterra Integro-Differential Equations
I. Parts and Arvid Pedas   **175**

30 Hybrid Laplace and Poisson Solvers. II: Robin BCs
Fred R. Payne   **181**

Preface

The international conferences with the generic title of Integral Methods in Science and Engineering (IMSE) aim to discuss up-to-date research where analytic or numerical integration constitutes a major method of solution.

IMSE1985 and IMSE1990 were held at the University of Texas at Arlington under the chairmanship of Fred Payne, to whom this volume is dedicated. They were followed by IMSE1993 (Tohoku University, Sendai, Japan), IMSE1996 (University of Oulu, Finland), IMSE1998 (Michigan Technological University, Houghton, MI, USA), IMSE2000 (Banff, AB, Canada), and IMSE2002 (University of Saint-Étienne, France). The IMSE conferences have now come to be regarded as important biennial occasions for scientists and engineers to communicate their latest results and to interact in the development and applications of a fundamental class of mathematical techniques.

The IMSE conferences have also acquired the enviable reputation of being some of the friendliest, socially most enjoyable professional meetings. IMSE2002, organized at the University "Jean Monnet" in Saint-Étienne, fully confirmed this reputation and set new standards of conviviality and camaraderie for future participants. Its success owes a lot to the excellent work done by the Local Organizing Committee:

Alain Largillier, *Chairman*;

Mario Ahues;

Olivier Titaud;

Arlette Bernard.

The organizers and the participants wish to acknowledge the financial support received from the conference sponsors:

Université "Jean Monnet" of Saint Étienne;

Région Rhône Alpes;

Saint Étienne Métropole;

Société de Mathématiques Appliquées et Industrielles (S.M.A.I.);

Air France.

As on previous occasions, advice and general guidance were provided by the International Steering Committee.

The next IMSE conference will be held in August 2004 at the University of Central Florida, Orlando, FL, USA. Details concerning this event will be posted on the conference web page, to be constructed shortly.

This volume contains four invited papers and thirty-eight contributed papers accepted after peer review. The papers are arranged in alphabetical order by (first) author's name.

The editors would like to record their thanks to the referees for their willingness to review the papers, and to the staff at Birkhäuser Boston, who have handled the publication process with patience and efficiency. A special mention is due to Olivier Titaud, who helped with the preparation of the camera-ready copy.

Tulsa, Oklahoma, USA   *Christian Constanda, IMSE Chairman*

The International Steering Committee of IMSE:

C. Constanda (University of Tulsa), *Chairman*

M. Ahues (University of Saint-Étienne)

B. Bertram (Michigan Technological University)

C. Corduneanu (University of Texas at Arlington)

A. Haji-Sheikh (University of Texas at Arlington)

A. Largillier (University of Saint-Étienne)

B. Limaye (Indian Institute of Technology, Bombay)

A. Mioduchowski (University of Alberta, Edmonton)

D. Mitrea (University of Missouri-Columbia)

Z. Nashed (University of Central Florida)

A. Nastase (Rhein.-Westf. Technische Hochschule, Aachen)

F.R. Payne (University of Texas at Arlington)

M.E. Pérez (University of Cantabria, Santander)

M.-C. Rivara (University of Chile, Santiago de Chile)

K. Ruotsalainen (University of Oulu)

P. Schiavone (University of Alberta, Edmonton)

S. Seikkala (University of Oulu)

Contributors

Asdin Aoufi: PECM, UMR CNRS 5146, 158, cours Fauriel, 42023 Saint-Étienne cedex 02, France
aoufi@emse.fr

Enrique Arias: Departamento de Informática, Escuela Politécnica Superior de Albacete, Universidad de Castilla-La Mancha, 02071-Albacete, Spain
earias@info-ab.uclm.es

Mario Ahues: Équipe d'Analyse Numérique, Université Jean Monnet de Saint-Étienne, 23 rue Dr. Paul Michelon, F-42023 Saint-Étienne, France
mario.ahues@univ-st-etienne.fr

Lahcene Bencheikh: Département de Génie des Procédés, Faculté des Sciences de l'Ingénieur, Université Ferhat Abbès, Sétif 19000, Algeria
bencheikh_lahcene@yahoo.com

Pamela N. Blair: Department of Mathematics, University of Strathclyde, 26 Richmond Street, Glasgow G1 1XH, UK
rs.pbla@maths.strath.ac.uk

Luis P. Castro: Department of Mathematics, University of Aveiro, 3810-193 Aveiro, Portugal
lcastro@mat.ua.pt

Ricardo Celorrio: Departamento de Matemática Aplicada, Universidad de Zaragoza, Centro Politécnico Superior, María de Luna, 3, E 50015 Zaragoza, Spain
celorrio@unizar.es

Roma Chakrabarti: School of Computing, Mathematical, and Information Sciences, University of Brighton, Lewes Road, Brighton BN2 4GJ, UK
r.chakrabarti@brighton.ac.uk

Jen Shi Chang: Nuclear Research Building #118, Engineering Physics Department, McMaster University, Hamilton, ON, Canada L8S 4M1
changj@mcmaster.ca

Loïc Chevallier: Observatoire de Lyon, 9 avenue Charles André, F-69561 Saint-Genis-Laval cedex, France
loic.chevallier@obs.univ-lyon1.fr

Igor Chudinovich: Department of Mechanical Engineering, University of Guanajuato, Salamanca, Mexico

chudynovich@salamanca.ugto.mx

Raimondas Čiegis: Department of Mathematical Modeling, Vilnius Gediminas Technical University, Sauletekio av. 11, LT-2040 Vilnius, Lithuania

rc@fm.vtu.lt

Christian Constanda: Department of Mathematical and Computer Sciences, University of Tulsa, 600 S. College Avenue, Tulsa, OK 74104-3189, USA

christian-constanda@utulsa.edu

Constantin Corduneanu: Department of Mathematics, University of Texas at Arlington, P.O. Box 19408, Arlington, TX 76019-0408, USA

cordun@uta.edu

Filomena D. d'Almeida: Faculdade de Engenharia da Universidade do Porto, Rua Dr. Roberto Frias 4200-464 Porto, Portugal

falmeida@fe.up.pt

Haroldo F. de Campos Velho: Laboratorio de Computaçao e Matematica Aplicada, Instituto Nacional de Pesquisas Espaciais, P.O. Box 515, 12245-970 Sao José dos Campos (SP), Brazil

haroldo@lac.inpe.br

Luc Giraud: CERFACS, 42, avenue G. Coriolis, 31057 Toulouse cedex 01, France

giraud@cerfacs.fr

Roger Godard: Department of Mathematics, Royal Military College of Canada, P.O. Box 17000, Stn Forces, Kingston, ON, Canada K7K 7B4

godard-r@rmc.ca

Glênio A. Gonçalves: Departamento de Engenharia Nuclear, Universidade Federal do Rio Grande do Sul, Av. Osvaldo Aranha 99–4o. andar, 90046-900 Porto Alegre (RS), Brazil

glenio_a@yahoo.com

Paul J. Harris: School of Computing, Mathematical, and Information Sciences, University of Brighton, Lewes Road, Brighton BN2 4GJ, UK

p.j.harris@bton.ac.uk

Seppo Heikkilä: Department of Mathematical Sciences, University of Oulu, P.O. Box 3000, FIN-90014, Oulu, Finland

sheikki@cc.oulu.fi

David Henwood: School of Computing, Mathematical, and Information Sciences, University of Brighton, Lewes Road, Brighton BN2 4GJ, UK

davidhenwood@talk21.com

Vicente Hernández: Departamento de Sistemas Informáticos y Computación, Universidad Politécnica de Valencia, 46022-Valencia, Spain

vhernand@dsic.upv.es

Markku Hihnala: Mathematics Division, Department of Electrical Engineering, Faculty of Technology, University of Oulu, PL 4500, FIN-90014 Oulu, Finland

markku.hihnala@ee.oulu.fi

Hiroshi Hirayama: Department of System Design Engineering, Kanagawa Institute of Technology, 1030 Shimo-Ogino, Atsugi-Shi, Kanagawa-Ken, 243-0292, Japan

hirayama@sd.kanagawa-it.ac.jp

Nancy Hitschfeld-Kahler: Department of Computer Science, Universidad de Chile, Casilla 2777, Santiago, Chile

nancy@dcc.uchile.cl

Jiri V. Horák: Department of Mathematical Analysis and Applied Mathematics, Faculty of Sciences, Palacky University, tr. Svobody 26, 771 46 Olomouc, Czech Republic

jiri.horak@upol.cz

Jacinto-Ibáñez: Departamento de Sistemas Informáticos y Computación. Universidad Politécnica de Valencia, 46022-Valencia, Spain

jjibanez@dsic.upv.es

Alexander O. Ignatyev: Institute for Applied Mathematics and Mechanics, R. Luxemburg Street 74, Donetsk-83114, Ukraine

mila@budinf.donetsk.ua

Alexei A. Ignatyev: Department of Mathematical Sciences, Kent State University, Kent, OH 44242, USA

aignatye@kent.edu

Rekha P. Kulkarni: Department of Mathematics, Indian Institute of Technology Bombay, Powai, Mumbai 400 076, India

rpk@math.iitb.ac.in

Wilson Lamb: Department of Mathematics, University of Strathclyde, 26 Richmond Street, Glasgow G1 1XH, UK

wl@maths.strath.ac.uk

Alain Largillier: Équipe d'Analyse Numérique, Université Jean Monnet de Saint-Étienne, 23 rue Dr. Paul Michelon, F-42023 Saint-Étienne, France

larg@anum.univ-st-etienne.fr

Balmohan V. Limaye: Department of Mathematics, Indian Institute of Technology Bombay, Powai, Mumbai 400 076, India

bvl@math.iitb.ac.in

Naglaa Madbouli: Department of Mathematics, American University in Cairo, P.O. Box 2511, 113 Sharia Kasr El Aini, Cairo, Egypt

nmadbouly@hotmail.com

Paul A. Martin: Department of Mathematics & Computer Sciences, Colorado School of Mines, Golden, CO 80401-1887, USA

pamartin@mines.edu

Desmond F. McGhee: Department of Mathematics, University of Strathclyde, 26 Richmond Street, Glasgow G1 1XH, UK

dfmcg@maths.strath.ac.uk

Sergey E. Mikhailov: Department of Mathematics, Glasgow Caledonian University, Cowcaddens Road, Glasgow G4 0BA, UK

s.mikhailov@gcal.ac.uk

Dorina Mitrea: Department of Mathematics, University of Missouri-Columbia, 202 Mathematical Sciences Building, Columbia, MO 65211, USA

dorina@math.missouri.edu

Radu Mitric: Faculty of Medical Bioengineering, University of Medicine and Pharmacy, Str. Universitatii nr. 16, Iasi, Romania

rmitric@yahoo.com

Andrey B. Muravnik: Moscow State Aviation Institute Department of Differential Equations, GSP-3, Volokolamskoe shosse 4, Moscow, A-80, 125993 Russia

muravnik@tmo-10.vrn.ru

Adriana Nastase: Aerodynamik des Fluges, Rhein.-Westf. Technische Hochschule, Templergraben 55, 52062 Aachen, Germany

nastase@lafaero.rwth-aachen.de

David Natroshvili: Department of Mathematics, georgian Technical University, 77, M. Kostava st., 0175 Tbilisi, Georgia

natrosh@hotmail.com

Inga Parts: Institute of Applied Mathematics, University of Tartu, Liivi 2, Tartu 50409, Estonia

inga.parts@ut.ee

Fred R. Payne: 1003 Shelley Court, Arlington, TX 76012, USA

frpdfi@airmail.net

Arvet Pedas: Institute of Applied Mathematics, University of Tartu, Liivi 2, Tartu 50409, Estonia

arvet.pedas@ut.ee

Eugenia Pérez: Departamento de Matematica Aplicada y C.C., E.T.S.I. Caminos, Canales y Puertos, Universidad de Cantabria, Av. de los Castros s/n, 39005 Santander, Spain

meperez@unican.es

Maria-Luisa Rapún: Dep. Matemática Aplicada, Universidad de Zaragoza, Centro Politécnico Superior, María de Luna, 3, E 50015 Zaragoza, Spain

mlrapun@unizar.es

Maria-Cecilia Rivara: Department of Computer Science, Universidad de Chile, Casilla 2777, Santiago, Chile

mcrivara@dcc.uchile.cl

Gary F. Roach: Department of Mathematics, University of Strathclyde, 26 Richmond Street, Glasgow, UK

gfr@maths.strath.ac.uk

Keijo Ruotsalainen: Division of Mathematics, Faculty of Technology, University of Oulu, Box 4500, 90014 Oulu, Finland

keijo.ruotsalainen@ee.oulu.fi

Bernard Rutily: Observatoire de Lyon, 9 avenue Charles André, F-69561 Saint-Genis-Laval cedex, France

rutily@obs.univ-lyon1.fr

Francisco-Javier Sayas: Departamento de Matemática Aplicada, Universidad de Zaragoza, Centro Politécnico Superior, María de Luna, 3, E 50015 Zaragoza, Spain

jsayas@unizar.es

Cynthia F. Segatto: Departamento de Engenharia Nuclear, Universidade Federal do Rio Grande do Sul, Av. Osvaldo Aranha 99–4o. andar, 90046-900 Porto Alegre (RS), Brazil

csegatto@mat.ufrgs.br

Seppo Seikkala: Mathematics Division, Department of Electrical Engineering, Faculty of Technology, University of Oulu, PL 4500, FIN-90014 Oulu, Finland

seppo.seikkala@ee.oulu.fi

Iain W. Stewart: Department of Mathematics, University of Strathclyde, 26 Richmond Street, Glasgow G1 1XH, UK

iws@maths.strath.ac.uk

Jianzhong Su: Department of Mathematics, University of Texas at Arlington, Arlington, TX 76019, USA

su@uta.edu

Tadie: Matematiske Institut, Universitetsparken 5, 2100 Copenhagen, Denmark

tad@math.ku.dk

Olivier Titaud: Équipe d'Analyse Numérique, Université Jean Monnet de Saint-Étienne, 23 rue Dr. Paul Michelon, F-42023 Saint-Étienne cedex 02, France

olivier.titaud@free.fr

Bao Loc Tran: Department of Mathematics, University of Texas at Arlington, Arlington, TX 76019, USA

bmt3814@exchange.uta.edu

Martin B. van Gijzen: CERFACS, 42, avenue G. Coriolis, 31057 Toulouse cedex 01, France

vangijzen@cerfacs.fr

Paulo B. Vasconcelos: Faculdade de Economia da Universidade do Porto, Rua Dr. Roberto Frias, 4200-464 Porto, Portugal

pjv@fep.up.pt

Marco T. Vilhena: Departamento de Engenharia Nuclear, Universidade Federal do Rio Grande do Sul, Av. Osvaldo Aranha 99–4o. andar 90046-900 Porto Alegre (RS), Brazil

vilhena@mat.ufrgs.br

Hui Wang: School of Computing, Mathematical, and Information Sciences, University of Brighton, Lewes Road, Brighton BN2 4GJ, UK

h.wang@brighton.ac.uk

Xiaoyi Xu: Department of Mathematics, Royal Military College of Canada, P.O. Box 17000, Stn Forces, Kingston, ON, Canada K7K 7B4

xu_xiaoyi@yahoo.com

Guozheng Yan: Department of Mathematics, Central China Normal University, Wuhan 430079, P.R. China

scip6003@yahoo.com

Bo Zhang: School of Mathematical and Information Sciences, Coventry University, Coventry CV1 5FB, UK

b.zhang@coventry.ac.uk

This volume is dedicated to Fred R. Payne,
the initiator of the IMSE conferences,
on the occasion of his retirement

1 A Finite Volume Scheme for a Nonlinear Reaction-Diffusion PDE Arising in Combustion

Asdin Aoufi

1.1 Introduction

The aim of this paper is to describe a finite volume scheme for the numerical discretization of a reaction-diffusion equation arising in the modelling of titanium carbide exothermic combustion synthesis. The paper is organized as follows. In section two, the mathematical modelling is formulated. Section three outlines the finite volume discretization principles. In section four, an L_∞ stability result of the numerical scheme for both the nonlinear parabolic equation and the kinetics equation is established. Section five analyses the stability of the implicit scheme. The last section presents some numerical results.

1.2 Governing Equations

We assume that the following hypotheses are fulfilled :
1. Ω is a bounded subset of $\mathbb{R}^d$ ($d \leq 3$) with polygonal boundary $\partial\Omega$.
2. The thermal conductivity, λ is strictly positive, continuous over $\mathbb{R}_+^* \times [0,1]$.
3. The heat capacity, C_p is strictly positive, continuous over $\mathbb{R}_+^* \times [0,1]$.
4. The sensible enthalpy $h_s(T,\xi) = \rho \int_{T_a}^T C_p(\theta,\xi)\,d\theta$ is defined by a bounded integral equation and is such that $\frac{\partial h_s(T,\xi)}{\partial \xi} = 0$ and $\frac{\partial h_s(T,\xi)}{\partial T} > 0$.

The mathematical modelling relates the temperature field $T(x,t)$ to the conversion rate $\xi(x,t)$ for each $x \in \Omega$ and $t \in [0,\tau[$ ($\tau > 0$) through the nonlinear parabolic equation expressing the enthalpy balance and the first order differential equation expressing the exothermic kinetics.

1.2.1 Enthalpy Balance

The enthalpy balance is written, for every $x \in \Omega$, while the nonlinear Robin boundary condition is defined on $\partial\Omega$ as

$$\frac{\partial h_s(T,\xi)}{\partial t} = \operatorname{div}\left(\lambda(T,\xi)\operatorname{grad}T\right) + \rho\Delta H_f\frac{d\xi}{dt}, \quad T(x,0) = T_a > 0,$$

$$-\lambda(T,\xi)\frac{\partial T}{\partial n} = \epsilon\,\sigma\,(T^4 - T_\infty^4).$$

1.2.2 Exothermic Chemical Kinetics

We assume that the reaction rate of the first step, irreversible exothermic reaction synthesis follows an Arrhenius law such that

$$\frac{d\xi\,(x,t)}{dt} = k(T)\,(1 - \xi)\,, \quad \xi(x,0) = 0.$$

1.3 Finite Volume Discretization

The finite volume method is particularly suited for the discretization of partial differential equations expressed in conservation form. In order to construct such a finite-volume discretization, we need to define the notion of *admissible mesh* according to Ref. [1] as follows. The elements of a mesh $\tau_h(\Omega)$ of Ω are called control volumes. For each $(P,Q) \in \tau_h(\Omega)^2$ with $P \neq Q$, we denote by $e_{PQ} = \overline{P} \cap \overline{Q}$ their common interface. m_{PQ} denotes the Lebesgue measure of e_{PQ} and n_{PQ} the unit vector normal to e_{PQ} oriented from P to Q. $\partial\tau_h(\Omega) = \{P \in \tau_h(\Omega) : \mathrm{meas}(\partial P \cap \partial\Omega) \neq 0\}$ is the set of external control volumes. The time step $\Delta t > 0$. We define by $N(P)$ the set of neighbors of P. For all $P \in \tau_h(\Omega)$, the measure of P is denoted by $m(P)$ in R^d. The size h of the mesh $\tau_h(\Omega)$ refers to the maximum of the diameter of all the control volumes P of the mesh. We assume that there exists $x_P \in P$ such that $\frac{x_Q - x_P}{|x_Q - x_P|} = n_{PQ}$ and note $d_{PQ} = |x_P - x_{P \cap Q}|$.

1.3.1 Discretization of the Enthalpy Balance

We integrate the enthalpy balance over the control-volume $[t_n, t_{n+1}] \times P$ and apply a simple Euler finite-difference scheme for the time discretization and the finite-volume method for the spatial terms. The key feature of the methodology is the application of Green's formula for the discretization of the Laplacian and the derivation of a consistent approximation of the fluxes over the edges of the boundaries of the control volumes. We define $S(T,\xi) = \rho\Delta H_f\,d\xi/dt$ and index by $i = 0$ (*resp.* $i = 1$) the explicit (*resp. implicit*) finite-volume scheme discretization

$$m(P)\,\frac{h_s(T_P^{n+1}, \xi_P^{n+1}) - h_s(T_P^n, \xi_P^n)}{\Delta t}$$
$$= \sum_{Q \in N(P)} \tau_{Q,P}^{n+i}(T_Q^{n+i} - T_P^{n+i}) + m(P)\,S(T_P^{n+i}, \xi_P^{n+i}).$$

According to [1], the exchange term $\tau_{Q,P}^{n+i}$ is computed by the following harmonic mean formula so that the heat flux $-\lambda\,(T,\xi)\,\nabla T$ is continuous over $P \cap Q$:

$$\tau_{Q,P}^{n+i} = \frac{\lambda_P^{n+i}\lambda_Q^{n+i}}{d_{QP}\lambda_P^{n+i} + d_{PQ}\,\lambda_Q^{n+i}}.$$

In order to solve numerically by an iterative method this set of nonlinear equations, we linearize $(T, \xi) \mapsto h_s(T, \xi)$ between (T_P^n, ξ_P^n) and (T_P^{n+1}, ξ_P^{n+1}) either at time t_n (*resp.* t_{n+1}) in the case of the explicit Euler scheme (resp. *implicit Euler scheme*). We notice from hypothesis $H4$, that $\dfrac{\partial h_s(T, \xi)}{\partial \xi} = 0$ and obtain

$$
\frac{m(P)}{\Delta t} \frac{\partial h_s(T_P^{n+i}, \xi_P^{n+i})}{\partial T}(T_P^{n+1} - T_P^n)]
$$
$$
= \sum_{Q \in N(P)} \tau_{Q,P}^{n+i} (T_Q^{n+i} - T_P^{n+i}) + m(P)\, S(T_P^{n+i}, \xi_P^{n+i}).
$$

1.3.2 Discretization of the Kinetics

Applying an Euler explicit/implicit scheme leads to :

$$
\frac{\xi_P^{n+1} - \xi_P^n}{\Delta t} = k(T_P^{n+i})\,(1 - \xi_P^{n+i}).
$$

1.4 Properties of the Explicit Scheme

1.4.1 Stability Condition for Reaction-Diffusion

Theorem 1. *Assuming that an admissible mesh $\tau_h(\Omega)$ of Ω is given, the stability condition over the time-step Δt of the explicit Euler scheme is given by*

$$
\Delta t \left(\sum_{Q \in N(P)} \frac{\tau_{Q,P}^n}{m(P)\dfrac{\partial h_s(T_P^n, \xi_P^n)}{\partial T}} \right) < 1.
$$

Proof. For every control volume $P \in \tau_h(\Omega)$ we can write

$$
T_P^{n+1} = \frac{\Delta t}{\dfrac{\partial h_s(T_P^n, \xi_P^n)}{\partial T}} S(T_P^n, \xi_P^n) \left(1 - \sum_{Q \in N(P)} \frac{\Delta t\, \tau_{Q,P}^n}{m(P)\dfrac{\partial h_s(T_P^n, \xi_P^n)}{\partial T}} \right) T_P^n
$$
$$
+ \sum_{Q \in N(P)} \left(\frac{\Delta t\, \tau_{Q,P}^n}{m(P)\dfrac{\partial h_s(T_P^n, \xi_P^n)}{\partial T}} \right) T_Q^n.
$$

From the hypotheses of section 1.2, $\dfrac{\partial h_s(T_P^n, \xi_P^n)}{\partial T} > 0$. Omitting the source term S, we notice that T_P^{n+1} is a linear barycentric combination of T_P^n and T_Q^n with positive weight. This implies that we have L_∞ stability.

Moreover, since $0 < S_M = \sup\{S(T,\xi) : T \geq 0, \xi \in [0,1]\} < \infty$ and $0 < A_m = \inf\{\frac{\partial h_s(T,\xi)}{\partial T} : T \geq 0, \xi \in [0,1]\} < \infty$, we obtain :

$$\|T^{n+1}\|_\infty \leq \|T^n\|_\infty + \Delta t \frac{S_M}{A_m}.$$

1.4.2 Stability Condition for the Kinetics Discretization

Theorem 2. *Assuming that an admissible mesh $\tau_h(\Omega)$ of Ω is given, the stability condition over the time-step Δt of the explicit Euler scheme is given by*

$$\Delta t \, \max \{k(T_P^n) : P \in \tau_h(\Omega)\} \leq 1.$$

Proof. For every control volume $P \in \tau_h(\Omega)$ we can write

$$\xi_P^{n+1} = \xi_P^n \left(1 - k(T_P^n)\, \Delta t\right) + k(T_P^n)\Delta t.$$

From the hypothesis, we notice that ξ_P^{n+1} is a barycentric combination of ξ_P^n and 1, therefore $\xi_P^{n+1} \in [\xi_P^n, 1]$. An easy induction shows that $\xi_P^n \in [0,1]$ $\forall n \geq 1$ and $\forall P \in \tau_h(\Omega)$.

1.5 Properties of the Implicit Scheme

In order to solve numerically the enthalpy balance and the kinetics we define an iterative scheme which solves in a decoupled procedure the discrete reaction-diffusion equation and the discrete kinetics equation in the following way. We define the two sequences $(T_P^{n+1,m})_{m \geq 0, P \in \tau_h(\Omega)}$ and $(\xi_P^{n+1,m})_{m \geq 0, P \in \tau_h(\Omega)}$ such that

$$\frac{m(P)}{\Delta t} \frac{\partial h_s(T_P^{n+1,m}, \xi_P^{n+1,m})}{\partial T}(T_P^{n+1,m+1} - T_P^n)$$

$$= \sum_{Q \in N(P)} \tau_{Q,P}^{n+1,m} (T_Q^{n+1,m+1} - T_P^{n+1,m+1}) + m(P)\, S(T_P^{n+1,m}, \xi_P^{n+1,m})$$

and

$$\frac{\xi_P^{n+1,m+1} - \xi_P^n}{\Delta t} = k(T_P^{n+1,m})\, (1 - \xi_P^{n+1,m+1}).$$

These two sequences are initialized as $T_P^{n+1,0} = T_P^n$ and $\xi_P^{n+1,0} = \xi_P^n$.

1.5.1 Properties of the Matrix

Theorem 3. *The sparse-matrix arising in the implicit finite-volume discretization of the enthalpy balance is an M-matrix and the numerical solution is always positive.*

Proof. At each iteration m of the nonlinear solver, the implicit scheme involves the inversion of a linear system $A^m T^{n+1,m} = B^m$, for the determination of the new temperature field $T^{n+1,m}$, where the sparse matrix A^m is defined as

$$- \left(\sum_{Q \in N(P)} \tau_{Q,P}^{n+1,m} \right) T_Q^{n+1,m+1}$$

$$+ \left(\frac{m(P)}{\Delta t} \frac{\partial h_s(T_P^{n+1,m}, \xi_P^{n+1,m})}{\partial T} + \sum_{Q \in N(P)} \tau_{Q,P}^{n+1,m} \right) T_P^{n+1,m+1}$$

$$= \frac{m(P)}{\Delta t} \frac{\partial h_s(T_P^{n+1,m}, \xi_P^{n+1,m})}{\partial T} T_P^n + m(P) S(T_P^{n+1,m}, \xi_P^{n+1,m}) \geq 0.$$

From this equation, we notice that the matrix A^m indexed by the iteration m is diagonally dominant, therefore invertible, and is an $M-matrix$, since the off-diagonal terms are negative and the right-hand side is positive. We conclude that the solution of the algebraic system is positive. In one dimension, the tridiagonal matrix is inverted by the classical direct *Thomas* algorithm. In upper dimensions, we use an S.O.R. iterative method for the inversion of the sparse matrix.

1.5.2 Stability for the Kinetics Discretization

Theorem 4. *The implicit scheme for the kinetics is unconditionally L_∞-stable, i.e., $\xi_P^n \in [0,1]$ $\forall n \geq 1$ and $\forall P \in \tau_h(\Omega)$.*

Proof. For every control volume $P \in \tau_h(\Omega)$ we can write

$$\xi_P^{n+1,m+1} = \frac{1}{1 + \Delta t\, k(T_P^{n+1,m})} \xi_P^n + \frac{\Delta t\, k(T_P^{n+1,m})}{1 + \Delta t\, k(T_P^{n+1,m})} 1.$$

It is clear that for each m, $\xi_P^{n+1,m+1}$ is a linear barycentric combination of ξ_P^n and 1, therefore $\xi_P^{n+1,m+1} \in [0,1]$. Since the function k is continuous, and assuming that the sequence $m \mapsto T_P^{n+1,m}$ converges, then $m \mapsto \xi_P^{n+1,m}$ converges towards a limit ξ_P^{n+1} such that $\xi_P^{n+1} \in [0,1]$.

It is worth mentioning that the computational cost of the kinetics solver is neglectible with respect to the reaction-diffusion solver and that no spurious oscillations are introduced by the numerical solver. This may not be the case if a high order finite-difference approximation is used for the temporal discretization.

1.6 Numerical Simulations

The finite-volume scheme has been implemented from the ground-up in our C^{++} code *Hephaïstos* and some results are presented in Fig. 1 and Fig. 2.

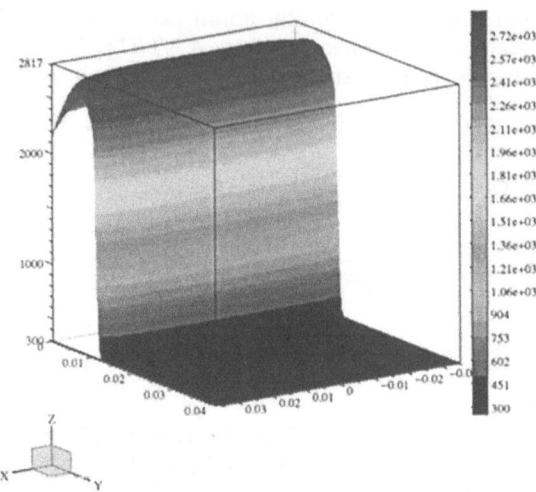

Fig. 1. Temperature profile inside the cylindrical sample of radius $R = 3$cm and height $H = 4$cm at time t = 10 sec . The combustion front propagates along the z-axis from $z = 0$ to $z = H$. Radiative heat losses are taken into account and explain the curvature of the front along the r-axis from $r = 0$ to $r = R$.

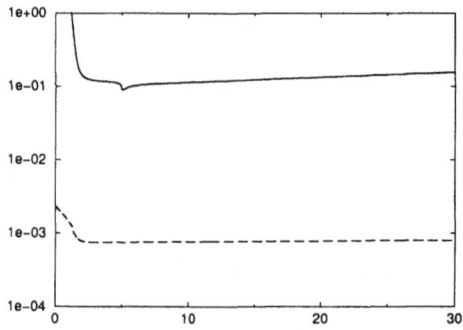

Fig. 2. CFL stability condition for the reaction-diffusion explicit finite-volume scheme (long-dashed), and CFL stability condition for the kinetics explicit numerical scheme (solid line).

References

1. R. Eymard, T. Gallouët, and R. Herbin, *Handbook of Numerical Analysis, Vol. VII*, North Holland, 2000.

2 High Performance Algorithms for Computing Nonsingular Jacobian-Free Piecewise Linearization of Differential Algebraic Equations

Enrique Arias, Vicente Hernández,
and Jacinto-Javier Ibáñez

2.1 Introduction

Numerical methods for solving differential algebraic equations (DAEs) have received considerable attention in recent years. Most of the available numerical methods are based on the approximation of a continuous model by a discrete model, and the computation of an approximate solution in a finite set of points. In the present work an extension of the methodology presented in [1] to consider the following differential algebraic equation (DAE) problem has been carried out:

$$M\dot{x}(t) = f(x(t), t), \quad x(t_0) = x_0,$$

where $M \in R^{N \times N}$, $x(t) \in R^N$, $t \in R$, f is a continuous Lipschitzian function on the domain $\Omega \subset R^N \times R$.

Notice that if M is the identity matrix, then the following ordinary differential equation (ODE) problem appears:

$$\dot{x}(t) = f(x(t), t), \quad x(t_0) = x_0.$$

In this work, sequential and parallel implementations, based on shared and distributed memory schemes, have been developed. The implementation presents the following features: good performance, portability and efficiency, by using Standard Linear Algebra Libraries like BLAS, LAPACK, PBLAS and ScaLAPACK, and PTHREADS and MPI for shared and distributed memory multiprocessors.

The paper is structured as follows. In the second section the above mentioned DAE's extension is described. Implementation details are treated in section three. The fourth section presents the experimental results. Finally some conclusions and future work guidelines are outlined.

2.2 Piecewise Linearization Methods for DAEs: Initial Value Problems

Let us consider the following initial value problem:

$$M\dot{x}(t) = f(x(t), t), \quad x(t_0) = x_0, \quad t \in I = [t_0, t_f].$$

An approximation to the analytic solution can be obtain by means of piecewise linearization methods by splitting the interval I into n subintervals, $I = (t_0, t_1] \cup \ldots \cup (t_{n-1}, t_n]$, where $t_n = t_f$, and on each subinterval, $I_i = (t_i, t_{i+1}]$, the previous problem is approximated by

$$My'(t) = g(y(t), t) \equiv F_i + J_i(y - y_i) + T_i(t - t_i), \quad y(t_i) = y_i, \qquad (2.1)$$

where

$$F_i = f(y_i, t_i), \quad J_i = \frac{\partial f}{\partial x}(y_i, t_i), \quad T_i = \frac{\partial f}{\partial t}(y_i, t_i), \quad i = 0, 1, \ldots, n-1,$$

with $y_0 = x_0$. The right-hand side of equation (2.1) corresponds to the first order Taylor polynomial approximation of $f(x, t)$ at (y_i, t_i) in the subinterval I_i.

Using the following steps, y_{i+1} is obtained from $y_i, i = 0, 1, \ldots, n-1$:

1. Compute the SVD of the matrix M, $M = USV^T$.

2. Make the change of variable $z = V^T y$ to obtain

$$Sz'(t) = U^T J_i V(z(t) - V^T y_i) + U^T T_i(t - t_i) + U^T F_i.$$

By updating $J_i \leftarrow U^T J_i V$, $T_i \leftarrow U^T T_i$, $F_i \leftarrow U^T F_i$, this expression can be written in the following way:

$$Sz'(t) = J_i(z(t) - z_i) + T_i(t - t_i) + F_i, z_i = V^T y_i.$$

3. By decoupling the previous equation

$$\begin{pmatrix} S_1 & 0 \\ 0 & 0 \end{pmatrix} \begin{pmatrix} z_1'(t) \\ z_2'(t) \end{pmatrix} = \begin{pmatrix} J_{i1} & J_{i2} \\ J_{i3} & J_{i4} \end{pmatrix} \left[\begin{pmatrix} z_1(t) \\ z_2(t) \end{pmatrix} - \begin{pmatrix} z_{1i} \\ z_{2i} \end{pmatrix} \right]$$
$$+ \begin{pmatrix} T_{i1} \\ T_{i2} \end{pmatrix} (t - t_i) + \begin{pmatrix} F_{i1} \\ F_{i2} \end{pmatrix}, \quad \begin{pmatrix} z_{1i} \\ z_{2i} \end{pmatrix} = \begin{pmatrix} V_1^T \\ V_2^T \end{pmatrix} y_i$$

the following pair of equations is obtained:

$$S_1 z_1'(t) = J_{i1}(z_1(t) - z_{1i}) + J_{i2}(z_2(t) - z_{2i}) + T_{i1}(t - t_i) + F_{i1},$$
$$0 = J_{i3}(z_1(t) - z_{1i}) + J_{i4}(z_2(t) - z_{2i}) + T_{i2}(t - t_i) + F_{i2},$$

$$z_{1i} = V_1^T y_i, \quad z_{2i} = V_2^T y_i.$$

4. Compute the SVD of J_{i4}, $J_{i4} = U_2 S_2 V_2^T$, and the pseudoinverse $J_{i4}^+ = V_2 S_2^+ U_2^T$.

5. From the second equation in step 3 we obtain

$$z_2(t) - z_{i2} = J_{i3}(z_1(t) - z_{i1}) + T_{i2}(t - t_i) + F_{i2}$$

where $J_{i3} = -J_{i4}^+ J_{i3}$, $T_{i2} = -J_{i4}^+ T_{i2}$, $F_{i2} = -J_{i4}^+ F_{i2}$.

6. Substituting in the first equation of step 3,

$$z_1'(t) = J_{i1}(z_1(t) - z_{i1}) + T_{i1}(t - t_i) + F_{i1}, z_{i1} = V_{i1}^T y_i,$$

where $J_{i1} = S_1^{-1}(J_{i1} + J_{i2}J_{i3})$, $T_{i1} = S_1^{-1}(T_{i1} + J_{i2}T_{i2})$, $F_{i1} = S_1^{-1}(F_{i1} + J_{i2}F_{i2})$. Then obtain $z_1(t)$ by solving this ODE initial value problem by using the methodology presented in [1]. After that, $z_2(t)$ is computed substituting in step 5

7. Finally, compute $x(t_{i+1}) \equiv y_{i+1} = V z_{i+1}$.

2.3 Implemented Algorithms

The algorithms developed in the present work have been implemented taking into account the following features:

• **Portability**: To achieve portable algorithms, Standard Linear Algebra Libraries, such as BLAS (I, II and III) (*Basic Linear Algebra Subroutines* [2]) and LAPACK (*Linear Algebra PACKage* [3]), have been used in the sequential and shared memory implementation and PBLAS (*Parallel Basic Linear Algebra Subroutines* [4]) and ScaLAPACK (*Scalable Linear Algebra PACKage* [5]) in the distributed memory implementation. The shared memory implementation employs the standard library PTHREADS [6] to deal with threads. The distributed memory implemention uses the standard message passing interface MPI (*Message Passing Interface* [7]).

• **Efficiency**: Good performance has been achieved by using block oriented algorithms, thus exploiting the memory hierarchy of current computers.

In the sequential and shared memory implementations, the next BLAS and LAPACK subroutines have been used:

• **BLAS**: _SCAL (it scales a matrix by a scalar), _AXPY (scalar times a vector plus a vector), and _GEMM (matrix-matrix multiplication).

• **LAPACK**: _GESV (it solves a general system of linear equations $AX = B$).

To improve performance, the algorithm is block-oriented. Operations performed by blocks reduce the data traffic between main memory and cache memory. In this way, better performance is achieved. So, LAPACK routines use an *ilaenv* routine to obtain optimal block size at run-time.

The shared memory implementation is based on a distribution by blocks of rows, where the number of local rows is computed in order to obtain an optimal data distribution. This implementation uses the above mentioned routines of BLAS and LAPACK.

On the other hand, distributed memory implementations use a 2-D cyclic distribution. This kind of distribution is supported by MPI and is the most general type of distribution based on the row and column block size. Depending on these values, other distributions can be obtained. Also, MPI supports the configuration of the network topology. So, this parameter is established to exploit benefits of our network topology. With respect to the routines, in the distributed memory implementation, BLAS and LAPACK routines are substituted by the corresponding PBLAS and ScaLAPACK routines in most cases. However, some steps of the algorithm can be done sequentially with local computations, as first order actualizations, thus improving performance.

The performance obtained in the parallel implementations are expressed in terms of:

• **Speed-up:** It is defined as the ratio of the time taken to solve a problem on a processor to the time required to solve the same problem on a parallel computer with p identical processors.

• **Efficiency:** It is a measure of the fraction of time for which a processor is usefully employed; it is defined as the ratio of the speed-up to the number of processors.

2.4 Experimental Results

In this section the test set and the experimental results are described.

The test battery comprises three cases for the ordinary differential equations (ODE) solver and two cases for the differential algebraic equations (DAE) solver.

• **Chemical Akzo Nobel Problem (CHEMAKZO)**: This case has been obtained from the IVPTestSet [8] collection. It is a stiff system of six nonlinear ordinary differential equations.

• **Chemical Kinetics 1 Problem (LSODE)**: This ODE case has been obtained from the ODEPACK test battery [9]. It is a 3-equations stiff dense problem.

• **Medical Akzo Nobel Problem (MEDAKZO)**: This case has been obtained from the IVPTestSet [8] collection. The problem consists of two partial differential equations. Semi-discretization of this system yields a stiff ODE.

• **Chemical Akzo Nobel Problem (CHEMAKZO2)**: This case has been obtained from the IVPTestSet [8] collection. It is a stiff system of six nonlinear differential algebraic equations. It is the DAE version of CHEMAKZO1.

• **Chemical Kinetics 2 Problem (LSODA)**: This DAE problem has been obtained from the ODEPACK test battery [9].

The experiments have been carried out in a cluster of 12 nodes IBM xSeries 330 SMP, interconnected by a Gigabit Ethernet network. Each node consists of two Pentium III-866 Mhz, 256 KB of cache memory, and 512 MB at 133 MHz of main memory. The cluster runs Red Hat Linux 7.1

(Seawolf) operating system, with Kernel 2.4.2-2 for multiprocessors, and the MPICH version of the message passing interface is used.

In order to study the accuracy of the method presented in this work, nonsingular Jacobian-free piecewise Linearization (NJF-PL) in comparison with other solvers (RADAU5 [10], PSIDE, ODEPACK [9]), let us consider the low dimension cases (CHEMAKXO, CHEMAKZO2, LSODE and LSODA). A high dimension problem (MEDAKZO) has been taken into account to analyze the performance of the parallel implementation.

The following parameters have been considered to run the test set into the parallel architectures:

• Block size: This parameter is very important in order to take advantage of the cache memory. The following block sizes have been considered: 8, 16, 32, 64 and 128.

• Number of processors or threads: The values considered are 2, 4, 8, and 16 processors, and 2 threads.

• Topology of the grid in ScaLAPACK: This parameter only affects the distributed memory implementation.

The following tables show the results of the new approximation with respect to other solvers.

	RADAU5	NJF-PL
$y(1)$	0.1161602E+00	0.1161602E+00
$y(2)$	0.1119418E-02	0.1119418E-02
$y(3)$	0.1621261E+00	0.1621262E+00
$y(4)$	0.3396981E-02	0.3396982E-02
$y(5)$	0.1646185E+00	0.1646185E+00
$y(6)$	0.1989533E+00	0.1989533E+00

	PSIDE	NJF-PL
$y(1)$	0.1150795E+00	0.1150795E+00
$y(2)$	0.1203831E-02	0.1203831E-02
$y(3)$	0.1611563E+00	0.1611563E+00
$y(4)$	0.3656156E-03	0.3656158E-03
$y(5)$	0.1708011E-01	0.1708012E-01
$y(6)$	0.4873531E-02	0.4873533E-02

Table 1. Results for CHEMAKZO (left) and CHEMAKZO-2 (right) problems.

	ODEPACK	NJF-PL
$y(1)$	0.9851726e+00	0.9851721E+00
$y(2)$	0.3386406E-04	0.3386395E-04
$y(3)$	0.1479357E-01	0.1479402E-01

	ODEPACK	NJF-PL
$y(1)$	0.9851712E+00	0.9851711E+00
$y(2)$	0.3386380E-04	0.3386377E-04
$y(3)$	0.1479493E-01	0.1479506E-01

Table 2. Results for the LSODE (left) and LSODA (right) problems.

NP	ET (secs)	S	E
1	1209.23509	1	1
2	609.54610	1.98	0.99
4	343.33289	3.52	0.88
8	237.91343	5.06	0.6325
16	146.23076	8.26	0.51625

NP	ET (secs)	S	E
1	1156.36614	1	1
2	617.44930	1.87	0.936

Table 3. Execution time (ET), speed-up (S) and efficiency (E) for the MEDAKZO problem in a distributed memory architecture, by using the MPI NJF-PL version (left), and PTHREADS NJF-PL version (right).

2.5 Conclusions and Future Tasks

In this paper, NJF-PL algorithms for the solution of DAE problems have been implemented using standard linear algebra libraries such as BLAS and LAPACK for sequential and shared memory implementations, and PBLAS and ScaLAPACK for the distributed memory implementations. Standard POSIX 1003.1c has been used in the shared memory case, and MPI in the distributed memory case for synchronization and communication purposes.

Experimental results show relatively good accuracy of the sequential version and that the parallel algorithms are suitable if the size of the problem is considerable. It is easy to see how performance increases according to the size of the problem.

New extensions of this methodology to implicit differential equations (IDEs), as well as parallel versions of this extension, are under development.

Also, the case of sparse systems has been considered. Then, algorithms that take into account the sparsity of the system are also under study.

Finally, new implementations based on shared memory platforms of all the algorithms will be carried out using the OpenMP library as a programming environment and the ATLAS software package as computational kernel.

References

1. E. Arias, I. Blanquer, V. Hernández, J.J. Ibañez, and P. Ruiz, "Nonsingular Jacobian Free Piecewise Linearization of Ordinary Differential Equation (ODE)", in *2nd NICONET Workshop*, Paris (France), December 1999.

2. J. Dongarra, J. Du Croz, S. Hammarling, and R.J. Hanson, "An Extended Set of FORTRAN Basic Linear Algebra Subroutines", in *ACMMathSoft*,1988.

3. E. Anderson et al., *LAPACK Users' Guide*, SIAM, 2nd ed., Philadelphia, 1994.

4. J. Choi, J. Dongarra, S. Ostrouchov, A. Petitet, and D. Walker, "A Proposal for a Set of Parallel Basic Linear Algebra Subprograms". Tech. Rep. UT-CS-95-292, Dept. of Computer Science, University of Tennessee, May 1995. *LAPACK Working Note 100*.

5. L.S. Blackford et al., *ScaLAPACK Users' Guide*, SIAM, Philadelphia, 1997.

6. J.M. Toribio, *Programación utilizando C threads*, Tech. Rep. SI9596-30, Universidad de Valladolid, February 1997.

7. W. Gropp, E. Lusk, and A. Skjellum, *Using MPI: Portable Parallel Programming with the Message-Passing Interface*, MIT Press, 1994.

8. W.M. Lioen, and J.J.B de Swart, *Test Set for Initial Value Problem Solvers*, Tech. Rep. NM-R9615, CWI, Amsterdam, 1996.

9. A.C. Hindmarsh, *Brief Description of ODEPACK - A systematized Collection of ODE Solvers*, available at http://www.netlib.org/odepack.

10. E. Hairer and G. Wanner, *Solving Ordinary Differential Equations II. Stiff and Differential-Algebraic Problems*, Springer Series in Computational Mathematics 14, Springer Verlag 1991, 2nd ed., 1996.

3 Multiple Scattering of Water Waves by N Floating Bodies Using Boundary Integral Equation Methods

Lahcene Bencheikh

3.1 Introduction

The use of boundary integral equation methods for solving exterior problems of scattering of time harmonic waves leads to a problem of nonuniqueness of the solutions of the relevant integral equations. This difficulty is inherent to the methods used. Various methods have been devised to overcome this problem. One of these methods consists in modifying the fundamental solution by adding to it a series of multipoles. With a mild condition on the coefficients of these multipoles, the problem of nonuniqueness is removed. In the context of water waves, Ursell [1,2] was the first to propose the modification of the fundamental solution for problems involving one floating cylindrical body. He proved uniqueness for the high frequencies. Sayer [3] did the same for the low frequencies. Later, Ursell [4] generalized the result for all frequencies. Martin [5] considered the case of one three-dimensional floating body. Later, Martin [6] treated the problem for two infinitely long floating cylinders. More recently, we extended Martin's work [6] to the case of two three-dimensional floating bodies [7]. Here, we generalize the case to N three-dimensional floating bodies. We found out that the conditions on some of the coefficients of the multipoles depend on the number, N, of the scatterers. It is worth noting that Martin [8] treated the case of N scattering bodies for three-dimensional acoustic waves. He found out that the conditions on all the coefficients depended on N.

3.2 Formulation of the Problem

Consider N rigid floating three-dimensional bodies in water of infinite depth. Let $N + 1$ Cartesian coordinates $(O_k x_k y_k z_k)\}_{k=1}^{N}$ and $(Oxyz)$ be such that the origins $O_k\}_{k=1}^{N}$ and O are all in the plane of the free undisturbed surface. The axes $(O_k y_k)\}_{k=1}^{N}$ and (Oy) are positively oriented downwards into the fluid. We denote by D the fluid domain, by $\partial D_k\}_{k=1}^{N}$ the wetted surfaces of the bodies and $\partial D = \bigcup_{k=1}^{N} \partial D_k$. The free undisturbed surface is denoted by F. The origin O_k is supposed to lie inside the body k $(k = 1, \ldots, N)$. We also denote by F_-^k the portion of the surface inside the body k and lying in the same plane as the free undisturbed surface. The domain bounded by ∂D_k and F^k is denoted D^k_-. Capital letters P, Q denote points of D; lower case letters p^k, q^k denote points of ∂D_k; P^k_-, Q^k_- denote points of D^k_- and $\partial/\partial n_{q^k}$ denotes the derivative, at the point q^k, in the normal direction from ∂D_k into D.

The N bodies are held fixed and a time harmonic wave of small amplitude is incident upon them. We assume that the water is incompressible and inviscid and that the motion is irrotational. It follows that there exists a velocity potential $\mathrm{Re}(\phi(P) \exp(-i\omega t))$ which should satisfy the following: Boundary value problem S: Find a function $\phi(P)$ such that

$$(\frac{\partial^2}{\partial x^2} + \frac{\partial^2}{\partial y^2} + \frac{\partial^2}{\partial z^2})\phi(P) = 0 \text{ for } P \in D, \tag{3.1}$$

$$\partial\phi(P)/\partial n_p = f(p) \text{ for } p \in \partial D, \tag{3.2}$$

$$K\phi + \partial\phi/\partial y \,|_{y=0} = 0 \text{ on } F, \tag{3.3}$$

$$r(\partial\phi/\partial r - iK\phi) \to 0 \text{ as } r \to \infty, \tag{3.4}$$

$$\partial\phi/\partial y \to 0 \text{ as } y \to \infty, \tag{3.5}$$

where r is the distance of the observation point measured from the origin O, $f(p)$ is the prescribed value of the velocity on the boundary ∂D and $K = \omega^2/g$, where ω is the frequency of the incident wave and g is gravity. We denote by ∂D^*_k the union of ∂D_k and its mirror image with respect to F^k_-. John [9] showed that if $\partial D^*_k \}^N_{k=1}$ is a convex, twice differentiable surface, then the boundary value problem S possesses exactly one solution for every value of K. The boundary value problem S is generally solved using boundary integral equation methods. These methods are based on integral representations from which boundary integral equations can be derived. In order to get integral representations, a fundamental solution is required. This is expressed as follows [5]:

$$G_0(P; Q) = \tfrac{1}{2}[R^2 + (y - \eta)^2]^{-1/2} + \tfrac{1}{2}[R^2 + (y + \eta)^2]^{-1/2}$$
$$+ K \int_0^\infty e^{-k(y+\eta)} J_0(kR)\frac{dk}{k - K}, \tag{3.6}$$

where $R^2 = (x - \xi)^2 + (z - \zeta)^2$, (x, y, z) and (ξ, η, ζ) are the Cartesian coordinates of the points P and Q, respectively, and $J_0(\cdot)$ denotes the Bessel function of the first kind and order zero. In the integral in equation (3.6) the k-contour passes below the pole at $k = K$. This fundamental solution, known as the Green's function, can be expanded as follows [5]:

$$G_0(P; Q) = \sum_{\sigma, \ell, m} \alpha^\sigma_{\ell m}(\mathbf{r}_P)\Phi^\sigma_{\ell m}(\mathbf{r}_Q) \text{ for } \mathrm{r_P} < \mathrm{r_Q}, \tag{3.7}$$

where the summation $\sum\limits_{\sigma, \ell, m}$ means $\sum\limits_{\sigma=1}^{2} \sum\limits_{\ell=0}^{\infty} \sum\limits_{m=0}^{\infty}$ and the functions $\alpha^\sigma_{\ell m}(\mathbf{r}_P)$,

$\Phi_{\ell m}^{\sigma}(\mathbf{r}_P)$ are as defined in [7]. The $\alpha_{\ell m}^{\sigma}$ are regular solutions of (1) and satisfy the condition (3.3), while the functions $\Phi_{\ell m}^{\sigma}$ are known as multipoles and satisfy equations (3.1) and (3.3)–(3.5). These multipoles are also singular at the origin with respect to which they are defined. The solution is sought in the following form:

$$\phi(P) = \int_{\partial D} G_0(P;q)\mu(q)ds_q. \tag{3.8}$$

The application of the boundary condition (3.2) gives

$$\pi\mu(p) + \int_{\partial D} \frac{\partial G_0(p;q)}{\partial n_p}\mu(q)ds_q = f(p). \tag{3.9}$$

This is the boundary integral equation corresponding to the integral representation(3.8). This integral equation will have a unique solution if the corresponding homogeneous integral equation posseses only the trivial solution. However, it is known [9] that the corresponding homogeneous integral equation of (3.9) possesses nontrivial solutions whenever K coincides with one of the eigenvalues of the corresponding 'interior wave-Dirichlet problem'. These values of K are known as the irregular values. In what follows, we are going to show how this difficulty can be overcame.

3.3 The Question of Uniqueness

We shall show how to avoid the problem of irregular values. This consists in modifying the fundamental Green's function by adding to it N infinite series of multipoles. In each origin O_k, we shall center one of the series of the multipoles. With a mild condition on the coefficients of these multipoles, we shall establish uniqueness for the corresponding modified integral equations. The modified Green's function can be written as

$$G_1(P;Q) = G_0(P;Q) + H(P;Q), \tag{3.10}$$

with

$$H(P;Q) = \sum_{k=1}^{N} \sum_{\sigma,\ell,m} \frac{1}{\pi K^{2m+1}} a_{\ell m}^{\sigma k} \Phi_{\ell m}^{\sigma}(\mathbf{r}_P^k)\Phi_{\ell m}^{\sigma}(\mathbf{r}_Q^k), \tag{3.11}$$

where $a_{\ell m}^{\sigma k}$ are the coefficients of the multipoles and $\mathbf{r}_P^k$ is the position vector of the point P with respect to the origin O_k. The factors $1/(\pi K^{2m+1})$ are added so as to make the conditions on the coefficients of the multipoles dimensionless. Consider now the integral representation (3.8) with this modified Green's function:

$$\phi(P) = \int_{\partial D} G_1(P;q)\mu(q)ds_q.$$

This representation satisfies equations (3.1) and (3.3)–(3.5). Implementing the boundary condition (3.2) gives

$$\pi\mu(p) + \int_{\partial D} \frac{\partial G_1(p;q)}{\partial n_p}\mu(q)ds_q = f(p). \tag{3.12}$$

Equation (3.12) will have a unique solution if and only if its corresponding homogeneous integral equation possesses the trivial solution only. This homogeneous integral equation will have a trivial solution if the corresponding "interior wave-Dirichlet problem", i.e. with a vanishing condition on ∂D, has the trivial solution only (see [6] for a detailed argument for the proof of this assertion). Let the potential

$$U(P_-^k) = \int_{\partial D} G_1(P_-^k;q)\lambda(q)ds_q \tag{3.13}$$

be defined in D_-^k, where $\lambda(q)$ is the unknown density, and with the condition $U(p) = 0$ on ∂D. Function (3.13) satisfies (3.1), in $D_-^k - \{O_k\}$, (3) on F_-^k. Let S_-^k be the inscribed hemisphere in D_-^k. Denote by D_N^k the interior of the hemisphere S_-^k. All we have to do is show that $U(P_-^k) \equiv 0$ in D_-^k for $k = 1, \ldots, N$. We have the following theorem.

Theorem 1. *If we choose* $\mathrm{Im}(a_{\ell m}^{\sigma k}) > 0$ *for* $\ell > 0$ *and* $\mathrm{Im}(a_{0m}^{\sigma k}) + (2 - N)|a_{0m}^{\sigma k}|^2 > 0$ *for* $\sigma = 1, 2$, $m = 0, 1, \ldots$, *and* $k = 1, \ldots, N$, *then* $U(P_-^k) \equiv 0$ *in* D_-^k *for* $k = 1, \ldots, N$.

Proof. Consider the case when the interior point is in D_-^s ($s = 1, \ldots, N$). Inserting (3.10) and (3.11) in (3.13) and making use of (3.7) with respect to O_s (since $r_P^s < r_q^s$), gives

$$U(P_-^s) = \sum_{\sigma,\ell,m} A_{\ell m}^{\sigma s}\alpha_{\ell m}^{\sigma}(\mathbf{r}_P^s) + \sum_{\sigma,\ell,m} \frac{1}{\pi K^{2m+1}}a_{\ell m}^{\sigma s}A_{\ell m}^{\sigma s}\Phi_{\ell m}^{\sigma}(\mathbf{r}_P^s)$$

$$+ \sum_{k=1,k\neq s}^{N}\sum_{\sigma,\ell,m} \frac{1}{\pi K^{2m+1}}a_{\ell m}^{\sigma k}A_{\ell m}^{\sigma k}\Phi_{\ell m}^{\sigma}(\mathbf{r}_P^k), \tag{3.14}$$

where $A_{\ell m}^{\sigma k} = \int_{\partial D} \Phi_{\ell m}^{\sigma}(\mathbf{r}_q^k)\lambda(q)ds_q$ for $\sigma = 1, 2$, $m = 0, 1, \ldots$, and $k = 1, \ldots, N$. We shall now make use of the following addition theorem for the multipoles in three dimensions.

Theorem 2. *If there is a point in the fluid having* $\mathbf{r}_P^k$ *as a position vector with respect to the origin* O_k, *then*

$$\Phi_{\ell m}^{\sigma}(\mathbf{r}_P^k) = \sum_{\upsilon,q,n} S_{\ell q m n}^{\sigma\upsilon}(\mathbf{a}_{ks})\alpha_{\ell m}^{\sigma}(\mathbf{r}_P^s) \text{ for } r_P^s < a_{ks}, \tag{3.15}$$

where $\mathbf{r}_P^k = \mathbf{r}_P^s + \mathbf{a}_{ks}$ *and* $\mathbf{a}_{ks} = \overrightarrow{O_k O_s}$.

The elements $S_{\ell qmn}^{\sigma v}(\mathbf{a}_{ks})$ as well as the proof of this theorem is given in [7]. Using (15) in the last group of summations of (14) leads to

$$U(P_-^s) = \sum_{\sigma,\ell,m} C_{\ell m}^{\sigma s} \alpha_{\ell m}^\sigma(\mathbf{r}_P^s) + \sum_{\sigma,\ell,m} \frac{1}{\pi K^{2m+1}} a_{\ell m}^{\sigma s} A_{\ell m}^{\sigma s} \Phi_{\ell m}^\sigma(\mathbf{r}_P^s),$$

where

$$C_{\ell m}^{\sigma s} = A_{\ell m}^{\sigma s} + \sum_{k=1, k\neq s}^{N} \sum_{v,q,n} \frac{1}{\pi K^{2n+1}} a_{qn}^{vk} A_{qn}^{vk} S_{q\ell nm}^{v\sigma}(\mathbf{a}_{ks}).$$

We now apply Green's theorem to $U(P_-^s)$ and $\overline{U(P_-^s)}$, the complex conjugate, in D_-^s/D_N^s and we make use of condition (3.3) and the vanishing condition on ∂D_s. We then sum over s and after some manipulations involving the use of some properties of $S_{\ell qmn}^{\sigma v}(\mathbf{a}_{ks})$, we get the following

$$\sum_{s=1}^{N} \sum_{\sigma,\ell,m} \frac{1}{K^{2m+1}} \mathrm{Im}(a_{\ell m}^{\sigma s})|A_{\ell m}^{\sigma s}|^2$$

$$+ \sum_{s=1}^{N} \sum_{\sigma=1}^{2} \sum_{m=0}^{\infty} \frac{1}{K^{2m+1}}[|B_{0m}^{\sigma s}|^2 - \sum_{k=1, k\neq s}^{N} \mathrm{Re}(B_{0m}^{\sigma s}\overline{Q}_{m\sigma}^{sk})] = 0, \quad (3.16)$$

where $B_{\ell m}^{\sigma s} = a_{\ell m}^{\sigma s} A_{\ell m}^{\sigma s}$ and $Q_{m\sigma}^{sk} = \sum_{v=1}^{2} \sum_{n=0}^{\infty} B_{0n}^{vk} \mathrm{Im}(S_{00mn}^{\sigma v}(\mathbf{a}_{sk}))/(\pi K^{2n+1})$.

Following Martin [8], $\mathrm{Re}(B_{0m}^{\sigma s}\overline{Q}_{m\sigma}^{sk})$ in (16) may be rewritten in the following way:

$$\mathrm{Re}(B_{0m}^{\sigma s}\overline{Q}_{m\sigma}^{sk}) = (|B_{0m}^{\sigma s}|^2 + |Q_{m\sigma}^{sk}|^2 - |B_{0m}^{\sigma s} - Q_{m\sigma}^{sk}|^2)/2. \quad (3.17)$$

If we use this expression given by (3.17) in (3.16) and after some calculations, this yields

$$\sum_{s=1}^{N} \sum_{\sigma=1}^{2} \sum_{\ell=1}^{\infty} \sum_{m=0}^{\infty} \frac{1}{K^{2m+1}} \mathrm{Im}(a_{\ell m}^{\sigma s})|A_{\ell m}^{\sigma s}|^2$$

$$+ \sum_{s=1}^{N} \sum_{\sigma=1}^{2} \sum_{m=0}^{\infty} \frac{1}{K^{2m+1}}\{\mathrm{Im}(a_{0m}^{\sigma s}) + (2-N)|a_{0m}^{\sigma s}|^2\}|A_{0m}^{\sigma s}|^2 + L_1 = 0,$$

where

$$L_1 = \frac{1}{2} \sum_{k=1, k\neq s}^{N} \sum_{s=1}^{N} \sum_{\sigma=1}^{2} \sum_{m=0}^{\infty} \frac{1}{K^{2m+1}}|B_{0m}^{\sigma s} - Q_{m\sigma}^{sk}|^2 \geq 0.$$

In order to deduce that all the $A_{\ell m}^{\sigma s}$ are zero, it is sufficient to choose the $a_{\ell m}^{\sigma s}$ such that $\text{Im}(a_{\ell m}^{\sigma s}) > 0$ for $\ell > 0$ and $\text{Im}(a_{0m}^{\sigma s}) + (2 - N)|a_{0m}^{\sigma s}|^2 > 0$ for $\sigma = 1, 2$, $m = 0, 1, \ldots$. This leads to $U(P_-^s) \equiv 0$ in D_-^s for $s = 1, \ldots, N$. This concludes the proof of Theorem 1.

It follows that the homogeneous version of (3.12) has only the trivial solution, so the nonhomogeneous modified integral equation (3.12) has a unique solution for all K. The irregular values are thus eliminated.

3.4 Conclusion

The problem of nonuniqueness of the solutions of integral equations resulting from the use of boundary integral equation methods has been studied. It was shown that the difficulty which arises with the use of these methods can be removed. This involved the addition of a series of multipoles to the fundamental solution. With some mild sufficient conditions on the coefficients of the multipoles the problem of irregular values was overcome. The condition on some of these coefficients depends on the number of scatterers unlike the result derived by Martin [8] for the analogous case in acoustics where all the coefficients had their condition depend on the number of scatterers.

References

1. F. Ursell, Short surface waves due to an oscillating immersed body, *Proc. Roy. Soc. London A* **220** (1953), 90–103.

2. F. Ursell, The transmission of surface waves under surface obstacles, *Proc. Cam. Phil. Soc.* **57** (1961), 638–668.

3. P. Sayer, An integral equation method for determining the fluid motion due to a cylinder heaving on water of finite depth, *Proc. Roy. Soc. London A* **372** (1980), 93–110.

4. F. Ursell, Irregular frequencies and the motion of floating bodies, *J.Fluid Mech.* **105** (1981), 143–156.

5. P.A. Martin, On the null-field equations for water wave radiation problems, *J. Fluid Mech.* **113** (1981), 315–332.

6. P.A. Martin, Integral equation methods for multiple scattering problems. II. Water waves, *Q. Jl. Mech. Appl. Math.* **38** Pt.1, (1985), 119–133.

7. M. Sidi and L. Bencheikh, Uniqueness for a problem of multiple scattering of water waves, *Proc. Roy. Soc. London A* (submitted for publication).

8. P.A. Martin, Multiple scattering and modified Green's functions, *J. Math. Anal. Appl.* (in press).

9. F. John, On the motion of floating bodies II, *Comm. Pure Appl. Math.* **3** (1950), 45–101.

4 An Application of Semigroup Theory to a Fragmentation Equation

Pamela N. Blair, Wilson Lamb, and Iain W. Stewart

4.1 Introduction

The process of fragmentation arises in many physical situations, including polymer degradation, droplet breakage and rock crushing and grinding. Under suitable assumptions, the evolution of the size distribution $c(x,t)$, where x represents particle size and t is time, may be described by the linear integro-differential equation

$$\frac{\partial c(x,t)}{\partial t} = -a(x)c(x,t) + \int_x^\infty a(y)b(x,y)c(y,t)dy, \ x > 0, t > 0; \quad (4.1)$$

see [1] for details. In (4.1), $a(x)$ describes the rate at which a particle of size x fragments and $b(x,y)$ represents the rate of production of particles of size x due to the break-up of particles of size $y > x$. For the total mass in the system to remain constant during fragmentation, b must satisfy the condition

$$\int_0^y x\, b(x,y)\, dx = y\,.$$

Here we consider the case

$$a(x) = x^{\alpha+1}\,, \ b(x,y) = (\nu+2)(x/y)^\nu\, y^{-1}\,, \quad (4.2)$$

where $\alpha \in \mathbb{R}$ and $-2 < \nu \le 0$. The restriction $\nu > -2$ is mathematical to ensure integrability whereas $\nu \le 0$ is a physical property of the system that is explained in [1]. With these rate functions, equation (4.1) becomes

$$\frac{\partial c(x,t)}{\partial t} = -x^{\alpha+1}c(x,t) + (\nu+2)\int_x^\infty (x/y)^\nu\, y^\alpha\, c(y,t)\, dy,$$

$$x > 0, \ t > 0. \quad (4.3)$$

The work of the first author has been supported by a Scholarship from the Carnegie Trust for the Universities of Scotland.

Our aim is to show that the questions of existence, uniqueness and mass conservation of solutions to (4.3) can be resolved by using the theory of semigroups of linear operators. For certain values of the parameters, the equation can be dealt with immediately by simply quoting previously established results. To cater for the remaining values, we use the idea of similar semigroups [2, p.43] to determine the required information on solutions. In the case when $\alpha > -1$, we also obtain an explicit formula for the semigroup associated with the equation. This yields a closed-form expression for the solution that coincides with that given in [1].

4.2 The Abstract Problem

Before semigroup theory can be applied, the initial-value problem associated with equation (4.3) is reformulated as an abstract Cauchy problem (ACP) of the form

$$\frac{d}{dt}c(t) = A_{\nu,\alpha}[c(t)], \ \ t > 0, \ \ c(0) = c_0, \tag{4.4}$$

where the operator $A_{\nu,\alpha}$ is defined on suitable functions f by

$$(A_{\nu,\alpha}f)(x) = -x^{\alpha+1}f(x) + (\nu+2)\int_x^\infty (x/y)^\nu \, y^\alpha \, f(y)\, dy, \ \ x > 0.$$

As solutions to (4.3) are expected to be mass-conserving, an appropriate Banach space in which to study (4.4) is

$$Y = \{f : ||f|| = \int_0^\infty x|f(x)|\, dx \ < \ \infty\,\}.$$

In this case, a natural domain for $A_{\nu,\alpha}$ is

$$D(A_{\nu,\alpha}) = \{f \in Y : x^{\alpha+1}f \in Y\,\},$$

since a routine calculation shows that $A_{\nu,\alpha}f \in Y$ whenever $f \in D(A_{\nu,\alpha})$.

We note that the solution of (4.4) is interpreted as a strongly continuously differentiable Y-valued function of t. Such a solution is sought in the form $c(t) = S_{\nu,\alpha}(t)c_0, t \geq 0$, where $\{S_{\nu,\alpha}(t)\}_{t\geq0}$ is a strongly continuous semigroup of linear operators on Y that is generated by some extension of $(A_{\nu,\alpha}, D(A_{\nu,\alpha}))$. This semigroup is said to be *substochastic* if $S_{\nu,\alpha}(t) \geq 0$ and $||S_{\nu,\alpha}(t)|| \leq 1$ for each $t \geq 0$. If, additionally, $||S_{\nu,\alpha}(t)f|| = ||f||$ for all $t \geq 0$ and $f \in Y_+$, where Y_+ is the cone of non-negative a.e. functions in Y, then $\{S_{\nu,\alpha}(t)\}_{t\geq0}$ is called a *stochastic* semigroup.

4.3 Existing Results

For the case $\alpha = -1$, the operator $A_{\nu,-1}$ is bounded on Y and hence is the infinitesimal generator of a uniformly continuous semigroup $\{S_{\nu,-1}(t)\}_{t\geq 0}$ on Y, where

$$S_{\nu,-1}(t) \doteq \sum_{k=0}^{\infty} A_{\nu,-1}^{k} t^{k} /k! \,, \ t \geq 0 \,.$$

As shown in [3, Lemma 2.4], this semigroup is stochastic and this, together with [2, p. 145, Proposition 6.2], establishes that the corresponding ACP has a unique, non-negative, mass-conserving solution $c(t) = S_{\nu,-1}(t)c_0$ for all $t \geq 0$ and $c_0 \in Y_+$.

When $\alpha > -1$, the functions a and b defined by (4.2) satisfy the constraints imposed in [4]. Hence, in this case, the semigroup $\{S_{\nu,\alpha}(t)\}_{t\geq 0}$ can be obtained by means of a truncation/limit procedure which takes the form

$$S_{\nu,\alpha}^{(n)}(t) = \sum_{k=0}^{\infty} (A_{\nu,\alpha} P_n)^k t^k /k! = I - P_n + \exp(A_{\nu,\alpha} t) P_n \,, \ t \geq 0, \quad (4.5)$$

$$S_{\nu,\alpha}(t)f = \lim_{n\to\infty} S_{\nu,\alpha}^{(n)}(t)f \,, \ f \in Y, t \geq 0 \,, \quad (4.6)$$

where

$$(P_n f)(x) = \begin{cases} f(x), & \text{if } 0 < x < n; \\ 0, & \text{if } x \geq n. \end{cases} \quad (4.7)$$

Theorem 1. *The semigroup $\{S_{\nu,\alpha}(t)\}_{t\geq 0}$ defined by (4.5)–(4.7) is stochastic and is generated by an extension $(\tilde{A}_{\nu,\alpha}, D(\tilde{A}_{\nu,\alpha}))$ of $(A_{\nu,\alpha}, D(A_{\nu,\alpha}))$. Hence the ACP*

$$\frac{d}{dt}c(t) = \tilde{A}_{\nu,\alpha}[c(t)] \,, \ t > 0, \ c(0) = c_0 \,,$$

has a unique, non-negative, mass-conserving solution for each non-negative $c_0 \in D(\tilde{A}_{\nu,\alpha})$.

Proof. See [4, Sections 2–5]

One drawback to the approach used in [4] is that it cannot be applied when $\alpha < -1$. However, in [5] an alternative strategy involving the Kato–Voigt perturbation theorem is used to analyze (4.4) for any $\alpha \neq -1$ but under the additional constraint that $\nu = 0$.

Theorem 2. *For each $\alpha \neq -1$ there exists a smallest substochastic semigroup $\{S_{0,\alpha}(t)\}_{t\geq 0}$ on Y generated by an extension of $(A_{0,\alpha}, D(A_{0,\alpha}))$. Furthermore, if $\alpha > -1$, then this semigroup is stochastic with generator $(\overline{A}_{0,\alpha}, D(\overline{A}_{0,\alpha}))$ (i.e. the closure of $(A_{0,\alpha}, D(A_{0,\alpha}))$), whereas if $\alpha < -1$, the generator is a proper extension of $(\overline{A}_{0,\alpha}, D(\overline{A}_{0,\alpha}))$. In the latter case,* *for each $f \in Y_+$, there exists $t > 0$ such that $||S_{0,\alpha}(t)f|| < ||f||$, i.e. a mass loss occurs.*

The proof of this assertion can be found in [5, Sections 4,5].

4.4 Extension of Results

We rely on the idea of similar semigroups to extend the results stated in Theorem 2 to the case when $\nu \neq 0$. Motivated by transformations used in [1], we introduce an operator, R_ν, defined on Y by

$$(R_\nu f)(x) = x^{-2\nu/(\nu+2)} f(x^{2/(\nu+2)}).$$

It is a straightforward matter to show that R_ν is a homeomorphism from Y onto Y. Moreover,

$$R_\nu A_{\nu,\alpha} R_\nu^{-1} = A_{0,\beta}, \quad \text{where } \beta = (2\alpha - \nu)/(\nu + 2).$$

Consequently, as pointed out in [1], it is possible to transform the original multiple fragmentation problem into an equivalent binary problem.

Theorem 3. *For each $\alpha \neq -1$ and $\nu \in (-2, 0]$ there exists a smallest substochastic semigroup $\{S_{\nu,\alpha}(t)\}_{t\geq0}$ on Y generated by an extension of $(A_{\nu,\alpha}, D(A_{\nu,\alpha}))$. Furthermore, if $\alpha > -1$, then this semigroup is stochastic with generator $(\overline{A}_{\nu,\alpha}, D(\overline{A}_{\nu,\alpha}))$, whereas if $\alpha < -1$, the generator is a proper extension of $(\overline{A}_{\nu,\alpha}, D(\overline{A}_{\nu,\alpha}))$. In the latter case, for each $f \in Y_+$, there exists $t > 0$ such that $||S_{\nu,\alpha}(t)f|| < ||f||$, i.e. a mass loss occurs.*

Proof. Since $S_{\nu,\alpha}(t) = R_\nu^{-1} S_{0,\beta}(t) R_\nu \; \forall t \geq 0$, where $\beta = (2\alpha - \nu)/(\nu + 2)$, and $\beta > -1 \Leftrightarrow \alpha > -1$, the result follows from Theorem 2.

Corollary 1. *If $\alpha > -1$ and $\nu \in (-2, 0]$, then the semigroups in Theorems 1 and 3 are identical.*

Proof. This follows since the closure of the operator $(A_{\nu,\alpha}, D(A_{\nu,\alpha}))$ (which exists from Theorem 3) can be shown to be the infinitesimal generator of the semigroup defined by (4.5)–(4.7).

4.5 Exact Solutions

Following an approach similar to that used in [3] for the case $\alpha = -1$, we now obtain an explicit formula for the semigroup $\{S_{\nu,\alpha}(t)\}_{t\geq0}$ when $\alpha > -1$. Again, it is convenient to apply a similarity transformation to simplify $A_{\nu,\alpha}$. Defining the operator $W_{\nu,\alpha}$ and Banach space $Y_{\nu,\alpha}$ by

$$(W_{\nu,\alpha}f)(x) = x^{-\nu/(\alpha+1)} f(x^{1/(\alpha+1)}),$$

$$Y_{\nu,\alpha} = \{f : ||f||_{\nu,\alpha} = \int_0^\infty x^{(\nu-\alpha+1)/(\alpha+1)} |f(x)| \, dx < \infty\},$$

it can be shown that $W_{\nu,\alpha}$ is a homeomorphism from Y onto $Y_{\nu,\alpha}$. Moreover, $W_{\nu,\alpha}A_{\nu,\alpha}W_{\nu,\alpha}^{-1} = A_\gamma$ where

$$(A_\gamma f)(x) = -xf(x) + \frac{2}{\gamma}\int_x^\infty f(y)\,dy,$$

with $\gamma = 2(\alpha+1)/(\nu+2)$ and $D(A_\gamma) = \{f \in Y_{\nu,\alpha} : xf \in Y_{\nu,\alpha}\}$.

Lemma 1. *If $\alpha > -1$, then the operator*

$$(\overline{A}_\gamma, D(\overline{A}_\gamma)) = (W_{\nu,\alpha}\overline{A}_{\nu,\alpha}W_{\nu,\alpha}^{-1}, W_{\nu,\alpha}[D(\overline{A}_{\nu,\alpha}])$$

generates a stochastic semigroup $\{S_\gamma(t)\}_{t\geq 0} = \{W_{\nu,\alpha}S_{\nu,\alpha}(t)W_{\nu,\alpha}^{-1}\}_{t\geq 0}$ on $Y_{\nu,\alpha}$. Moreover,

$$S_\gamma(t)f = \lim_{n\to\infty}(I - P_{n^{\alpha+1}} + \exp(A_\gamma t)P_{n^{\alpha+1}})f, \quad f \in Y_{\nu,\alpha}, \qquad (4.8)$$

where $P_{n^{\alpha+1}}$ is defined by (4.7).

Proof. This follows from Theorem 3, Corollary 1 and formulae (4.5)–(4.7).

An explicit formula for $S_\gamma(t)$ can now be obtained using (4.8) and the power series definition of $\exp(A_\gamma t)$. First we require a formula for positive iterates of the operator A_γ.

Lemma 2. *For each $k = 1, 2, 3, \ldots$,*

$$(A_\gamma^k f)(x) = (-x)^k f(x)$$
$$+ \frac{2(-1)^{k-1}}{\gamma}\sum_{j=1}^k \binom{k}{j}(1 - \frac{2}{\gamma})_{j-1}\, x^{k-j}\int_x^\infty \frac{(y-x)^{j-1}}{(j-1)!}\, f(y)\,dy,$$

where $(c)_k = c(c+1)\ldots(c+k-1)$.

This assertion is proved by induction.

Lemma 3. *For each $t > 0$ and $f \in Y_{\nu,\alpha}$,*

$$(S_\gamma(t)f)(x) = e^{-xt}\left(f(x) + \frac{2t}{\gamma}\int_x^\infty {}_1F_1\left(1 - \frac{2}{\gamma}, 2; t(x-y)\right)f(y)\,dy\right),$$

where ${}_1F_1$ denotes the confluent hypergeometric function.

Proof. This can be deduced from (4.8) and Lemma 2.

Theorem 4. *For each $\alpha > -1$, $\nu \in (-2, 0]$, $t > 0$ and $f \in Y$,*

$$(S_{\nu,\alpha}(t)f)(x) = e^{-tx^{\alpha+1}}[f(x)$$
$$+ (\nu+2)t\int_x^\infty {}_1F_1\left(\frac{\alpha-\nu-1}{\alpha+1}, 2; t(x^{\alpha+1} - y^{\alpha+1})\right)\left(\frac{x}{y}\right)^\nu y^\alpha f(y)\,dy].$$

Moreover, the domain $D(A_{\nu,\alpha})$ is positively invariant under the semigroup $\{S_{\nu,\alpha}(t)\}_{t \geq 0}$.

Proof. The formula given for $S_{\nu,\alpha}(t)$ follows from Lemma 3 and the fact that $S_{\nu,\alpha}(t) = W_{\nu,\alpha}^{-1} S_\gamma(t) W_{\nu,\alpha}$. The invariance property can be established by a direct calculation.

The invariance of $D(A_{\nu,\alpha})$ leads immediately to the following more satisfactory version of Theorem 2.

Corollary 2. *If $\alpha > -1$ and $\nu \in (-2, 0]$, then the ACP (4.4) has the unique, non-negative mass-conserving solution $c(t) = S_{\nu,\alpha}(t) c_0$ for each non-negative c_0 in $D(A_{\nu,\alpha})$.*

References

1. E.D. McGrady and R.M. Ziff, "Shattering" transition in fragmentation, *Phys. Rev. Lett.* **58** (1987), 892–895.

2. K.-J. Engel and R. Nagel, *One-parameter semigroups for linear evolution equations*, Springer-Verlag, New York, 1999.

3. W. Lamb and A.C. McBride, On a continuous coagulation and fragmentation equation with a singular fragmentation kernel, in *Recent contributions to evolution equations*, Lect. Notes, Marcel Dekker (to appear).

4. D.J. McLaughlin, W. Lamb, and A.C. McBride, A semigroup approach to fragmentation models, *SIAM J. Math. Anal.* **28** (1997), 1158–1172.

5. J. Banasiak, On an extension of the Kato–Voigt perturbation theorem for substochastic semigroups and its application, *Taiwanese J. Math.* **5** (2001), 169–191.

5 Solution of a Sommerfeld Diffraction Problem with a Real Wave Number

Luis P. Castro

5.1 Introduction and Formulation of the Problem

We consider a problem of wave diffraction by a half-plane with general boundary and transmission conditions of first and second kind. The problem is taken in the framework of Bessel potential spaces and several Wiener–Hopf operators are introduced in order to translate the conditions initially stated. Similar problems can be found in the work of E. Meister and F.-O. Speck (see, e.g., [1]). In the present work, the main difference is the possibility to consider a real wave number. The class in study contains, as a particular case, the Rawlins' Problem [1] which was already considered by K. Rottbrand also in the limiting case of a wave number with a null imaginary part [2]. The study is carried out with the help of some factorization techniques, certain projectors and a representation due to Laplace-type integrals. As a consequence, the exact solution of the problem is obtained in a form that is still valid for the limiting case of a real wave number.

We will use the Bessel potential spaces $H^\sigma(\mathbb{R})$, with $\sigma \in \mathbb{R}$, formed by the tempered distributions φ such that $\|\varphi\| = \|\mathcal{F}^{-1}(1+\xi^2)^{\sigma/2} \cdot \mathcal{F}\varphi\|_{L^2(\mathbb{R})}$ is finite (where $\mathcal{F}$ represents the Fourier transformation). Moreover, we denote by $\widetilde{H}^\sigma(\mathbb{R}_+)$ the closed subspace of $H^\sigma(\mathbb{R})$ defined by the distributions with support contained in $\overline{\mathbb{R}_+}$ and $H^\sigma(\mathbb{R}_+)$ will denote the space of generalized functions on $\mathbb{R}_+$ which have extensions into $\mathbb{R}$ that belong to $H^\sigma(\mathbb{R})$. In particular, we shall denote by $L^2(\mathbb{R}_+)$ and $L^2_+(\mathbb{R})$, the spaces $H^0(\mathbb{R}_+)$ and $\widetilde{H}^0(\mathbb{R}_+)$, respectively. All those definitions can be extended to the multi-index case by taking the product topology.

From the mathematical point of view, the problem is how to find elements $u \in L^2(\mathbb{R}^2)$, with $u_{|\mathbb{R}^2_\pm} \in H^1(\mathbb{R}^2_\pm)$, so that

$$(\Delta + k^2)\, u \;=\; 0 \qquad \text{in} \qquad \mathbb{R}^2_\pm, \tag{5.1}$$

$$\begin{cases} a_0 u_0^+ + b_0 u_0^- = h_0 \\ a_1 u_1^+ + b_1 u_1^- = h_1 \end{cases} \qquad \text{on} \qquad \mathbb{R}_+, \tag{5.2}$$

$$\begin{cases} a_0' u_0^+ + b_0' u_0^- = 0 \\ a_1' u_1^+ + b_1' u_1^- = 0 \end{cases} \qquad \text{on} \qquad \mathbb{R}_-, \tag{5.3}$$

where $\mathbb{R}^2_\pm$ represents the upper/lower half-plane, $k \in \mathbb{C}$ (with $\operatorname{Re} ek > 0$ and $\operatorname{Im} mk > 0$, for start), $u_0^\pm = u_{|y=\pm 0}$, $u_1^\pm = (\partial u/\partial y)_{|y=\pm 0}$ and a_0, b_0, a_1,

b_1, a'_0, b'_0, a'_1, b'_1 are given complex numbers so that $a_0 b_1 + a_1 b_0$, $a'_0 b'_1 + a'_1 b'_0$, $-a_0 b'_0 + b_0 a'_0$, $-a_1 b'_1 + b_1 a'_1$, $a_1 b'_0 + b_1 a'_0 \neq 0$ and (for $l = 0, 1$) the elements $h_l \in H^{-l+1/2}(\mathbb{R}_+)$ are arbitrarily given.

Please note that we have just formulated a two-dimensional situation due to the assumption that the wave propagates perpendicular to the edge $x = y = 0$, $z \in \mathbb{R}$.

The problem to find an element u in the above conditions can be considered in view of a solution of the single equation $\mathcal{U}u = h$, described by the use of a linear operator $\mathcal{U} : D(\mathcal{U}) \to H^s(\mathbb{R}_+)$, $s = (1/2, -1/2)$, if we define $D(\mathcal{U})$ as the subspace of $H^1(\mathbb{R}_+^2) \times H^1(\mathbb{R}_-^2)$ whose functions fulfill the Helmholtz equation (5.1) and the homogeneous transmission conditions (5.3) whereas the action $\mathcal{U}u = [h_0, h_1]^T$ results from the nonhomogeneous conditions (5.2). In this sense, from an operator theoretical point of view [1, 3, 4, 5], we will say that operator $\mathcal{U}$ is *associated* to the problem and refer to the latter as the *Problem $\mathcal{U}$*.

5.2 Relations with Wiener–Hopf Operators

In the first instance, it is necessary to understand the structure of the operator $\mathcal{U}$. For that we will construct relations, in the form of explicit operator matrix identities, between the operator $\mathcal{U}$ and Wiener–Hopf operators.

Let $t(\xi) = (\xi^2 - k^2)^{1/2}$, $\xi \in \mathbb{R}$, denote the branch of the square root that tends to $+\infty$ as $\xi \to +\infty$ with branch cuts along $\pm k \pm i\nu$, $\nu \geq 0$. It is known [1] that a function $u \in L^2(\mathbb{R}^2)$, with $u_{|\mathbb{R}_\pm^2} \in H^1(\mathbb{R}_\pm^2)$, satisfies the Helmholtz equation (5.1) if and only if it is representable by

$$u(x,y) = \mathcal{F}_{\xi \mapsto x}^{-1} e^{-t(\xi)y} \mathcal{F}_{x \mapsto \xi} u_0^+(x) \chi_+(y) + \mathcal{F}_{\xi \mapsto x}^{-1} e^{t(\xi)y} \mathcal{F}_{x \mapsto \xi} u_0^-(x) \chi_-(y) \tag{5.4}$$

for $(x, y) \in \mathbb{R}^2$, where χ_+, χ_- denote the characteristic functions of the positive and negative half-line, respectively.

Taking into account the representation formula (5.4) and defining $Z = \{(\phi, \psi) \in [H^{1/2}(\mathbb{R})]^2 : a'_0 \phi + b'_0 \psi \in \widetilde{H}^{1/2}(\mathbb{R}_+), \mathcal{F}^{-1} t \cdot \mathcal{F}(-a'_1 \phi + b'_1 \psi) \in \widetilde{H}^{-1/2}(\mathbb{R}_+)\}$, we have that the trace operator $\mathcal{T}_0 : D(\mathcal{U}) \to Z$, given by $\mathcal{T}_0 u = u_0 := (u_0^+, u_0^-)^T$, is continuously invertible by the Poisson operator $\mathcal{K} : u_0 \mapsto u$ defined by (5.4). More than that, a direct computation yields

$$\mathcal{U} = \mathcal{W}_{\Psi, \mathbb{R}_+} \, \mathcal{C} \, \mathcal{T}_0, \tag{5.5}$$

where $\mathcal{C}$ is the convolution operator $\mathcal{C} = \mathcal{F}^{-1} \begin{bmatrix} a'_0 & b'_0 \\ -a'_1 t & b'_1 t \end{bmatrix} \cdot \mathcal{F} : Z \to \widetilde{H}^s(\mathbb{R}_+)$ and $\mathcal{W}_{\Psi, \mathbb{R}_+} = r_{\mathbb{R} \to \mathbb{R}_+} \mathcal{F}^{-1} \Psi \cdot \mathcal{F} : \widetilde{H}^s(\mathbb{R}_+) \to H^s(\mathbb{R}_+)$, with

$$\Psi = \frac{1}{a'_0 b'_1 + b'_0 a'_1} \begin{bmatrix} a_0 b'_1 + b_0 a'_1 & (-a_0 b'_0 + b_0 a'_0) t^{-1} \\ (-a_1 b'_1 + b_1 a'_1) t & a_1 b'_0 + b_1 a'_0 \end{bmatrix},$$

with $r_{\mathbb{R}\to\mathbb{R}_+} : H^s(\mathbb{R}) \to H^s(\mathbb{R}_+)$ being the restriction operator and $s = (1/2, -1/2)$.

Thus, (5.5) shows an operator equivalence relation (in the sense of [3, 5]) between $\mathcal{U}$ and the Wiener–Hopf operator $W_{\Psi,\mathbb{R}_+}$ because $C\mathcal{T}_0$ is continuously invertible by $\mathcal{K}C^{-1}$. We summarize in the next theorem what we have just demonstrated.

Theorem 1. *The operator $\mathcal{U}$ is equivalent [3, 5] to the Wiener–Hopf operator $W_{\Psi,\mathbb{R}_+}$, i.e. the two operators coincide up to homeomorphic linear transformations, cf. (5.5).*

We are now interested to relate the former operators $\mathcal{U}$ and $W_{\Psi,\mathbb{R}_+}$ with operators that act between L^2 spaces. For this purpose, let us denote $t_\pm(\xi) = \xi \pm k$, $\xi \in \mathbb{R}$, and introduce $\mathcal{E}_1 = r_{\mathbb{R}\to\mathbb{R}_+}\mathcal{F}^{-1} \text{diag}[t_-^{-1/2}, t_-^{1/2}] \cdot \mathcal{F}l_0 : [L^2(\mathbb{R}_+)]^2 \to H^s(\mathbb{R}_+)$, $\mathcal{E}_2 = l_0 r_{\mathbb{R}\to\mathbb{R}_+}\mathcal{F}^{-1} \text{diag}[t_+^{1/2}, t_+^{-1/2}] \cdot \mathcal{F} : \tilde{H}^s(\mathbb{R}_+) \to [L_+^2(\mathbb{R})]^2$, where $l_0 : [L^2(\mathbb{R}_+)]^2 \to [L_+^2(\mathbb{R})]^2$ is the zero extension operator. These new operators help us to define the following operator acting between L^2-spaces:

$$W_{\Theta,\mathbb{R}_+} = \mathcal{E}_1^{-1} W_{\Psi,\mathbb{R}_+} \mathcal{E}_2^{-1}. \tag{5.6}$$

Because $\mathcal{E}_1$ and $\mathcal{E}_2$ are bounded invertible operators, identity (5.6) shows an operator equivalence relation:

Theorem 2. *Let $\zeta := t_-/t_+$. The Wiener–Hopf operator $W_{\Psi,\mathbb{R}_+}$ is equivalent to $W_{\Theta,\mathbb{R}_+} = r_{\mathbb{R}\to\mathbb{R}_+}\mathcal{F}^{-1}\Theta \cdot \mathcal{F} : [L_+^2(\mathbb{R})]^2 \to [L^2(\mathbb{R}_+)]^2$, where*

$$\Theta = \frac{1}{a_0'b_1' + b_0'a_1'} \begin{bmatrix} (a_0b_1' + b_0a_1')\zeta^{1/2} & -a_0b_0' + b_0a_0' \\ -a_1b_1' + b_1a_1' & (a_1b_0' + b_1a_0')\zeta^{-1/2} \end{bmatrix}. \tag{5.7}$$

Taking into account (5.6) and (5.7), a straightforward computation leads us to the following result.

Theorem 3. *The Wiener–Hopf operator $W_{\Psi,\mathbb{R}_+}$ is equivalent to $W_{\Phi,\mathbb{R}_+} = r_{\mathbb{R}\to\mathbb{R}_+}\mathcal{F}^{-1}\Phi \cdot \mathcal{F} : [L_+^2(\mathbb{R})]^2 \to [L^2(\mathbb{R}_+)]^2$, where*

$$\Phi = \begin{bmatrix} 1 & \eta\zeta^{1/2} \\ \zeta^{-1/2} & 1 \end{bmatrix}, \qquad \eta = \frac{(a_0b_1' + b_0a_1')(a_1b_0' + b_1a_0')}{(-a_0b_0' + b_0a_0')(-a_1b_1' + b_1a_1')}.$$

Namely, one has $W_{\Phi,\mathbb{R}_+} = \mathcal{E}_3 W_{\Theta,\mathbb{R}_+} \mathcal{E}_4$, with

$$\mathcal{E}_3 = r_{\mathbb{R}\to\mathbb{R}_+}\mathcal{F}^{-1} \text{diag}\left[\frac{a_1b_0' + b_1a_0'}{-a_0b_0' + b_0a_0'}, 1\right] \cdot \mathcal{F}l_0,$$

$$\mathcal{E}_4 = l_0 r_{\mathbb{R}\to\mathbb{R}_+}\mathcal{F}^{-1}(a_0'b_1' + b_0'a_1') \begin{bmatrix} 0 & (-a_1b_1' + b_1a_1')^{-1} \\ (a_1b_0' + b_1a_0')^{-1} & 0 \end{bmatrix} \cdot \mathcal{F}.$$

From the above operator relations we derive that

$$\mathcal{U} = \mathcal{W}_{\Psi,\mathbb{R}_+} C T_0 = \mathcal{E}_1 \mathcal{W}_{\Theta,\mathbb{R}_+} \mathcal{E}_2 C T_0 = \mathcal{E}_1 \mathcal{E}_3^{-1} \mathcal{W}_{\Phi,\mathbb{R}_+} \mathcal{E}_4^{-1} \mathcal{E}_2 C T_0 \qquad (5.8)$$

and that $\mathcal{W}_{\Psi,\mathbb{R}_+} C \left[u_0^+, u_0^- \right]^T = [h_0, h_1]^T$ or, equivalently (assuming the existence of $\mathcal{W}_{\Phi,\mathbb{R}_+}^{-1}$),

$$\begin{bmatrix} u_0^+ \\ u_0^- \end{bmatrix} = C^{-1} \mathcal{E}_2^{-1} \mathcal{E}_4 \mathcal{W}_{\Phi,\mathbb{R}_+}^{-1} \mathcal{E}_3 \mathcal{E}_1^{-1} \begin{bmatrix} h_0 \\ h_1 \end{bmatrix}.$$

Corollary 1. *The operator $\mathcal{U}$ belongs to the same regularity classes [5, 6] as the Wiener–Hopf operators $\mathcal{W}_{\Psi,\mathbb{R}_+}$, $\mathcal{W}_{\Theta,\mathbb{R}_+}$ and $\mathcal{W}_{\Phi,\mathbb{R}_+}$ do. I.e., they are invertible, one-sided invertible, Fredholm, semi-Fredholm, one-sided regularizable, generalized invertible or normally solvable, only at the same time. Moreover, if we have the inverse (generalized, one-sided inverse) of one of these operators, then (from (5.8)) we will have the inverse (generalized, one-sided inverse) of the other operators.*

5.3 Factorization Procedures and Corresponding Solution of the Problem

Let us take into account the functions $\gamma_\pm = \frac{t_- \mp t_+}{\sqrt{\pm 2k}}$. By an application of the Daniele method [7], for $\eta \notin [1, +\infty[$, one finds a *canonical generalized factorization* of Φ relative to L^2 [8, 9, 10] given by $\Phi = \Phi_- \Phi_+$, with

$$\Phi_\pm = (1 - \eta)^{1/4} \begin{bmatrix} \cosh(\mu \log \gamma_\pm) & \eta^{1/2} \zeta^{1/2} \sinh(\mu \log \gamma_\pm) \\ \eta^{-1/2} \zeta^{-1/2} \sinh(\mu \log \gamma_\pm) & \cosh(\mu \log \gamma_\pm) \end{bmatrix}$$
$$(5.9)$$

and $\mu = \frac{1}{\pi i} \log \frac{1 + \sqrt{\eta}}{1 - \sqrt{\eta}}$. It is worthwhile to point out that the work of F.-O. Speck [9] about generalized factorizations also includes the above class of matrix functions. For further explanations to understand how to arrive at (5.9) we also refer to [10].

For shortness, we will use the notation $\Phi_\pm^{-1} = [\phi_{mn}^\pm]$, $m, n = 1, 2$.

Taking into account Theorem 1, Theorem 2, Theorem 3, formulas (5.9) and a decomposition of the representation formula (5.4) into an even-symmetric and an odd-symmetric part with respect to the second variable ($u^e = u_1^e + u_2^e$ and $u^o = u_1^o + u_2^o$, respectively), we obtain the following way to present the solution of the problem. Here we use the notation $\overline{h_0} = \mathcal{F} l^{even} h_0$ and $\overline{h_1} = \mathcal{F} l^{odd} h_1$ for the Fourier transforms of the corresponding even and odd extensions of the data.

Theorem 4. *For $y \neq 0$ and $\eta \notin [1, +\infty[$, the representation formula takes the form*

$$u(x, y) = \mathcal{F}_{\xi \mapsto x}^{-1} \left\{ [\chi_+(y), \chi_-(y)] e^{-t(\xi)|y|} \mathcal{F} C^{-1} \begin{bmatrix} a_0' u_0^+ + b_0' u_0^- \\ a_1' u_1^+ + b_1' u_1^- \end{bmatrix} \right\}.$$

This is equivalent to (respectively, in the upper and lower half-planes)

$$u_+(x,y)$$

$$= \chi_+(y)\mathcal{F}^{-1}_{\xi \mapsto x}\left\{e^{-yt(\xi)}\left(\eta_1\frac{\phi^+_{11}(\xi)}{t^{1/2}_-(\xi)} + \eta_2\frac{\phi^+_{21}(\xi)}{t^{1/2}_+(\xi)}\right)Fl_0r_{\mathbb{R}\to\mathbb{R}_+}\mathcal{F}^{-1}\theta(\xi)\right.$$

$$\left. + e^{-yt(\xi)}\left(\eta_1\frac{\phi^+_{12}(\xi)}{t^{1/2}_-(\xi)} + \eta_2\frac{\phi^+_{22}(\xi)}{t^{1/2}_+(\xi)}\right)Fl_0r_{\mathbb{R}\to\mathbb{R}_+}\mathcal{F}^{-1}\vartheta(\xi)\right\},$$

$$u_-(x,y)$$

$$= \chi_-(y)\mathcal{F}^{-1}_{\xi \mapsto x}\left\{e^{yt(\xi)}\left(\eta_3\frac{\phi^+_{11}(\xi)}{t^{1/2}_-(\xi)} + \eta_4\frac{\phi^+_{21}(\xi)}{t^{1/2}_+(\xi)}\right)Fl_0r_{\mathbb{R}\to\mathbb{R}_+}\mathcal{F}^{-1}\theta(\xi)\right.$$

$$\left. + e^{yt(\xi)}\left(\eta_3\frac{\phi^+_{12}(\xi)}{t^{1/2}_-(\xi)} + \eta_4\frac{\phi^+_{22}(\xi)}{t^{1/2}_+(\xi)}\right)Fl_0r_{\mathbb{R}\to\mathbb{R}_+}\mathcal{F}^{-1}\vartheta(\xi)\right\},$$

where $\theta = \eta_5\,\phi^-_{11}\,t^{1/2}_-\,\overline{h_0} + \phi^-_{12}t^{-1/2}_-\,\overline{h_1}$, $\vartheta = \eta_5\,\phi^-_{21}\,t^{1/2}_-\,\overline{h_0} + \phi^-_{22}t^{-1/2}_-\,\overline{h_1}$ *and*
$\eta_1 = -b'_0(a_1b'_0 + b_1a'_0)^{-1}$, $\eta_2 = b'_1(b_1a'_1 - a_1b'_1)^{-1}$, $\eta_3 = a'_0(a_1b'_0 + b_1a'_0)^{-1}$,
$\eta_4 = a'_1(-a_1b'_1 + b_1a'_1)^{-1}$, $\eta_5 = (a_1b'_0 + b_1a'_0)(-a_0b'_0 + b_0a'_0)^{-1}$.

Theorem 5. *Taking the Cauchy singular integral operator on the real line* $S_{\mathbb{R}}$, *let* $P_+ = \frac{1}{2}(I + S_{\mathbb{R}})$ *denote one of the generated Cauchy projectors [8]. For* $y \neq 0$ *and* $\eta \notin [1, +\infty[$ *(from the above result), the solution of Problem* $\mathcal{U}$ *is given by*

$$u(x,y) = \frac{1}{2\pi}\int_{-\infty}^{+\infty} P_+\left[e^{-i\xi x}\frac{e^{-|y|t}}{t^{1/2}_+}(\chi_+(y)\eta_2 + \chi_-(y)\eta_4)\,\phi^+_{21}\right]\theta\,d\xi$$

$$+ \frac{1}{2\pi}\int_{-\infty}^{+\infty} P_+\left[e^{-i\xi x}\frac{e^{-|y|t}}{t^{1/2}_+}(\chi_+(y)\eta_2 + \chi_-(y)\eta_4)\,\phi^+_{22}\right]\vartheta\,d\xi$$

$$+ \frac{1}{2\pi}\int_{-\infty}^{+\infty} P_+\left[e^{-i\xi x}\frac{e^{-|y|t}}{t^{1/2}_-}(\chi_+(y)\eta_1 + \chi_-(y)\eta_3)\,\phi^+_{11}\right]\theta\,d\xi$$

$$+ \frac{1}{2\pi}\int_{-\infty}^{+\infty} P_+\left[e^{-i\xi x}\frac{e^{-|y|t}}{t^{1/2}_-}(\chi_+(y)\eta_1 + \chi_-(y)\eta_3)\,\phi^+_{12}\right]\vartheta\,d\xi.$$

In the last result, an *inner product shifting* of the projector P_+ was used: $\langle f, P_+g\rangle := \langle(\xi^2 + 1)^{-\sigma/2}f, (\xi^2 + 1)^{\sigma/2}P_+g\rangle_{L^2(\mathbb{R})} = \langle P_+f, g\rangle$, for $f \in \mathcal{F}H^{-\sigma}$, $g \in \mathcal{F}H^{\sigma}$ and within $-1/2 < \sigma < 1/2$. This was possible due to the symmetry of the intermediate spaces [11] involved in the factorization and also due to the boundedness of P_+ for smoothness orders with $|\sigma| < 1/2$.

We reach our final goal because the above projections are still valid for a real wave number. To observe this we only need to interpret them in the form of Laplace-type transforms (cf. below). Namely, we are dealing with P_+-projections of elements of the form $\delta_{x,y,k}(\xi) = c_0e^{-i\xi x}\omega_{y,k}(\xi)$, where c_0 denotes a constant and ω is a function of negative order containing the elements $t_\pm^{1/2}$ (considering, once more, the branch cuts going from $\mp k$ to $\mp k \mp i\infty$). Thus, we have functions of the following type (respectively for $x < 0$ or $x \geq 0$)

$$P_+ \delta_{x,y,k}(\xi)$$

$$= \begin{cases} \frac{c_0 e^{-ikx}}{2\pi i} \int_0^\infty e^{\rho x} \left[\omega_{y,k}^+(k + i\rho) - \omega_{y,k}^-(k + i\rho) \right] \frac{d\rho}{\rho - i(k - \xi)}, \\ -\delta_{x,y,k}(\xi) - \frac{c_0 e^{ikx}}{2\pi i} \int_0^\infty e^{-\rho x} \left[\omega_{y,k}^+(-k - i\rho) - \omega_{y,k}^-(-k - i\rho) \right] \frac{d\rho}{\rho - i(k + \xi)}, \end{cases}$$

where the plus and minus signs on ω denote this function when taken on the right, or respectively, left bank of the branch cut. In this process, for $k = \operatorname{Re} ek > 0$, a radiation condition (like (37) in [12]) is also considered.

References

1. E. Meister and F.-O. Speck, Modern Wiener–Hopf methods in diffraction theory, in *Ordinary and partial differential equations*, Pitman Res. Notes Math. Ser. **216**, Longman Sci. Tech., Harlow, 1989, 130–171.

2. K. Rottbrand, Rawlins' problem for half-plane diffraction: Its generalized eigenfunctions with real wave numbers, *Math. Methods Appl. Sci.* **20** (1997), 989–1014.

3. L.P. Castro and F.-O. Speck, Relations between convolution type operators on intervals and on the half-line, *Integral Equations Oper. Theory* **37** (2000), 169–207.

4. L.P. Castro, A relation between convolution type operators on intervals in Sobolev spaces, *Appl. Anal.* **74** (2000), 393–412.

5. L.P. Castro, Regularity of convolution type operators with PC symbols in Bessel potential spaces over two finite intervals, *Math. Nach.* **161** (2003) (in press).

6. L.P. Castro and F.-O. Speck, Regularity properties and generalized inverses of delta-related operators, *Z. Anal. Anwend.* **17** (1998), 577–598.

7. V.G. Daniele, On the solution of two coupled Wiener–Hopf equations, *SIAM J. Appl. Math.* **44** (1984), 667–680.

8. S.G. Mikhlin and S. Prössdorf, *Singular integral operators*, Springer-Verlag, Berlin, 1986.

9. F.-O. Speck, Sommerfeld diffraction problems with first and second kind boundary conditions, *SIAM J. Math. Anal.* **20** (1989), 396–407.

10. A.B. Lebre and A.F. dos Santos, Generalized factorization for a class of non-rational 2×2 matrix functions, *Integral Equations Oper. Theory* **13** (1990), 671–700.

11. L.P. Castro and F.-O. Speck, On the characterization of the intermediate space in generalized factorizations, *Math. Nach.* **176** (1995), 39–54.

12. S.N. Chandler-Wilde, The impedance boundary value problem for the Helmholtz equation in a half-plane, *Math. Methods Appl. Sci.* **20** (1997), 813–840.

6 A Mixed BEM Applied to Scattering of Thermal Waves in Composite Materials

Ricardo Celorrio, Maria-Luisa Rapún,
and Francisco-Javier Sayas

6.1 Statement of the Problem

Let $\Omega_1, \ldots, \Omega_d$ denote a finite number of simply connected domains strictly contained in $\mathbb{R}^2_- := \{(x_1, x_2) \,|\, x_2 < 0\}$, with nonintersecting closures and such that $\overline{\Omega}_k \cap \Pi = \emptyset$ for all k, being $\Pi := \{(x_1, 0) \,|\, x_1 \in \mathbb{R}\}$. The boundaries $\Gamma_k := \partial\Omega_k$ are assumed to be parameterizable $\mathcal{C}^2$–curves. Normals are directed towards the exterior of Ω_k for each k and the normal derivative on Π is directed towards the exterior of Ω (see Fig. 1).

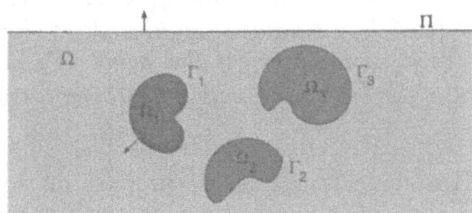

Fig. 1. Geometry of the problem.

We are looking for a solution of the problem

$$\Delta u + \lambda^2 u = 0 \quad \text{in } \Omega := \mathbb{R}^2_- \setminus \left(\cup_{k=1}^d \overline{\Omega}_k \right), \tag{6.1}$$

$$\Delta u + \mu_k^2 u = 0 \quad \text{in } \Omega_k, \quad k = 1, \ldots, d, \tag{6.2}$$

where $\lambda, \mu_k \in (1 + \imath)\mathbb{R}^+$ (i.e., λ^2, μ_k^2 are purely imaginary numbers). The transmission conditions on the inner boundaries are

$$u|_{\Gamma_k}^{\text{int}} - u|_{\Gamma_k}^{\text{ext}} = g_k^0, \quad k = 1, \ldots, d, \tag{6.3}$$

$$\nu_k \, \partial_n u|_{\Gamma_k}^{\text{int}} - \nu \, \partial_n u|_{\Gamma_k}^{\text{ext}} = g_k^1, \quad k = 1, \ldots, d, \tag{6.4}$$

This work has been partly supported by MCYT/FEDER Proj. BFM2001–2521.

where

$$\nu, \nu_k > 0, \quad g_k^0 = u_{\text{inc}}|_{\Gamma_k}, \quad g_k^1 = \nu \, \partial_n u_{\text{inc}}|_{\Gamma_k}, \quad u_{\text{inc}}(x) := \exp(\imath \lambda x_2),$$

u_{inc} being the so-called incident wave. We also demand the boundary condition

$$\partial_n u|_\Pi = 0, \tag{6.5}$$

and the Sommerfeld radiation condition at infinity ([1] Chapter 7),

$$\lim_{r \to \infty} r^{1/2} \left(\partial_r u - \imath \lambda u \right) = 0. \tag{6.6}$$

This condition has to be satisfied uniformly in all available directions (∂_r denotes the radial derivative).

For $r \in \mathbb{R}$ we consider the space

$$H_{loc}^r(\overline{\Omega}) := \{ v \mid v \phi \in H^r(\Omega), \forall \phi \in \mathcal{D}(\overline{\Omega}) \}.$$

Theorem 1. *Given $g_k^0 \in H^{1/2}(\Gamma_k)$ and $g_k^1 \in H^{-1/2}(\Gamma_k)$, the transmission problem (6.1)–(6.6) has a unique solution $u|_\Omega \in H_{loc}^1(\overline{\Omega})$ and $u|_{\Omega_k} \in H^1(\Omega_k)$, $k = 1, \ldots, d$. Moreover, if the curves Γ_k are smooth, $g_k^0 \in H^{1/2+s}(\Gamma_k)$ and $g_k^1 \in H^{-1/2+s}(\Gamma_k)$ with $s > 0$, then $u|_\Omega \in H_{loc}^{1+s}(\overline{\Omega})$ and $u|_{\Omega_k} \in H^{1+s}(\Omega_k)$.*

Proof. Taking Π as reflection axis, we propose an equivalent problem with reflected data that has a unique solution and the adequate regularity properties ([1] Chapter 7).

Physical motivation. Photothermal techniques are suitable means of inspecting composite materials. One of these techniques endeavors to determine internal properties of the material by observing its behavior when it is heated on one of its surfaces by a defocused laser beam modulated at a given frequency ω. After a sufficiently long time, the temperature distribution becomes time-harmonic and can be expressed as

$$T(\mathbf{x}, t) = \text{Re}(v(\mathbf{x}) \exp(-\imath \omega t)).$$

Taking as unknown

$$u := v - u_{\text{inc}} \quad \text{in } \Omega, \quad u := v \quad \text{in each } \Omega_k,$$

u is a solution of (6.1)–(6.6). The boundary condition (6.5) models an adiabatic situation whereas the transmission conditions (6.3) and (6.4) model the continuity of the temperature and heat flux (see [2]).

6.2 Boundary Integral Formulation

6.2.1 Potentials, Operators, and Some Notation

Let $\mathbf{x}_k : \mathbb{R} \to \Gamma_k$ be regular 1-periodic parameterizations of the boundaries of the obstacles.

The function $\Phi^\rho(\mathbf{x}, \mathbf{y}) := \frac{i}{4} H_0^{(1)}(\rho|\mathbf{x}-\mathbf{y}|)$, where $H_0^{(1)}$ is the Hankel function of first kind and order zero, the fundamental solution of the Helmholtz equation. To a given density $\eta : [0, 1] \to \mathbb{C}$ we associate the single-layer potentials

$$S_k^\rho \eta := \int_0^1 \Phi^\rho(\,\cdot\,, \mathbf{x}_k(t))\eta(t)dt : \mathbb{R}^2 \to \mathbb{C},$$

$$\widetilde{S}_k^\rho \eta := \int_0^1 \Big(\Phi^\rho(\,\cdot\,, \mathbf{x}_k(t)) + \Phi^\rho(\,\cdot\,, \widetilde{\mathbf{x}}_k(t)) \Big)\eta(t)dt : \mathbb{R}^2 \to \mathbb{C},$$

where $\widetilde{\mathbf{x}} := (x_1, -x_2)$ the reflected point of $\mathbf{x} = (x_1, x_2)$. Given densities $\varphi_k : \Gamma_k \to \mathbb{C}$, which we group in the vector $\varphi = (\varphi_1, \ldots, \varphi_d)$, we define the potential $\widetilde{S}^\rho \varphi := \sum_{k=1}^d \widetilde{S}_k^\rho \varphi_k$. We also consider the double-layer potentials

$$\mathcal{D}_k^\rho \eta := \int_0^1 \partial_{n_k(t)} \Phi^\rho(\,\cdot\,, \mathbf{x}_k(t))\eta(t)\, dt : \mathbb{R}^2 \to \mathbb{C},$$

($\mathbf{n}_k(t)$ is the normal vector at $\mathbf{x}_k(t)$), and the following boundary integral operators:

$$V_{ij}^\rho \eta := \int_0^1 \Phi^\rho(\mathbf{x}_i(\cdot), \mathbf{x}_j(t))\eta(t)dt,$$

$$\widetilde{V}_{ij}^\rho \eta := \int_0^1 \Big(\Phi^\rho(\mathbf{x}_i(\cdot), \mathbf{x}_j(t)) + \Phi^\rho(\mathbf{x}_i(\cdot), \widetilde{\mathbf{x}}_j(t)) \Big)\eta(t)dt,$$

$$K_{ij}^\rho \eta := \int_0^1 |\mathbf{x}_j'(t)|\partial_{n_j(t)} \Phi^\rho(\mathbf{x}_i(\cdot), \mathbf{x}_j(t))\eta(t)dt, \qquad (6.7)$$

$$\widetilde{J}_{ij}^\rho \eta := \int_0^1 |\mathbf{x}_i'(\cdot)|\partial_{n_i(\cdot)} \Big(\Phi^\rho(\mathbf{x}_i(\cdot), \mathbf{x}_j(t)) + \Phi^\rho(\mathbf{x}_i(\cdot), \widetilde{\mathbf{x}}_j(t)) \Big)\eta(t)dt.$$

We will be working in the frame of the 1-periodic Sobolev spaces H^s, for $s \in \mathbb{R}$ (see [3], Chapter 5, for the definition and properties of these spaces). We recall that the space H^0 can be identified with $L^2(0, 1)$ and its inner product admits an extension to represent the reciprocal duality of the spaces H^s and H^{-s} for all $s \in \mathbb{R}$. We also consider the product space

$$\mathbf{H}^s := \underbrace{H^s \times \cdots \times H^s}_{d \text{ times}}.$$

We rename the 1-periodic data functions

$$g_k^0 := g_k^0(\mathbf{x}_k(\,\cdot\,)) : \mathbb{R} \to \mathbb{C}, \qquad g_k^1 := |\mathbf{x}_k'(\,\cdot\,)|g_k^1(\mathbf{x}_k(\,\cdot\,)) : \mathbb{R} \to \mathbb{C},$$

and set $g^0 := (g_1^0, \ldots, g_d^0)$, $g^1 := (g_1^1, \ldots, g_d^1)$.

6.2.2 Boundary Integral Equations

We propose a mixed formulation using an exterior single layer potential, combined with a direct representation formula for the interior problems. As unknowns we will consider a set of densities for the exterior problems $\psi \in \mathbf{H}^{-1/2}$ and the Cauchy data for the interior problems:

$$\zeta_k := u|_{\Gamma_k}^{\text{int}} \circ \mathbf{x}_k, \quad \xi_k := |\mathbf{x}_k'| \partial_n u|_{\Gamma_k}^{\text{int}} \circ \mathbf{x}_k, \quad k = 1, \dots, d,$$

$$u = \begin{vmatrix} \widetilde{\mathcal{S}}^\lambda \psi & \text{in } \Omega, \\ -\mathcal{D}_k^{\mu_k} \zeta_k + \mathcal{S}_k^{\mu_k} \xi_k & \text{in } \Omega_k, \quad k = 1 \dots, d. \end{vmatrix}$$

By definition, u satisfies (6.1), (6.2), and (6.6), and since $\widetilde{\mathcal{S}}^\lambda \psi(\widetilde{\mathbf{x}}) = \widetilde{\mathcal{S}}^\lambda \psi(\mathbf{x})$ for all $\mathbf{x} \in \mathbb{R}^2$ it follows readily that $\partial_n u|_\Pi = 0$, i.e., (6.5) holds.

Now, we collect the operators defined in (6.7) using the matrices

$$\widetilde{V}^\lambda := (\widetilde{V}_{ij}^\lambda), \quad \widetilde{J}^\lambda := (\widetilde{J}_{ij}^\lambda), \quad V^\mu := \text{diag}(V_{ii}^{\mu_i}), \quad K^\mu := \text{diag}(K_{ii}^{\mu_i}),$$

and introduce the operator $N := \text{diag}(\nu_i I)$ to express (6.3) and (6.4) by

$$\zeta - \widetilde{V}^\lambda \psi = g^0, \quad N\zeta + \nu(\tfrac{1}{2}I - \widetilde{J}^\lambda) = g^1. \tag{6.8}$$

These are consequences of the well-known jump relations of the single layer potential (see [4]). Finally, from (6.8) and the integral identity satisfied by the Cauchy data of the interior problem [4]:

$$(\tfrac{1}{2}I + K^\mu)\zeta^- - V^\mu \xi^- = 0,$$

we obtain the following system of boundary integral equations:

$$\mathcal{H} \begin{bmatrix} \psi \\ \xi \\ \zeta \end{bmatrix} := \begin{bmatrix} -\widetilde{V}^\lambda & 0 & I \\ 0 & -V^\mu & \tfrac{1}{2}I + K^\mu \\ \nu(\tfrac{1}{2}I - \widetilde{J}^\lambda) & N & 0 \end{bmatrix} \begin{bmatrix} \psi \\ \xi \\ \zeta \end{bmatrix} = \begin{bmatrix} g^0 \\ 0 \\ g^1 \end{bmatrix}. \tag{6.9}$$

Theorem 2. *[5] For all $s \in \mathbb{R}$, the operator $\mathcal{H} : \mathbf{H}^s \times \mathbf{H}^s \times \mathbf{H}^{s+1} \to \mathbf{H}^{s+1} \times \mathbf{H}^{s+1} \times \mathbf{H}^s$ is a bounded isomorphism. Moreover, there exists a bounded isomorphism $V_0 = \text{diag}(\Lambda)$ with $\Lambda : H^s \to H^{s+1}$ independent of λ and $\{\lambda_k\}_{k=1}^d$ such that*

$$\mathcal{H} - \begin{bmatrix} -V_0 & 0 & I \\ 0 & -V_0 & \tfrac{1}{2}I \\ \tfrac{\nu}{2}I & N & 0 \end{bmatrix} \quad \text{is compact.}$$

6.3 A Galerkin Method

We construct a uniform mesh in $[0,1]$ with nodes $x_i := ih$, $i = 0,\ldots,N$ (with $h = 1/N$), and take a space of piecewise constant functions

$$S_h^0 := \{u_h \in H^0 \mid u_h|_{(x_i-h/2, x_i+h/2)} \in \mathbb{P}_0, \ \forall i\},$$

and a space of continuous piecewise linear functions

$$S_h^1 := \{u_h \in C^0 \mid u_h|_{(x_i, x_{i+1})} \in \mathbb{P}_1, \ \forall i\},$$

where $\mathbb{P}_m$ is the space of polynomials of degree less than or equal to m. Now, we introduce the product spaces $\mathbf{S}_h^0 := S_h^0 \times \cdots \times S_h^0$ and $\mathbf{S}_h^1 := S_h^1 \times \cdots \times S_h^1$, and take $\mathbf{S}_h^0 \times \mathbf{S}_h^0 \times \mathbf{S}_h^1$ as discrete space for a Galerkin scheme to approximate (6.9), i.e.,

$$
\left|
\begin{aligned}
&\text{find } \psi_h, \ \xi_h \in \mathbf{S}_h^0, \ \zeta_h \in \mathbf{S}_h^1, \ \text{such that} \\[4pt]
&(-\widetilde{V}^\lambda \psi_h, r_h) + (\zeta_h, r_h) && = (g^0, r_h) && \forall r_h \in \mathbf{S}_h^0, \\[4pt]
&(-V^\mu \xi_h, r_h) + ((\tfrac{1}{2}I + K^\mu)\zeta_h, r_h) && = 0 && \forall r_h \in \mathbf{S}_h^0, \\[4pt]
&(\nu(\tfrac{1}{2}I - \widetilde{J}^\lambda)\psi_h, t_h) + (N\xi_h, t_h) && = (g^1, t_h) && \forall t_h \in \mathbf{S}_h^1.
\end{aligned}
\right.
$$

$$(6.10)$$

As is commonly used in the literature, we say that a Petrov–Galerkin method is stable in a certain norm when the operator mapping the exact solution to the numerical one is bounded uniformly in the discretization parameter (see [6]).

Lemma 3. *The Petrov–Galerkin method $\{S_h^0; S_h^1\}$ is stable for $I : H^t \to H^r$ for all t,r such that $-3/2 < r \le t < 1/2$.*

Sketch of the proof. The result is first proven with $t = r = 0$. After that, by Céa's lemma in H^0, some inverse inequalities and properties of the orthogonal projection operators onto S_h^0 and S_h^1 (see [7] Chapter 2) we achieve the result (see [5] for the complete proof).

Lemma 4. *For all $s \in (-3/2, 1/2)$, the method $\{S_h^0; S_h^0\}$ is stable for the operator $\Lambda : H^s \to H^{s+1}$ introduced in Theorem 2.*

Proof. It is similar to the previous one taking into account that the operator Λ is elliptic in $H^{-1/2}$ (see [5]) and consequently the Galerkin method $\{S_h^0; S_h^0\}$ is $H^{-1/2}$-stable.

Theorem 5. *For all $s \in (-3/2, 1/2)$ the scheme (6.10) is convergent in $\mathbf{H}^s \times \mathbf{H}^s \times \mathbf{H}^{s+1}$. Moreover, for all r,t such that $-2 \le r \le t \le 1$, $r < 1/2$ and $-3/2 < t$, there holds*

$$\|\psi - \psi_h\|_r + \|\xi - \xi_h\|_r + \|\zeta - \zeta_h\|_{r+1} \le Ch^{t-r} (\|\psi\|_t + \|\xi\|_t + \|\zeta\|_{t+1}).$$

$$(6.11)$$

Sketch of the proof. As S_h^0 and S_h^1 satisfy the approximation property in H^s and H^{s+1} for $s \in (-3/2, 1/2)$ (see [7] Chapter 2), convergence is equivalent to the stability of the method for the principal part of $\mathcal{H}$ (see [6]). Moreover, the principal part of $\mathcal{H}$ can be decoupled as a family of operators

$$\begin{bmatrix} -\Lambda & 0 & I \\ 0 & -\Lambda & \frac{1}{2}I \\ \frac{\nu}{2}I & \nu_k I & 0 \end{bmatrix}$$

and we can use Lemmas 3 and 4 together with the standard theory of approximation of mixed problems (see [8]), with very slight adaptations, to prove stability in the desired norms.

The bound (6.11) can be proven by standard Aubin–Nitsche duality arguments. The complete proof can be seen in [5].

References

1. G. Chen and J. Zhou, *Boundary element methods*, Computational Mathematics and Applications, Academic Press, Ltd., London, 1992.

2. A. Salazar, A. Sánchez–Lavega and J.M. Terrón, Multiple scattering effects of thermal waves by two subsurface cylinders, *J. Appl. Phys*, **87** (2000), 2600–2607.

3. J. Saranen and G. Vainikko, *Periodic Integral and Pseudodifferential Equations with Numerical Approximation*, Springer-Verlag, Berlin, 2002.

4. M. Costabel and E. Stephan, A direct boundary integral equation method for transmission problems, *J. Math. Anal. Appl.* **106** (1985), 367–413.

5. R. Celorrio, M.-L. Rapún and F.-J. Sayas, Boundary integral formulation and solution of scattering of thermal waves (in preparation).

6. R. Kress, *Linear integral equations*. Second edition, Springer-Verlag, New York, 1999.

7. S. Prossdorf and B. Silbermann, *Numerical analysis for integral and related operator equations*. Akademie Verlag, Berlin, 1991.

8. F. Brezzi and M. Fortin, *Mixed and hybrid finite element methods*, Springer–Verlag, Berlin, 1991.

7 Stellar Atmosphere Modeling

Loïc Chevallier

7.1 Introduction

In this note, the formalism used to compute a simple stellar atmosphere model is described. The definition of a stellar atmosphere is strongly connected with the radiation field crossing the boundary layers of a star: the atmosphere is composed of those layers where photons interact with matter for the last time before leaving the star. The goal of a model is to calculate the radiation field at all frequencies, depths and directions consistent with the properties of matter (temperature, number densities, pressure, *etc.*) at all depths.

7.2 General Description

7.2.1 General Assumptions

The atmosphere is modelled as a finite slab of gas, with spatial coordinate $z \in [0, z^*]$ and angular coordinate $\mu = \cos\theta \in [-1, 1]$. We suppose azimuthal symmetry (Fig. 1). The atmosphere is static, in a steady state, in hydrostatic and radiative equilibrium.

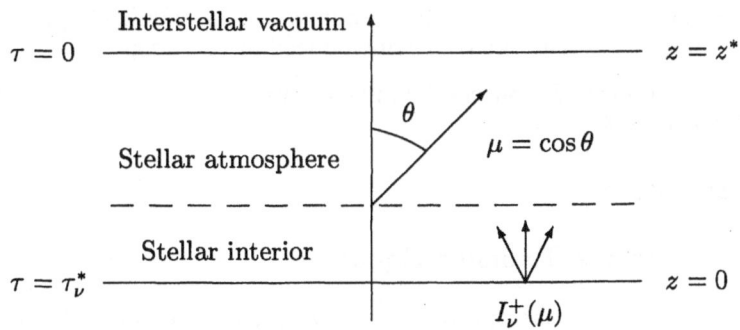

Fig. 1. Schematic representation of a plane-parallel atmosphere.

The author is indebted to B. Rutily for fruitful discussions on this topic.

7.2.2 Chemical Composition

A stellar atmosphere is composed with many kinds of particles, namely atoms, ions, free electrons, possibly molecules or even dust grains. To simplify the description of the model, we suppose that the atmosphere is composed of a single chemical species (*e.g.*, hydrogen, helium) whose atoms can be ionized several times when losing their electrons. Distinct energy states of this atom and associated ions are denoted by the subscript $i \in [\![1, N]\!]$, where $i = N$ refers to completely ionized atoms. The average number of particles in state i per unit volume (number density) is denoted by n_i.

7.2.3 Unknown Quantities

The basic unknown quantities of this model are the specific intensity of the radiation field $I(z, \mu, \nu)$, the complete set of number densities $n_i(z)$ for $i \in [\![1, N]\!]$, the number density of free electrons $n_e(z)$ and the temperature $T(z)$. The variables are the depth $z \in [0, z^*]$, the cosine of the inclination angle from the outer normal to the atmosphere $\mu \in [-1, 1]$, and the frequency $\nu \in]0, +\infty[$.

7.2.4 Input Parameters

Input parameters of a model are:

- the surface gravity $g := GM/R^2$, where M and R are the mass and the radius of the star respectively, and G is the gravitational constant,

- the effective temperature T_{eff}, as defined by $L = 4\pi R^2 \sigma T_{\text{eff}}^4$, where L is the luminosity of the star and σ is the Stefan–Boltzmann constant,

- the chemical composition (abundances of chemical species with respect to hydrogen), not introduced in this single species model.

The gas pressure P_0 on the boundary layer $z = z^*$ is also required in models with $P_0 \neq 0$.

7.3 Equations

7.3.1 Radiative Transfer Equation

This equation yields the specific intensity $I(z, \mu, \nu)$ of the radiation field for given $n_i(z)$, $n_e(z)$ and $T(z)$. It describes the distribution of photons in space (z), direction (μ) and frequency (ν) as a result of their interaction with matter. It is usually written in terms of the optical depth τ rather than the geometrical depth z. For a given frequency ν, the relation between z and τ is given by a bijection τ_ν from $[0, z^*]$ to $[0, \tau_\nu^*]$, where $\tau_\nu^* := \tau_\nu(0) > 0$ (see [1] for details). Then $I_\nu(\tau, \mu) := I(\tau_\nu^{-1}(\tau), \mu, \nu)$.

The radiative transfer equation is, for all $\nu \in]0, +\infty[$, $\tau \in]0, \tau_\nu^*[$ and $\mu \in [-1, 1]$,

$$\mu \frac{\partial I_\nu}{\partial \tau}(\tau, \mu) = I_\nu(\tau, \mu) - \tfrac{1}{2} \varpi_\nu(\tau) \int_{-1}^{1} I_\nu(\tau, \mu') \, d\mu' - S_\nu^*(\tau),$$

with boundary conditions $\begin{cases} I_\nu(0, \mu) = 0 & \mu < 0 \\ I_\nu(\tau_\nu^*, \mu) = I_\nu^+(\mu) & \mu > 0 \end{cases}$.

$S_\nu^*(\tau)$ is the primary source function and $\varpi_\nu(\tau) \in [0, 1]$ is the albedo at depth τ.[1] It is usual to introduce, instead of $\varpi_\nu(\tau)$, the destruction probability of photons $\varepsilon_\nu(\tau) = 1 - \varpi_\nu(\tau)$. The primary source function and the albedo depend on $n_i(z)$, $n_e(z)$ and $T(z)$. We have supposed isotropic and monochromatic (*i.e.*, with no change of frequency) scattering to simplify the presentation. Once the radiative transfer equation is solved for the variable τ, its solution is written in terms of the z-variable via the bijection τ_ν^{-1}.

7.3.2 Chemical Structure Equations

This system of $N + 1$ equations yields all number densities $n_i(z)$ for $i \in [\![1, N]\!]$, together with the electron number density $n_e(z)$, for given $I(z, \mu, \nu)$ and $T(z)$. We introduce the total number density of particles

$$n(z) := \sum_{i \in [\![1, N]\!]} n_i(z) + n_e(z),$$

the volumic mass

$$\rho(z) := \sum_{i \in [\![1, N]\!]} m_i n_i(z) + m_e n_e(z),$$

and the pressure $P(z) := n(z) k T(z)$, assuming that the atmosphere is a perfect gas. m_i and m_e are particles i and electron mass respectively, and k is the Boltzmann constant.

The chemical structure equations are, for all $z \in [0, z^*]$,

$$\begin{cases} n_i(z) \displaystyle\sum_{j \in [\![1, N]\!] \setminus \{i\}} P_{i,j}(z) = \displaystyle\sum_{j \in [\![1, N]\!] \setminus \{i\}} n_j(z) P_{j,i}(z) & (i \in [\![1, N-1]\!]), \\[2mm] \displaystyle\sum_{i \in [\![1, N]\!]} q_i \, n_i(z) = n_e(z), \\[2mm] \dfrac{dP}{dz}(z) = -g\rho(z), \text{ with boundary condition } P(z^*) = P_0. \end{cases}$$

The first $N - 1$ equations are the statistical equilibrium equations. They describe the change of state of particles i, due to their radiative and collisional interactions. $P_{i,j}(z)$ is the rate of change from state i to state j. It depends on $I(z, \mu, \nu)$, $n_e(z)$ and $T(z)$. The equation for $i = N$ is unnecessary since it can be deduced from the first $N - 1$ equations. The second equation is the charge conservation equation. It means that the atmosphere is electrically neutral (q_i is the electrical charge of particles i when expressed in units of electron charge). The last equation is the hydrostatic equilibrium equation. It means that the atmosphere is static, since the pressure forces (left-hand side) are equal to the gravity forces (right-hand side). We have neglected radiation and turbulent pressure in this equation.

7.3.3 Radiative Equilibrium Equation

This equation yields the temperature $T(z)$ for given $I(z, \mu, \nu)$, $n_i(z)$ and $n_e(z)$. For all $z \in [0, z^*]$,

$$F_r(z) := 2\pi \int_{-1}^{1} \int_{0}^{+\infty} I(z, \mu, \nu)\, \mu\, d\mu d\nu = 4\pi\sigma T_{\text{eff}}^4,$$

where $F_r(z)$ is the integrated (on frequency) radiative flux at z.

This equation involves the coefficients of the radiative transfer equation at z, which depend on $T(z')$ at every point z' of the atmosphere. Making this dependence explicit in the radiative equilibrium equation leads to a nonlinear integral equation for $T(z)$. This equation is very difficult to solve because of the highly nonlinear dependence of $I(z, \mu, \nu)$ on the temperature.

In a stellar atmosphere in radiative equilibrium, energy is carried out by radiation only (there is no convection) and the total flux is conserved within the atmosphere. It is thus equal to the flux $4\pi\sigma T_{\text{eff}}^4$ leaving the star.

7.4 Starting a Model

As seen in Sec. 3, all quantities are coupled and the system of equations they satisfy has to be solved iteratively. To start the iterative process, a first idealized model is computed by assuming that the atmosphere is in local thermodynamic equilibrium.

7.5 Validation

A model is declared valid if the calculated emerging monochromatic flux $F_r(0, \nu) := 2\pi \int_{-1}^{1} I(z, \mu, \nu)\mu\, d\mu$ fits the observed one for all $\nu > 0$. The model then defines the boundary conditions of a stellar interior model.

References

1. B. Rutily, Multiple scattering theory and integral equations, this volume, 211–232.

8 Time-Dependent Bending of a Plate with Mixed Boundary Conditions

Igor Chudinovich and Christian Constanda

8.1 The Mathematical Model

Consider a homogeneous and isotropic elastic plate of thickness $h_0 = \text{const} > 0$, which occupies a region $\tilde{S} \times [-h_0/2, h_0/2]$ in $\mathbb{R}^3$, where S is a domain in $\mathbb{R}^2$ with a simple, closed boundary ∂S. In the transverse shear deformation model proposed in [1] it is assumed that the displacement vector at (x, x_3), $x = (x_1, x_2) \in \mathbb{R}^2$, at time $t \geq 0$, is of the form

$$\left(x_3 u_1(x,t), x_3 u_2(x,t), u_3(x,t) \right)^{\mathrm{T}},$$

where the superscript T denotes matrix transposition. Then the vector $u = (u_1, u_2, u_3)^{\mathrm{T}}$ satisfies the equation of motion

$$B(\partial_t^2 u)(x,t) + (Au)(x,t) = q(x,t), \quad (x,t) \in G = S \times (0, \infty);$$

here $B = \mathrm{diag}\{\rho h^2, \rho h^2, \rho\}$, $h^2 = h_0^2/12$, ρ is the plate density, $\partial_t = \partial/\partial t$,

$$A = \begin{pmatrix} -h^2\mu\Delta - h^2(\lambda + \mu)\partial_1^2 + \mu & -h^2(\lambda + \mu)\partial_1\partial_2 & \mu\partial_1 \\ -h^2(\lambda + \mu)\partial_1\partial_2 & -h^2\mu\Delta - h^2(\lambda + \mu)\partial_2^2 + \mu & \mu\partial_2 \\ -\mu\partial_1 & -\mu\partial_2 & -\mu\Delta \end{pmatrix},$$

$\partial_\alpha = \partial/\partial x_\alpha$, $\alpha = 1, 2$, λ and μ are the Lamé constants satisfying $\lambda + \mu > 0$ and $\mu > 0$, and q is a combination of the forces and moments acting on the plate and its faces $x_3 = \pm h_0/2$.

In what follows we work with three-component distributions; however, for simplicity, we use the same symbols for their spaces and norms as in the scalar case.

We denote by $H_{m,p}(\mathbb{R}^2)$, $m \in \mathbb{R}$, $p \in \mathbb{C}$, the space that coincides with $H_m(\mathbb{R}^2)$ as a set but is equipped with the norm

$$\|u\|_{m,p} = \left\{ \int\limits_{\mathbb{R}^2} (1 + |p|^2 + |\xi|^2)^m |\tilde{u}(\xi)|^2 \, d\xi \right\}^{1/2},$$

where $\tilde{u}$ is the distributional Fourier transform of $u \in \mathcal{S}'(\mathbb{R}^2)$. Next, $\mathring{H}_{m,p}(S)$ is the subspace of $H_{m,p}(\mathbb{R}^2)$ consisting of all $u \in H_{m,p}(\mathbb{R}^2)$ with supp $u \subset \bar{S}$, and $H_{m,p}(S)$ is the space of the restrictions to S of all $v \in H_{m,p}(\mathbb{R}^2)$. The norm of $u \in H_{m,p}(S)$ is defined by

$$\|u\|_{m,p;S} = \inf_{v \in H_{m,p}(\mathbb{R}^2):\, v|_S=u} \|v\|_{m,p}.$$

Also, $H_{-m,p}(\mathbb{R}^2)$ is the dual of $H_{m,p}(\mathbb{R}^2)$ with respect to the duality generated by the inner product $(\cdot,\cdot)_0$ in $L^2(\mathbb{R}^2)$; the dual of $\mathring{H}_{m,p}(S)$ is $H_{-m,p}(S)$. Let γ be the trace operator that maps $H_{1,p}(S)$ continuously to the space $H_{1/2,p}(\partial S)$, which coincides as a set with $H_{1/2}(\partial S)$ but is equipped with the norm

$$\|f\|_{1/2,p;\partial S} = \inf_{u \in H_{1,p}(S):\, \gamma u=f} \|u\|_{1,p;S}.$$

The continuity of γ from $H_{1,p}(S)$ to $H_{1/2,p}(\partial S)$ is uniform with respect to $p \in \mathbb{C}$. Finally, $H_{-1/2,p}(\partial S)$ is the dual of $H_{1/2,p}(\partial S)$ with respect to the duality generated by the inner product $(\cdot,\cdot)_{0;\partial S}$ in $L^2(\partial S)$.

We fix $\kappa > 0$ and introduce the complex half-plane $\mathbb{C}_\kappa = \{p = \sigma + i\tau \in \mathbb{C} : \sigma > \kappa\}$. Consider the space $H^{\mathcal{L}}_{m,k,\kappa}(S)$, m, $k \in \mathbb{R}$, of all $\hat{u}(x,p)$, $x \in S$, $p \in \mathbb{C}_\kappa$, such that $U(p) = \hat{u}(\cdot,p)$ is a holomorphic mapping from $\mathbb{C}_\kappa$ to $H_m(S)$ (which implies that $U(p)$ also belongs to $H_{m,p}(S)$ for every $p \in \mathbb{C}_\kappa$) and for which

$$\|\hat{u}\|^2_{m,k,\kappa;S} = \sup_{\sigma > \kappa} \int_{-\infty}^{\infty} (1 + |p|^2)^k \|U(p)\|^2_{m,p;S}\, d\tau < \infty.$$

The norm on $H^{\mathcal{L}}_{m,k,\kappa}(S)$ is defined by this equality. In what follows, we use the symbol $\hat{u}(x,p)$ when we want to emphasize that this is a distribution in $H_{m,p}(S)$, and the symbol $U(p)$ when we need to regard it as a mapping from $\mathbb{C}_\kappa$ to $H_m(S)$. The space $H^{\mathcal{L}}_{\pm 1/2,k,\kappa}(\partial S)$ and its norm $\|\cdot\|_{\pm 1/2,k,\kappa;\partial S}$ are introduced similarly.

Let $H^{\mathcal{L}^{-1}}_{m,k,\kappa}(G)$ and $H^{\mathcal{L}^{-1}}_{\pm 1/2,k,\kappa}(\Gamma)$, $\Gamma = \partial S \times (0,\infty)$, be the spaces of the inverse Laplace transforms u and f of all $\hat{u} \in H^{\mathcal{L}}_{m,k,\kappa}(S)$ and $\hat{f} \in H^{\mathcal{L}}_{\pm 1/2,k,\kappa}(\partial S)$, with norms

$$\|u\|_{m,k,\kappa;G} = \|\hat{u}\|_{m,k,\kappa;S}, \quad \|f\|_{\pm 1/2,k,\kappa;\partial G} = \|\hat{f}\|_{\pm 1/2,k,\kappa;\partial S}.$$

We assume that ∂S is a C^2-curve consisting of two arcs ∂S_ν, $\nu = 1,2$, such that $\partial S = \overline{\partial S}_1 \cup \overline{\partial S}_2$, $\partial S_1 \cap \partial S_2 = \emptyset$, and mes $\partial S_\nu > 0$, $\nu = 1,2$. Let S^+ and S^- be the interior and exterior domains into which ∂S divides $\mathbb{R}^2$, and let $G^\pm = S^\pm \times (0,\infty)$ and $\Gamma_\nu = \partial S_\nu \times (0,\infty)$, $\nu = 1,2$. We denote by $\gamma^\pm$ the trace operators corresponding to $S^\pm$. For simplicity, we use the same symbols for the trace operators in the spaces of originals and in those of their Laplace transforms. Thus, $\gamma^\pm$ also denote the trace operators mapping $H_{1,k,\kappa}^{\mathcal{L}^{-1}}(G^{\pm})$ continuously to $H_{1/2,k,\kappa}^{\mathcal{L}^{-1}}(\Gamma)$ for any $k \in \mathbb{R}$. Also, we denote by π_{ν}, $\nu = 1,2$, the operators of restriction from Γ to Γ_{ν} (and from ∂S to ∂S_{ν}), and write $\gamma_{\nu}^{\pm} = \pi_{\nu}\gamma^{\pm}$, $\nu = 1,2$. Finally, let $\pi^{\pm}$ be the operators of restriction from $\mathbb{R}^2 \times (0,\infty)$ to $G^{\pm}$ (or from $\mathbb{R}^2$ to $S^{\pm}$).

We introduce the subspace $\mathring{H}_{1/2,p}(\partial S_{\nu})$ of $H_{1/2,p}(\partial S)$ consisting of all $\varphi \in H_{1/2,p}(\partial S)$ such that $\operatorname{supp}\varphi \in \overline{\partial S}_{\nu}$, $\nu = 1,2$. Let $H_{1/2,p}(\partial S_{\nu})$ be the space of the restrictions from ∂S to ∂S_{ν} of the elements of $H_{1/2,p}(\partial S)$. The norm of $\varphi \in H_{1/2,p}(\partial S_{\nu})$ is defined by

$$\|\varphi\|_{1/2,p;\partial S_{\nu}} = \inf_{f \in H_{1/2,p}(\partial S):\pi_{\nu}f = \varphi} \|f\|_{1/2,p;\partial S}, \quad \nu = 1,2.$$

Let l_{ν}, $\nu = 1,2$, be extension operators from ∂S_{ν} to ∂S which map $H_{1/2,p}(\partial S_{\nu})$ to $H_{1/2,p}(\partial S)$ continuously and uniformly with respect to p and satisfy

$$\|l_{\nu}f\|_{1/2,p;\partial S} \leq c\|f\|_{1/2,p;\partial S_{\nu}} \quad \forall f \in H_{1/2,p}(\partial S_{\nu}).$$

Also, let $l^{\pm}$ be operators of extension from ∂S to $S^{\pm}$ which map $H_{1/2,p}(\partial S)$ to $H_{1,p}(S^{\pm})$ continuously and uniformly with respect to p.

By $\mathring{H}_{-1/2,p}(\partial S_{\nu})$ and $H_{-1/2,p}(\partial S_{\nu})$, $\nu = 1,2$, we denote the duals of $H_{1/2,p}(\partial S_{\nu})$ and $\mathring{H}_{1/2,p}(\partial S_{\nu})$, respectively, with respect to the duality generated by the inner product in $[L^2(\partial S_{\nu})]^3$; their norms are $\|\cdot\|_{-1/2,p;\partial S}$ and $\|\cdot\|_{-1/2,p;\partial S_{\nu}}$. The corresponding spaces $H_{\pm 1/2,k,\kappa}^{\mathcal{L}^{-1}}(\Gamma_{\nu})$ and $\mathring{H}_{\pm 1/2,k,\kappa}^{\mathcal{L}^{-1}}(\Gamma_{\nu})$, $\nu = 1,2$, and their norms $\|\cdot\|_{\pm 1/2,k,\kappa,\Gamma_{\nu}}$ and $\|\cdot\|_{\pm 1/2,k,\kappa,\Gamma}$, $k \in \mathbb{R}$, $\kappa > 0$, are introduced in the usual way.

In what follows, we denote by

$$a_{\pm}(u,v) = 2 \int\limits_{S^{\pm}} E(u,v)\,dx$$

the sesquilinear form of the internal energy density, where

$$2E(u,v) = h^2 E_0(u,v) + h^2 \mu (\partial_2 u_1 + \partial_1 u_2)(\partial_2 \bar{v}_1 + \partial_1 \bar{v}_2)$$
$$+ \mu[(u_1 + \partial_1 u_3)(\bar{v}_1 + \partial_1 \bar{v}_3) + (u_2 + \partial_2 u_3)(\bar{v}_2 + \partial_2 \bar{v}_3)],$$
$$E_0(u,v) = (\lambda + 2\mu)\big[(\partial_1 u_1)(\partial_1 \bar{v}_1) + (\partial_2 u_2)(\partial_2 \bar{v}_2)\big]$$
$$+ \lambda\big[(\partial_1 u_1)(\partial_2 \bar{v}_2) + (\partial_2 u_2)(\partial_1 \bar{v}_1)\big].$$

The classical mixed dynamic problems (DM$^{\pm}$) consist in finding $u \in C^2(G^{\pm}) \cap C^1(\bar{G}^{\pm})$ that satisfies

$$B(\partial_t^2 u)(x,t) + (Au)(x,t) = 0, \quad (x,t) \in G^+ \text{ or } G^-,$$
$$u(x,0+) = (\partial_t u)(x,0+) = 0, \quad x \in S^+ \text{ or } S^-,$$
$$u^{\pm}(x,t) = f_1(x,t), \quad (x,t) \in \Gamma_1, \quad (Tu)^{\pm}(x,t) = g_2(x,t), \quad (x,t) \in \Gamma_2,$$

where T is the moment-force boundary operator defined by

$$
\begin{pmatrix}
h^2(\lambda+2\mu)n_1\partial_1 + h^2\mu n_2\partial_2 & h^2\mu n_2\partial_1 + h^2\lambda n_1\partial_2 & 0 \\
h^2\lambda n_2\partial_1 + h^2\mu n_1\partial_2 & h^2\mu n_1\partial_1 + h^2(\lambda+2\mu)n_2\partial_2 & 0 \\
\mu n_1 & \mu n_2 & \mu(n_1\partial_1 + n_2\partial_2)
\end{pmatrix},
$$

$n = (n_1, n_2)$ is the outward unit normal to ∂S the superscripts $\pm$ denote the limiting values of the corresponding functions as $(x, t) \to \Gamma$ from inside $G^\pm$ (or $x \to \partial S$ from inside $S^\pm$), and f_1 and g_2 are given functions.

We call $u \in H^{\mathcal{L}^{-1}}_{1,0,\kappa}(G^\pm)$ a weak solution of the corresponding problem (DM$^\pm$) if it satisfies

$$
\int\limits_0^\infty \{a_\pm(u,v) - (B^{1/2}\partial_t u, B^{1/2}\partial_t v)_{0;S^\pm}\}\, dt = \pm \int\limits_0^\infty (g_2, v)_{0;\partial S_2}\, dt,
\tag{8.1}
$$

$$
\gamma_1^\pm u = f_1 \quad \forall v \in C_0^\infty(\bar{G}^\pm) \text{ such that } \gamma_1^\pm v = 0.
$$

8.2 Solvability of the Problems

In what follows we use the same symbol c for all positive constants that occur in various estimates and are independent of the functions in those estimates and of $p \in \mathbb{C}_\kappa$ (but may depend on κ). The following assertion holds.

Theorem 1. *For every* $\kappa > 0$, $f_1 \in H^{\mathcal{L}^{-1}}_{1/2,1,\kappa}(\Gamma_1)$, *and* $g_2 \in H^{\mathcal{L}^{-1}}_{-1/2,1,\kappa}(\Gamma_2)$, *problems* (8.1) *have a unique solution* $u \in H^{\mathcal{L}^{-1}}_{1,0,\kappa}(G^\pm)$. *Furthermore, if* $f_1 \in H^{\mathcal{L}}_{1/2,k,\kappa}(\Gamma_1)$ *and* $g_2 \in H^{\mathcal{L}}_{-1/2,k,\kappa}(\Gamma_2)$, *then* $u \in H^{\mathcal{L}}_{1,k-1,\kappa}(G^\pm)$ *and for every* $k \in \mathbb{R}$,

$$
\|u\|_{1,k-1,\kappa;G^\pm} \le c\big(\|f_1\|_{1/2,k,\kappa;\Gamma_1} + \|g_2\|_{-1/2,k,\kappa;\Gamma_2}\big).
$$

8.3 Retarded Plate Potentials

Consider a matrix $D(x,t)$ of fundamental solutions for the equation of motion; that is, a (3×3)-matrix such that

$$
B(\partial_t^2 D)(x,t) + AD(x,t) = \delta(x,t)I, \quad (x,t) \in \mathbb{R}^2 \times \mathbb{R},
$$
$$
D(x,t) = 0, \quad (x,t) \in \mathbb{R}^2 \times (-\infty, 0),
$$

where δ is the Dirac delta distribution and I is the identity (3×3)-matrix. Clearly, its Laplace transformation $\hat{D}(x,p)$ satisfies

$$
Bp^2\hat{D}(x,p) + A\hat{D}(x,p) = \delta(x)I, \quad x \in \mathbb{R}^2.
$$

The explicit form of $\hat{D}(x, p)$ can be found in [2].

Let α, $\beta \in C^2(\partial S \times \mathbb{R})$ be functions with compact support in $\bar{\Gamma}$, and let $\hat{\alpha}$ and $\hat{\beta}$ be their Laplace transforms. We define the single-layer and double-layer potentials

$$(V_p\hat{\alpha})(x, p) = \int_{\partial S} \hat{D}(x - y, p)\hat{\alpha}(y, p)\, ds_y, \quad x \in \mathbb{R}^2, \ p \in \mathbb{C}_0,$$

$$(W_p\hat{\beta})(x, p) = \int_{\partial S} (T_y\hat{D}(y - x, p))^{\mathrm{T}}\hat{\beta}(y, p)\, ds_y, \quad x \in S^+ \cup S^-, \ p \in \mathbb{C}_0,$$

where T_y is the boundary operator T acting with respect to y.

Since $\hat{D}(x, p)$ has a polynomial growth with respect to $p \in \mathbb{C}_\kappa$, $\kappa > 0$, we may now define the retarded single-layer and double-layer potentials

$$(V\alpha)(x, t) = (\mathcal{L}^{-1}V_p\hat{\alpha})(x, t)$$

$$= \int_0^\infty \int_{\partial S} D(x - y, t - \tau)\alpha(y, \tau)\, ds_y\, d\tau, \quad (x, t) \in \mathbb{R}^2 \times (0, \infty),$$

$$(W\beta)(x, t) = (\mathcal{L}^{-1}W_p\hat{\beta})(x, t)$$

$$= \int_0^\infty \int_{\partial S} (T_yD(y - x, t - \tau))^{\mathrm{T}}\beta(y, \tau)\, ds_y\, d\tau, \quad (x, t) \in G^+ \cup G^-.$$

We consider the boundary operators V_0, $W^\pm$, and N defined by

$$V_0\alpha = \gamma^\pm\pi^\pm V\alpha, \quad W^\pm\beta = \gamma^\pm\pi^\pm W\beta, \quad N\beta = T^\pm\pi^\pm W\beta.$$

These operators can be extended by continuity to much wider classes of densities [2].

8.4 Integral Representations of the Solutions

We now consider four representations for the solutions of problems (DM$^\pm$) in terms of retarded potentials and comment on the unique solvability of the corresponding systems of boundary integral equations. We begin with the representation

$$u(x, t) = (V\alpha)(x, t), \quad (x, t) \in G^+ \text{ or } G^-, \tag{8.2}$$

which yields the system of boundary equations

$$\begin{aligned}
(\pi_1 V_0\alpha)(x, t) &= f_1(x, t), \quad (x, t) \in \Gamma_1, \\
(\pi_2 T^\pm V_0\alpha)(x, t) &= g_2(x, t), \quad (x, t) \in \Gamma_2,
\end{aligned} \tag{8.3}$$

where T is the Poincaré–Steklov operator [2].

Theorem 2. *For every $\kappa > 0$, $k \in \mathbb{R}$, $f_1 \in H^{\mathcal{L}^{-1}}_{1/2,k,\kappa}(\Gamma_1)$, and $g_2 \in H^{\mathcal{L}^{-1}}_{-1/2,k,\kappa}(\Gamma_2)$, system (8.3) has a unique solution $\alpha \in H^{\mathcal{L}^{-1}}_{-1/2,k-2,\kappa}(\Gamma)$, in which case u defined by (8.2) belongs to $H^{\mathcal{L}^{-1}}_{1,k-1,\kappa}(G^{\pm})$. If $k \geq 1$, then u is the solution of $(DM^{\pm})$.*

The remaining representations, namely,

$$u(x,t) = (W\beta)(x,t), \quad (x,t) \in G^+ \text{ or } G^-,$$
$$u(x,t) = (V\alpha_1)(x,t) + (W\beta_2)(x,t), \quad (x,t) \in G^+ \text{ or } G^-,$$
$$u(x,t) = (W\beta_1)(x,t) + (V\alpha_2)(x,t), \quad (x,t) \in G^+ \text{ or } G^-,$$

are handled similarly.

A fuller version of the above results, including all the detailed proofs, will appear elsewhere.

References

1. C. Constanda, *A mathematical analysis of bending of plates with transverse shear deformation*, Pitman Res. Notes Math. Ser. **215**, Longman/Wiley, Harlow-New York, 1990.

2. I. Chudinovich and C. Constanda, Nonstationary integral equations for elastic plates, *C.R. Acad. Sci. Paris Sér. I* **329** (1999), 1115–1120.

9 Finite-Difference Schemes for a Nonlinear Parabolic Problem with Nonlocal Boundary Conditions

Raimondas Čiegis

9.1 Introduction

Consider the nonlinear parabolic equation

$$\frac{\partial u}{\partial t} = \frac{\partial}{\partial x}\left(p(x)\frac{\partial u}{\partial x}\right) - q(x,t)u + f(u,x,t),\qquad(9.1)$$

for $(x,t) \in Q_T = (0,1)\times(0,T], \ 0 < T \leq \infty$, subject to the initial condition

$$u(x,0) = u_0(x), \ \ x \in [0,1]$$

and the nonlocal boundary conditions

$$u(0,t) = \gamma_0\left(\alpha_0(t)u(c_0(t),t) + \int_0^1 \beta_0(x,t)u(x,t)\,dx\right) + g_0(t),$$

$$u(1,t) = \gamma_1\left(\alpha_1(t)u(c_1(t),t) + \int_0^1 \beta_1(x,t)u(x,t)\,dx\right) + g_1(t).$$

We assume that all coefficients are sufficiently smooth functions and that there exist constants p_1, p_2 such that

$$0 < p_0 \leq p(x) \leq p_1, \ \ q(x,t) \geq 0.$$

Also we assume that f is Lipschitz continuous on compact sets.

The existence and uniqueness of solutions of problems with nonlocal boundary conditions are investigated in [1,2]. Numerical solution of such problems is considered in [3-6]. Linear problems are investigated in [3,5,6] and only paper [4] deals with the analysis of Galerkin methods for a nonlinear parabolic problem with nonlocal boundary conditions. In [5,6] the convergence of finite-difference approximations is proved for stability regions much larger than in [3, 4].

The purpose of this paper is to extend results of [6] for the nonlinear case. The rest of the paper is organized as follows. In Section 2 the finite-difference scheme is formulated and a brief review of main results from [6] is given. Section 3 deals with the convergence of the nonlinear finite-difference scheme.

9.2 Finite-Difference Scheme

In this section we formulate a finite-difference scheme which approximates the differential problem with nonlocal boundary conditions. Let ω_h and ω_τ be partitions of the space and time coordinates, respectively:

$$\omega_h = \left\{ x_i : \ x_i = ih, \ i = 1, 2, \ldots, N - 1, \ h = \frac{1}{N} \right\},$$

$$\omega_\tau = \left\{ t^n : \ t^n = n\tau, \ n = 1, 2, \ldots, M, \ \tau = \frac{T}{M} \right\}.$$

For functions $U_i^n = U(x_i, t^n)$ defined on the mesh $\omega_h \times \omega_\tau$, we introduce the notation

$$U_t = \frac{U^n - U^{n-1}}{\tau}, \quad U_x = \frac{U_{i+1} - U_i}{h}, \quad U_{\bar{x}} = \frac{U_i - U_{i-1}}{h}.$$

Then the finite-difference approximation is defined by

$$
\begin{aligned}
U_t &= \left(p_{i+0.5}\, U_x^n\right)_{\bar{x}} - q_i^n U_i^n + f_i^n(U^n), \quad x \in \omega_h, \\
U_0^n &= \gamma_0\left(\sigma K_0^n(U^n) + (1 - \sigma)K_0^{n-1}(U^{n-1})\right) + g_0^n, \\
U_N^n &= \gamma_1\left(\sigma K_1^n(U^n) + (1 - \sigma)K_1^{n-1}(U^{n-1})\right) + g_1^n, \\
U_i^0 &= u_0(x_i), \quad x \in \omega_h \cup \{x_0, x_N\};
\end{aligned}
\tag{9.2}
$$

here we used the notation

$$K_j^n(U^n) = \alpha_j^n U^n(c_0(t^n)) + \sum_{l=1}^{N} \beta_j^n(x_l)\frac{U_{l-1}^n + U_l^n}{2}, \quad j = 0, 1.$$

Thus, the difference equation is approximated by the implicit Euler method and nonlocal boundary conditions can be approximated explicitly, if $\sigma = 0$, or implicitly, if $\sigma = 1$. By expending the truncation error in Taylor series we get $|\psi^n| \leq C(\tau + h^2)$ for $u \in C^{4,2}([0,1] \times [0,T])$.

Ekolin [3] investigated problem (9.1) for $q = 0$ and $\alpha_j = 0$. He proved the convergence of the backward Euler method (9.2) and $\sigma = 1$ under the assumption that inequalities

$$\gamma_j \int_0^1 |\beta_j(x)|\, dx < 1, \quad j = 0, 1,$$

hold, whereas the analysis of the Crank–Nicolson method required that

$$\sum_{j=0}^{1} \gamma_j \left(\int_0^1 |\beta_j(x)|^2 \, dx \right)^{1/2} \leq \frac{\sqrt{3}}{2}.$$

Fairweather and Lopez-Marcos [4] investigated Crank–Nicolson Galerkin approximation of (9.1) with integral nonlocal conditions and proved convergence under the assumption that

$$\gamma_j^2 \int_0^1 |\beta_j(x)|^2 \, dx \leq \mu^2 < 1.$$

In our previous papers [5,6] we have developed a new technique for investigation of finite-difference problems with nonlocal boundary conditions. It is based on one generalization of the maximum principle. We review very briefly essential steps of this analysis.

Two auxiliary problems are formulated, namely,

$$\Phi_{jt} = \left(p_{i+0.5} \, \Phi_{jx}^n \right)_{\bar{x}} - \tilde{q}_i^n \Phi_{ji}^n, \quad j = 1, 2,$$
$$\Phi_{j0}^n = \delta_{j1}, \quad \Phi_{jN}^n = \delta_{j2}, \quad n \geq 1,$$
$$\Phi_j^0(x_i) = 0, \quad x \in \omega_h \cup \{x_0, x_N\};$$

here the coefficient $\tilde{q}$ is such that

$$0 \leq \tilde{q}_i^n \leq q_i^n, \quad x \in \omega_h,$$

and δ_{jk} is the Kronecker symbol. Let us define linear functionals

$$\tilde{K}_j^n(V^n) = |\alpha_j^n| V^n(c_0(t^n)) + \sum_{l=1}^{N} |\beta_j^n(x_l)| \frac{V_{l-1}^n + V_l^n}{2}, \quad j = 0, 1.$$

The stability region D of parameters (γ_0, γ_1) depends on the matrix

$$\mathbf{K} = (k_{ij}), \quad 0 \leq i, j \leq 1,$$

where

$$k_{ij} = \max_{1 \leq n \leq M} \left(\sigma \tilde{K}_i^n(\Phi_j^n) + (1 - \sigma) \tilde{K}_i^{n-1}(\Phi_j^{n-1}) \right).$$

D is given by

$$D(\mathbf{K}) = \left\{ (\gamma_0, \gamma_1) : \; \theta > 0, \; 0 < \gamma_j < \frac{1}{k_{jj}}, \; j = 0, 1 \right\},$$

where

$$\theta = 1 - \gamma_0 k_{00} - \gamma_1 k_{11} + \gamma_0 \gamma_1 \det \mathbf{K}.$$

Then we have the following result [6].

Theorem 1. *Let $u \in C^{4,2}([0,1] \times [0,T])$ be the solution of linear problem (9.1) with the right-hand side $f(x,t)$. If $(\gamma_0, \gamma_1) \in D(\mathbf{K})$, then there exists a unique solution of finite-difference scheme (9.2) such that*

$$\|u^n - U^n\|_{C(\omega_h)} \leq C(\tau + h^2), \quad n = 1, 2, \ldots, K.$$

Remark. It is interesting to note that the following two properties hold:

1. Explicit approximations of nonlocal boundary conditions give larger stability regions than the implicit approximations.

2. In many cases stability regions of the parabolic problem monotonically decrease with respect to the time coordinate and converge to the stability region of the stationary problem.

9.3 Analysis of the Nonlinear Difference Scheme

To analyze the nonlinear finite-difference scheme (9.2), we use the framework developed in [7]. Let us define the neighborhood of the exact solution

$$B(u, \delta) = \left\{ v : \ \|v - u\|_C \leq \delta \right\}.$$

Next we assume that instead of function f we have a modified function $\tilde{f}$, which satisfies the equality

$$\tilde{f}(v, x, t) = f(v, x, t), \ \ \forall v \in B(u, \delta)$$

and $\tilde{f}$ is a globally Lipschitz function, that is,

$$|\tilde{f}(v_1, x, t) - \tilde{f}(v_2, x, t)| \leq L|v_1 - v_2|.$$

It is obvious that u is also a solution of the modified problem (9.1).

In order to be able to use such an assumption we must guarantee that the finite-difference solution always belongs to the discrete neighborhood of the solution

$$B_h(u^n, \delta) = \left\{ v : \ \|V^n - u^n\|_{C(\omega_h)} \leq \delta \right\}.$$

This part of the proof can be done by using mathematical induction and taking into account the fact that $u \in C^{4,2}$ and therefore the inequalities

$$|V^n - u^n| \leq |V^n - u^{n-1}| + |u^n - u^{n-1}| \leq |V^n - u^{n-1}| + \frac{\delta}{2}$$

are valid for sufficiently small $\tau \leq \tau_0$.

We find a solution of (9.2) by using the following iterative method:

$$\overset{s}{U}_t = \left(p_{i+0.5}\,\overset{s}{U}_x\right)_{\bar{x}} - q_i^n\overset{s}{U}_i + \tilde{f}_i^n(\overset{s-1}{U}), \quad x \in \omega_h,$$

$$\overset{s}{U}_0 = \gamma_0\big(\sigma K_0^n(\overset{s}{U}) + (1-\sigma)K_0^{n-1}(U^{n-1})\big) + g_0^n,$$

$$\overset{s}{U}_N = \gamma_1\big(\sigma K_1^n(\overset{s}{U}) + (1-\sigma)K_1^{n-1}(U^{n-1})\big) + g_1^n,$$

$$\overset{0}{U}_i = U_i^{n-1}, \quad x \in \omega_h \cup \{x_0, x_N\}.$$

Using the maximum principle from [6], we prove that $\{\overset{s}{U}\}$ is a bounded fundamental system. Since $\overset{s}{U}$ belongs to a finite-dimensional space, then $\overset{s}{U} \to U^n$.

It remains to prove that nonlinear finite-difference scheme (9.2) is stable. The global error of the discrete solution $Z_i^n = u_i^n - U_i^n$ satisfies the following problem:

$$Z_t = \left(p_{i+0.5}\,Z_x^n\right)_{\bar{x}} - q_i^n Z_i^n + + \tilde{f}'(\tilde{U})Z_i^n + \psi_i^n, \quad x \in \omega_h,$$

$$Z_0^n = \gamma_0\big(\sigma K_0^n(Z^n) + (1-\sigma)K_0^{n-1}(Z^{n-1})\big) + \psi_0^n, \tag{9.3}$$

$$Z_N^n = \gamma_1\big(\sigma K_1^n(Z^n) + (1-\sigma)K_1^{n-1}(Z^{n-1})\big) + \psi_N^n,$$

$$Z_i^0 = 0, \quad x \in \omega_h \cup \{x_0, x_N\}.$$

We have two cases. If the nonlinearity of the right-hand side of (9.1) is weak (or, equivalently, if the sink term is strong), i.e.:

$$q_i^n - L \geq 0, \quad x \in \omega_h,$$

then we can apply the results from [6]. Thus Theorem 1 is valid for the nonlinear finite-difference scheme with slightly changed stability region $D(\mathbf{K})$.

Otherwise the effect of the nonlinear term must be investigated directly. Let us introduce an auxiliary function V_i^n, which is defined by

$$Z_i^n = (1+\tau\tilde{L})^n V_i^n, \quad \tilde{L} = (1+\tau L)L.$$

Substituting this expression into (9.3), we see that V satisfies the problem

$$\frac{1}{1+\tau\tilde{L}}V_t = \left(p_{i+0.5}\,V_x^n\right)_{\bar{x}} - q_i^n V_i^n + + (\tilde{f}'(\tilde{U}) - L)V_i^n + \frac{\psi_i^n}{(1+\tau\tilde{L})^n},$$

$$V_0^n = \gamma_0\big(\sigma K_0^n(V^n) + \frac{1-\sigma}{1+\tau\tilde{L}}K_0^{n-1}(V^{n-1})\big) + \frac{\psi_0^n}{(1+\tau\tilde{L})^n},$$

$$V_0^N = \gamma_1\big(\sigma K_1^n(V^n) + \frac{1-\sigma}{1+\tau\tilde{L}}K_1^{n-1}(V^{n-1})\big) + \frac{\psi_N^n}{(1+\tau\tilde{L})^n},$$

$$V_i^0 = 0, \quad x \in \omega_h \cup \{x_0, x_N\}.$$

Again we can apply the results of [6] and prove that for $(\gamma_0, \gamma_1) \in D(\tilde{\mathbf{K}})$ the inequality

$$\|V^n\| \leq C(\tau + h^2)$$

is valid. Therefore using it we obtain the main stability estimate

$$\|Z^n\| \leq e^{2t^n L} C(\tau + h^2),$$

which shows that under the hypothesis made above the solution of nonlinear finite-difference scheme (9.2) converges to the solution of the differential problem with nonlocal boundary conditions.

Remark. The stability region $D(\mathbf{K})$ is only sufficient for the convergence of the discrete solution, but for stationary problems it is proved in [5] that $(\gamma_0, \gamma_1) \in D(\mathbf{K})$ is also a necessary condition.

References

1. W.A. Dey, Extensions of property of the heat equation to linear thermoelasticity and other theories, *Quart. Appl. Math.* **40** (1982), 319–330.

2. A. Friedman, Monotonic decay of solutions of parabolic equations with nonlocal boundary conditions, *Quart. Appl. Math.* **44** (1986), 401–407.

3. G. Ekolin, Finite difference method for a nonlocal boundary value problem for the heat equation, *BIT* **31** (1991), 245–261.

4. G. Fairweather and J.C. Lopez-Marcos, Galerkin methods for a semilinear parabolic problem with nonlocal boundary conditions, *Adv. Comput. Math.* **6** (1996), 243–262.

5. R. Čiegis, A. Štikonas, O. Štikonienė, and O. Suboč, Stationary problems with nonlocal boundary conditions, *Math. Modelling Anal.* **6** (2001), 178–191.

6. R. Čiegis, A. Štikonas, O. Štikonienė, and O. Suboč, Monotone finite-difference scheme for parabolic problem with nonlocal boundary conditions, *Differential Equations* **38** (2002), 968–975.

7. R. Čiegis and M. Meilūnas, On the difference scheme for a nonlinear diffusion-reaction type problem, *Liet. Mat. Rink.* **33** (1993), 16–29.

10 Absolute Stability for Neutral Systems with Infinite Delay

Constantin Corduneanu

10.1 Introduction

The results of this paper complete those recently obtained by the author [4]. The feedback term is considered in a more general case than in [4]. The method of reducing the problem of absolute stability to the investigation of asymptotic behaviour for nonlinear integral equations is used again in this paper.

10.2 Formulation of the Problem

Let us consider the nonlinear neutral differential system

$$\frac{d}{dt}\left[x(t) - \int_0^t C(t-s)x(s)ds\right] = (Ax)(t) + (b\varphi)(\sigma(t)),$$

$$\sigma = <a, x>, \qquad (10.1)$$

where the operator A is formally given by

$$(Ax)(t) = \sum_{j=0}^{\infty} A_j x(t - t_j) + \int_0^t B(t-s)x(s)ds, \qquad (10.2)$$

under the assumption that

$$t_j \geq 0, \quad j \geq 0, \quad \sum_{j=0}^{\infty} |A_j| < \infty, \quad \int_0^{\infty} |B(t)|dt < \infty, \qquad (10.3)$$

while b stands for an operator represented by

$$(b\xi)(t) = \sum_{j=0}^{\infty} b_j \xi(t - t_j) + \int_0^{\infty} c(t-s)\xi(s)ds. \qquad (10.4)$$

Conditions similar to (10.3) will be imposed on b:

$$t_j \geq 0, \quad j \geq 0, \quad \sum_{j=0}^{\infty} |b_j| < \infty, \quad \int_0^{\infty} |c(t)|dt < \infty. \qquad (10.5)$$

The meaning of the notation used above is as follows: $x : \mathbb{R}_+ \to \mathbb{R}^n$, $a \in \mathbb{R}^n$, $\varphi : \mathbb{R} \to \mathbb{R}$, $A_j \in L(\mathbb{R}^n, \mathbb{R}^n)$, $b_j \in \mathbb{R}^n$, $j \geq 0$, $B : \mathbb{R}_+ \to L(\mathbb{R}^n, \mathbb{R}^n)$, $c : \mathbb{R}_+ \to \mathbb{R}$. Finally, $C : \mathbb{R}_+ \to L(\mathbb{R}^n, \mathbb{R}^n)$, and will be subject to further conditions.

The matrix or vector norms are the Euclidean ones, and $< \cdot, \cdot >$ denotes the scalar product in $\mathbb{R}^n$. Let us notice that the nature of the operators A and C requires for (10.1) an initial condition of the form

$$x(t) = h(t), \quad t < 0, \quad x(0) = x^0 \in \mathbb{R}^n. \tag{10.6}$$

The assumptions on $h(t)$ will be specified below.

The problem of absolute stability for the system (10.1) can be formulated as follows: show that the solution $x = 0$, under condition (10.6), is globally asymptotically stable for each (scalar) function φ belonging to a certain class (for instance, $\varphi : \mathbb{R} \to \mathbb{R}$, continuous with $\sigma\varphi(\sigma) > 0$ for $\sigma \neq 0$, and subject to further restrictions).

The condition

$$\sigma\varphi(\sigma) > 0 \quad \text{for} \quad \sigma \neq 0, \tag{10.7}$$

together with continuity, imply that $\varphi(0) = 0$. This property justifies the fact that $x = 0$ is a solution of the system (10.1) on the half-axis $\mathbb{R}_+$.

10.3 Reduction of the Problem to an Integral Equation of Volterra Type

Let us consider the linear auxiliary system associated to (10.1), namely

$$\frac{d}{dt}\left[x(t) - \int_0^t C(t - s)x(s)ds\right] = (Ax)(t) + f(t), \tag{10.8}$$

where $f : \mathbb{R}_+ \to \mathbb{R}^n$ is (at least!) locally integrable. As shown in [4], if $C(t)$ is differentiable, the system (10.8) can be reduced to the (non-neutral!) form

$$\dot{x}(t) = (\bar{A}x)(t) + f(t), \tag{10.9}$$

where

$$(\bar{A}x)(t) = C(0)x(t) + (Ax)(t) + \int_0^t [B(t - s) + \dot{C}(t - s)]x(s)ds, \tag{10.10}$$

which is of the same nature as the operator A given by (10.2). In order to satisfy the second condition in (10.3), we will assume that $|\dot{C}(t)| \in L^1(\mathbb{R}_+, \mathbb{R})$.

A formula of variation of parameters for the system (10.9), under initial conditions (10.6), has been provided in several papers under various assumptions: [2],[3],[4],[5].

We shall use here a result which is obtained in [6]:

Lemma. *Consider the system (10.8), under initial conditions (10.7), and assume that the following hypotheses are satisfied:*
1) The operator A is such that conditions (10.3) are verified, and

$$det[sI - \bar{A}^*(s)] \neq 0 \quad for \quad \mathrm{Re}\, s \geq 0, \tag{10.11}$$

where $\bar{A}^*(s)$ is defined by

$$\bar{A}^*(s) = C(0) + \sum_{j=0}^{\infty} A_j e^{-t_j s} + \int_0^{\infty} [B(t) + \dot{C}(t)] e^{-ts} dt.$$

2) $C(t)$ is locally absolutely continuous on $\mathbb{R}_+$, and

$$|\dot{C}(t)| \in L^1(\mathbb{R}_+, \mathbb{R}). \tag{10.12}$$

3)
$$f \in L^p(\mathbb{R}_+, \mathbb{R}^n), \quad p \geq 1. \tag{10.13}$$

4)
$$h \in L^p(\mathbb{R}_-, \mathbb{R}^n), \quad p \geq 1. \tag{10.14}$$

Then the unique solution of the problem can be represented by the formula

$$x(t) = \bar{X}(t)x^0 + (Yh)(t) + \int_0^t \bar{X}(t-s)f(s)ds, \tag{10.15}$$

where $\bar{X}(t)$ is defined by

$$\dot{\bar{X}}(t) = (A\bar{X})(t), \quad t > 0, \quad \bar{X}(0) = I, \quad \bar{X}(t) = O, \quad t < 0,$$

and $(Yh)(t)$ is represented by the series

$$(Yh)(t) = \sum_{j=0}^{\infty} \int_{-t_j}^0 \bar{X}(t - t_j - u) A_j h(u) du.$$

Moreover,
$$x \in L^p(\mathbb{R}_+, \mathbb{R}^n), \quad p \geq 1. \tag{10.16}$$

The proof of the lemma is given in [6].

Remark. The condition (10.4) on the operator $\bar{A}$ is equivalent to

$$|\bar{X}(t)| \in L^p(\mathbb{R}_+, \mathbb{R}), \quad p \geq 1. \tag{10.17}$$

Condition (10.17) implies $|\dot{\bar{X}}(t)| \in L^p(\mathbb{R}_+, \mathbb{R})$, which easily leads to the asymptotic stability property

$$\lim_{t \to \infty} |\bar{X}(t)| = 0. \tag{10.18}$$

By using the lemma, we can achieve the reduction of our problem on absolute stability for the system (10.1), to an integral equation of Volterra type. Indeed, the system (10.1) can be rewritten in the equivalent form

$$\dot{x}(t) = (\bar{A}x)(t) + (b\varphi)(\sigma(t)),\tag{10.19}$$

which leads to the formula (taking into account the initial conditions (10.6))

$$x(t) = \bar{X}(t)x^0 + (Yh)(t) + \int_0^t \bar{X}(t-s)(b\varphi)(\sigma(s))ds.$$

Substituting this into the second equation of the system (10.1), we obtain the Volterra integral equation

$$\sigma(t) = < a, \bar{X}(t)x^0 + (Yh)(t) > + \int_0^t < a, \bar{X}(t-s)(b\varphi)(\sigma(s)) > ds.\tag{10.20}$$

The equation (10.20) can be reduced to the classical form if we conveniently transform the convolution integral. Such a transformation has been used in [2], and we are not going to repeat the details. As a result of such transformation, the equation (10.20) takes the form

$$\sigma(t) = f(t) + \int_0^t k(t-s)\varphi(\sigma(s)ds,\tag{10.21}$$

where

$$f(t) = < a, \bar{X}(t)x^0 + (Yh)(t) + (Y_0h)(t) >,\tag{10.22}$$

and

$$k(t) = < a, \widetilde{X}(t) >,\tag{10.23}$$

where

$$\widetilde{X}(t) = \sum_{j=0}^{\infty} \bar{X}(t-t_j)b_j + \int_0^t \bar{X}(t-u)c(u)du,\tag{10.24}$$

and

$$(Y_0h)(t) = \sum_{j=0}^{\infty} \int_{-t_j}^0 \bar{X}(t-t_j-s)b_j\varphi(< a, h(s) >)ds.\tag{10.25}$$

The legitimacy of these transformations is discussed in [2],[3], where the case of the space $L^1(\mathbb{R}_+, \mathbb{R}^n)$ is emphasized. For the case of spaces $L^p(\mathbb{R}_+, \mathbb{R}^n)$, $p > 1$, the discussion can be conducted on the same lines as in [2]. It is necessary to assume one more restriction on φ, namely

$$\sigma\varphi(\sigma) \le L\sigma^2, \quad \sigma \in \mathbb{R},\tag{10.26}$$

where $L > 0$ is a constant.

Therefore, we shall concentrate on the bahaviour of solutions to the equation (10.21), with the purpose of showing that any solution of (10.21) on $\mathbb{R}_+$, satisfies

$$\lim_{t \to \infty} \sigma(t) = 0. \tag{10.27}$$

The property (10.27) easily implies the absolute stability for the system (10.1), as we shall see by using formula (10.15).

10.4 Equation (10.21)

Let us consider the equation (10.21), and check some of the properties enjoyed by the term f, given by (10.22), and the kernel k given by (10.23). These properties will be necessary in applying to (10.21) results available in the literature: [1],[7],[8],

First, we can show that $f \in L^1(\mathbb{R}_+, \mathbb{R})$, $p \geq 1$. Indeed $|\bar{X} \in L^p(\mathbb{R}_+, \mathbb{R})$, as seen above (a consequence of the lemma; see the remark to it). Again, according to the lemma, $|Yh| \in L^p(\mathbb{R}_+, \mathbb{R})$, $p \geq 1$, when the conditions 1), 2) and 4) are verified. Finally, to obtain the property $|Y_0 h| \in L^p(\mathbb{R}_+, \mathbb{R})$ one has to remark that in (10.25), $|\bar{X} \in L^p(\mathbb{R}_+, \mathbb{R})$, that φ satisfies $|\varphi(\sigma)| \leq L|\sigma|$, according to (10.26), while $h \in L^p(\mathbb{R}_-, \mathbb{R})$. Moreover, b_j are such that condition (10.5) holds true, which implies that $\varphi(<a, h>)$ is integrable on $\mathbb{R}_-$.

Second, in regard to the kernel $k(t)$ given by (10.23), we notice that it suffices to prove that $|\widetilde{X}(t)| \in L^1(\mathbb{R}_+, \mathbb{R})$, in order to apply results available in the literature [1], [8].

Based on conditions (10.3) and the fact that $|\bar{X}(t)| \in L^p(\mathbb{R}_+, \mathbb{R})$, as well as on the invariance of any $L^p(\mathbb{R}_+, \mathbb{R})$ with respect to the operator A, one sees that $|\widetilde{X}(t)| \in L^1(\mathbb{R}_+, \mathbb{R})$, $p \geq 1$.

As an example of application of the results for integral equations to the case of equation (10.21), we shall consider the following case (see [1], Theorem 2.2).

Consider the integral equation (10.21) under the following assumptions:
1) $f, f' \in L^1(\mathbb{R}_+, \mathbb{R})$;
2) $k, k' \in L^1(\mathbb{R}_+, \mathbb{R})$;
3) $\varphi : \mathbb{R} \to \mathbb{R}$ is continuous, bounded and such that (10.7) and (10.26) hold true;
4) there exists $q \geq 0$ such that

$$\mathrm{Re}\{(1 + i\omega q)\tilde{k}(i\omega)\} \leq 0, \quad \omega \in \mathbb{R}, \tag{10.28}$$

where $\tilde{k}(i\omega)$ is the Fourier transform of the kernel k, that is,

$$\tilde{k}(i\omega) = \int_0^\infty k(t)e^{-i\omega t}dt. \tag{10.29}$$

Then, there exists a solution $\sigma(t)$ of (10.21), continuous on $\mathbb{R}_+$, and such that (10.27) is verified. Moreover, any continuous solution of (10.21), defined on $\mathbb{R}_+$ satisfies (10.27).

Based on the above result, we can state the following theorem related to (10.21), in the case f and k are given by (10.22) and (10.23), i.e., in case (10.21) is equivalent to the system (10.1), with initial conditions (10.6).

Theorem. *Consider the neutral system (10.1), under the following assumptions:*

1) The operator A, given by (10.2), satisfies the conditions (10.3).

2) The operator C, given by (10.4) satisfies the conditions (10.5).

3) $C(t)$ is locally absolutely continuous on $\mathbb{R}_+$, and

$$|\dot{C}(t)| \in L^1(\mathbb{R}_+, \mathbb{R}). \tag{10.30}$$

4) The operator $\bar{A}$ satisfies the (stability) condition (10.11), with $\bar{A}^*$ given by (10.12).

5) The same as Condition 3) in the theorem quoted from [1].

6) The same as Condition 4) in the above quoted result from [1].

7) $h \in L^1(\mathbb{R}_-, \mathbb{R}^n)$.

Then, there exists a solution $x(t)$ of the system (10.1), under initial conditions (10.6), such that

$$\lim_{t \to \infty} |x(t)| = 0, \tag{10.31}$$

for each $x^0 \in \mathbb{R}^n$ and $h \in L^1(\mathbb{R}_-, \mathbb{R}^n)$. Moreover, any solution $x(t)$ of the system (10.1), under condition (10.6), defined on $\mathbb{R}_+$ will satisfy (10.31).

In other words, the system (10.1) is absolutely stable.

Proof. We shall check all the conditions in the theorem quoted above (see [1], Theorem 2.2).

Condition 1) requires that both f and f' be in $L^1(\mathbb{R}_+, \mathbb{R}^n)$. Indeed, if one takes (10.22) into account, in which $a \in \mathbb{R}^n$ is a constant, we must only show that $\bar{X}(t)$, $(Yh)(t)$, and $(Y_0 h)(t)$ are in L^1. This property follows for $\bar{X}(t)$ due to the assumption 4) in this theorem, and from $\dot{\bar{X}}(t) = (A\bar{X})(t)$ on $\mathbb{R}_+$. Of course, we have to keep in mind that L^1 is invariant with respect to $\bar{A}$. Since $(Yh)(t)$ is also a solution of $\dot{x}(t) = (\bar{A}x)(t)$, and according to the lemma it is in L^1, when h is in L^1. A similar argument is valid in case of the term $(Y_0 h)(t)$, also representing a solution of $\dot{x}(t) = (\bar{A}x)(t)$ on $\mathbb{R}_+$. Consequently, f given by (10.22) satisfies the condition 1) of our theorem.

Let us check now the validity of condition 2) in the theorem, for k given by (10.23). From (10.24) we easily derive $\widetilde{X}(t) \in L^1(\mathbb{R}_+, \mathbb{R}^n)$. The formula (10.24) shows that $\widetilde{X}(t)$ is a solution of $\dot{x}(t) = (\bar{A}x)(t) + c(t)$, which implies $\dot{\widetilde{X}}(t) \in L^1(\mathbb{R}_+, \mathbb{R}^n)$ if we rely on $\widetilde{X}(t) \in L^1(\mathbb{R}_+, \mathbb{R}^n)$. The last inclusion is a consequence of (10.24), $\bar{X}(t) \in L^1(\mathbb{R}_+, \mathbb{R}^n)$ and (10.5). Hence, $k(t)$ from (10.23) satisfies condition 2) in the above quoted theorem.

Condition 3) in the reference theorem is the same as condition 5) in the statement above, while condition 4) is condition 6) in our theorem.

Finally, condition 3) in our statement has the role to provide to the operator $\bar{A}$ the same properties we imposed on A, while condition 6) in our statement is used in representing the solution of (10.1) by means of the variation of parameter formula (10.15).

In order to conclude the proof of the theorem, it suffices to prove that the property (10.27) of $\sigma(t)$ implies the property (10.31) for $x(t)$ for arbitrary $x^0 \in \mathbb{R}^n$ and $h \in L^1(\mathbb{R}_-, \mathbb{R}^n)$, which amounts to absolute stability for the solution of (10.1). This property for $x(t)$ is a consequence of the representation formula

$$x(t) = \bar{X}(t)x^0 + (Yh)(t) + (Y_0h)(t) + \int_0^t \tilde{X}(t-s)\varphi(\sigma(s))ds, \quad (10.32)$$

which is derived easily from (10.19), as shown, for example, in [2] and [3]. We have already noticed that the first, second, and third terms on the right-hand side above tend to zero (in $\mathbb{R}^n$) as $t \to \infty$ (because they belong to L^1 together with their first derivatives). Concerning the integral on the right-hand side in (10.32), we see that it is the convolution product of a function in $L^1(\mathbb{R}_+, \mathbb{R}^n)$ and of a scalar function $\varphi(\sigma, (t))$, which tends to zero as $t \to \infty$.

On the basis of this discussion, we can claim that the proof of the theorem is complete.

Remark. The frequency domain condition (10.28), which is also part of our condition 5), can be improved. Indeed, as seen in various sources (see, for instance [1], [8]) one can obtain the basic result about $\sigma(t)$ if $\tilde{k}(i\omega)$ satisfies the inequality (weaker than (10.28))

$$\text{Re}\{(1 + i\omega q)\tilde{k}(i\omega)\} \le L^{-1}, \quad \omega \in \mathbb{R}.$$

We preferred the stronger variant (10.28), in order to avoid further technicalities (which have been carried out elsewhere).

In concluding the paper we notice that, with adequate transformations, the method of reduction to integral equations will work in case the relation $\sigma = < a, x >$ is replaced by more complex formulas.

References

1. C. Corduneanu, *Integral Equations and Stability of Feedback Systems*, Academic Press, New York-London, 1973.

2. C. Corduneanu, Asymptotic behaviour for some systems with infinite delay, in *VII. Internationale Konferenz über nichtlineare Schwingungen*, Band I, 1, Akademie-Verlag, Berlin, 1977, 155–160.

3. C. Corduneanu, *Recent contributions to the theory of differential systems with infinite delay*, Vander, Louvain, 1976.

4. C. Corduneanu, Absolute stability for neutral differential systems, *European J. Control* **8** (2002), 209–212.

5. C. Corduneanu, *Functional Equations with Causal Operators*, Taylor & Francis, London-New York, 2002.

6. C. Corduneanu and N. Luca, The stability of some feedback systems with delay, *J. Math. Anal. Appl.* **51** (1975), 377–393.

7. Ch.A. Desoer and M. Vidyasagar, *Feedback Systems: Input-Output Properties*, Academic Press, New York, 1975.

8. K.S. Narendra and J.H. Taylor, *Frequency Domain Criteria for Absolute Stability*, Academic Press, New York, 1973.

11 Large Scale Acoustic Simulations on Clusters of SMPs

Luc Giraud and Martin B. van Gijzen

11.1 Introduction

Finite element codes are usually parallelized either at a low level by exploiting fine grain loop parallelism or at a much higher level by exploiting the coarse grain parallelism of a mesh splitting in a domain decomposition type approach. The advantage of the first technique is its simplicity, in particular if the code already exists and, even better, is already vectorized. This approach is usually the preferred method if the target machine is a computer with a global address space, on which the cost of communication between computing entities (in this setting commonly denoted by threads) is usually relatively low. This is in particular the case if all the processors of the target computer physically share the same memory; this type of platform is usually referred to as Symmetric Multi-Processors (SMP). The second strategy, based on mesh splitting, is much more involved. Turning a sequential single-domain code into a parallel multi-domain code might require a complete redesign and at least imposes to add new communication subroutines at many places in the existing code. This, however, is a necessary step to exploit parallelism on platforms where the computing entities do not share any address space (in this setting commonly denoted by processes). This is typically the case on distributed memory computers.

In the recent years, new computer architectures have appeared that combine disjoint memory address space between groups of processors and a global memory address space within each group of processors. This kind of computer is usually called "Cluster of SMPs". This physical memory organization perfectly matches the requirements of parallel algorithms that can exploit two levels of parallelism. The outer/coarser is implemented between the SMPs and the inner/finer within each SMP. The corresponding parallel programming paradigms are message passing at the coarser level and loop level parallelism at the finer. This two-level parallelism has received considerable attention, see [1] and its references.

In this paper we investigate the parallelization of an existing, fully vectorized finite element code on a cluster of SMPs. Through numerical examples from ocean acoustics we show the merits of mixing the two programming models in relation to the numerical performance of the algorithms that are being used. We consider two test cases: a time-dependent problem that is solved with an explicit time integration method and a stationary example that is solved with a preconditioned iterative solution method.

11.2   Description of the Clusters of SMPs

Our two target machines are the two clusters currently in use at CER-FACS. The first is a Compaq ES40 Alphaserver with 10 nodes (SMPs), of which two could be used for the experiments. Each of the nodes has four processors with a peak performance of 2 Gflops. The memory per node is 4 Gb. Each processor has a primary cache memory of 64 Kb and a secondary cache of 8Mb. The second machine is a cluster of eight Pentium bi-processor PC's. The processor speed is 933 Mflops and the memory per node is 1 Gb. The size of the primary cache memory is 16 Kb and of the secondary cache 256 Kb.

The compilers on both machines support the OpenMP directives [2] to create parallel threads. The message passing library is based on MPI-CH [3] optimized for the interconnecting network. We refer to [1] for more detailed information about the clusters.

11.3   The Model Problem: Reflection of Sound Against an Object Buried in the Sediment

We will study the combined coarse grain/fine grain parallelization by means of examples from ocean acoustics. The examples model reflection of sound against an object that is half buried in the sediment. This problem has clear applications, for example to determine whether it is possible to detect under certain propagation conditions a buried container with a sound source (SONAR). In order to study this problem we define a (2D) domain of 200 m by 200 m. The lower 20 m of the domain consists of sediment and the upper 180 m of sea water. The reflecting object has a diameter of 10 m and is half buried in the sediment. It is located at 50 m from the left edge of the domain.

For our model problem we assume the following realistic parameters:
- sound speed c: 1500 m/s in the water, 1800 m/s in the sediment;
- density ρ: 1000 kg/m^3 in the water, 2000 kg/m^3 in the sediment;
- damping τ: 10^{-7} $kg/(m^3 s)$ in the sediment, 0 in the water.

The SONAR is located on the left edge at a depth of 100 m and transmits a short LFM-pulse of 50 ms that propagates through the domain. The pulse has a centre frequency of 1 kHz and a bandwidth of 1 kHz.

The above problem is mathematically described by the wave equation plus boundary conditions. Appropriate boundary conditions for this problem are pressure release at the surface, symmetry at the left edge of the domain and radiation conditions at the other two edges. The mathematical problem can be turned into a computational problem by discretization in space with the finite element method. We refer to [4] for the exact mathematical formulation and for the discretization. The process results in the matrix equation

$$\mathbf{M\ddot{p} + C\dot{p} + Kp + f = 0}. \tag{11.1}$$

In this equation $\mathbf{K}$ corresponds to a discretized Laplace operator. The matrix is sparse and SPD. The matrix $\mathbf{M}$, usually called mass matrix, is diagonal and SPD. The damping matrix $\mathbf{C}$ is diagonal and complex. The vector $\mathbf{p}$ contains the numerical approximations of the nodal values of the acoustic pressure. In order to have sufficient gridpoints per wavelength we need, for the given frequency of the transmitted pulse, at least 1000 gridpoints in each direction. Hence the number of unknowns is 10^6.

The system is only discretized in space, not in time. The vectors $\ddot{\mathbf{p}}$ and $\dot{\mathbf{p}}$ are the second, resp. first derivative of the acoustic pressure with respect to time. To integrate the system in time, a time stepping scheme like the explicit Newmark method can be applied. We will consider this problem as our first test case for the combined parallelization approach. It will be discussed in the next section.

The generation of very low frequency sound by a ship can be modeled by a source at the surface. If the frequency is low enough (lower than 1 Hz) we can assume the problem to be stationary. In that case equation (11.1) simplifies to the (real) linear system of equations

$$\mathbf{Kp} + \mathbf{f} = \mathbf{0}. \tag{11.2}$$

The mesh for this problem can be taken coarser than for the time integration problem. For the stationary problem we have taken 200 gridpoints in each direction which yields a system of order 40,000. Solving a linear system requires other numerical techniques then for explicit time integration. For this reason we will consider this problem as our second test case. It will be discussed in Section 5.

11.4 Explicit Time Integration

Equation (11.1) can be integrated in time with an explicit Newmark time integration method (see [4], page 452). This method is composed of three different types of operations: multiplication with (inverses of) diagonal matrices, vector updates and matrix-vector multiplication (specifically with $\mathbf{K}$). All three operations are well parallelizable, although the matrix-vector multiplication requires special attention.

In the finite element method, the matrix $\mathbf{K}$ is assembled from element matrices $\mathbf{K_e}$. This fact can be exploited in the matrix-vector multiplication which can be performed elementwise. The element-by-element (EBE) matrix-vector multiplication $\mathbf{Kv} = \mathbf{w}$ can be described by

$$\mathbf{Kv} = \sum_{e=1}^{n_e} \mathbf{K_e v_e} = \sum_{e=1}^{n_e} \mathbf{w_e} = \mathbf{w}.$$

The advantage of the above approach is that one avoids the assembly and storage of the global matrix $\mathbf{K}$. For this reason this matrix-free method was popular in the time that computer memory was smaller and more expensive than today. The (EBE) matrix-vector multiplication can be vectorized by making a multi-color ordering of the elements that are not connected. Operations with element matrices of the same color can be done in vector mode. See [5] for the details. The parallelization with OpenMP is a trivial task; one only has to add the appropriate directives around vectorized loops to force them to be performed in parallel.

Coarse grain parallelization can be extracted by making a domain decomposition. All element matrices that correspond to a domain are assigned to the same processor. Since nodal points at the edges of an internal subdomain boundary are shared with other subdomains, communication between processors has to be performed to exchange these nodal values. Further details can be found in [6].

It is important to note that the parallelization, both for the domain decomposition and for the multi-color ordering is extracted by changing the order of the operations. The result of the matrix-vector multiplication must, apart from round-off effects, be the same irrespective which parallelization technique is used.

We have integrated equation (11.1) 0.2s in time with 4,000 timesteps. The elapsed times and speed-ups on the Compaq are tabulated in Table 1.

Procs	1	2	4	8
Nodes	1	1	1	2
OpenMP [s]	3400	1788	939	-
Speed-up	*1*	*1.90*	*3.62*	-
MPI [s]	3400	1716	828	329
Speed-up	*1*	*1.98*	*4.11*	*10.33*
Combined [s]	3400	-	-	398
Speed-up	*1*	-	-	*8.54*

Table 1. Explicit time integration: elapsed time and speed-up on the Compaq.

Although parallelization with OpenMP requires little effort, the parallel performance for this example is close to optimal (i.e. a speed-up of p on p processors). Domain decomposition with MPI gives even better than optimal speed-ups. This can be explained by a better use of the fast cache memories due to the smaller size of the data. See [1] for more detail on the effect of the cache memory. For the above example it does not pay off to combine OpenMP and MPI. The pure MPI implementation is faster.

11.5　Solution of a Linear System

The solution of system (11.2) can be determined with the well-known conjugate gradient method [7]. This method is composed of the following operations: matrix-vector multiplication, vector updates, inner products and preconditioning. Here we will give special attention to the parallelization of the preconditioning operation.

A preconditioner is an easily invertible approximation $\mathbf{P}$ to the matrix $\mathbf{K}$. It is applied to speed up the convergence of the CG-method. A popular way to obtain a preconditioner is to construct it in factorized form, $\mathbf{P} = \mathbf{C}\mathbf{C}^T$ if $\mathbf{K}$ is symmetric as in our case. Here, $\mathbf{C}$ is a lower triangular matrix. The preconditioner is applied by making a back and forward substitution. Hughes et al. [5] have proposed a preconditioner of this structure that is composed of a product of factorized element matrices. Vectorization, and hence parallelization with OpenMP, is performed in a way similar to the EBE matrix-vector product with a multi-color ordering [5]. In a domain decomposition setting, however, there is no straightforward way to parallelize the back and forward substitution. A simple solution is to construct and apply local preconditioners, this is per subdomain [6]. Nodal values at the interfaces of subdomains can simply be added together. Note that this procedure changes the preconditioner depending on the number of subdomains. Table 2 gives the results (elapsed times, number of iterations and speed-ups) on the Compaq and Table 3 on the clusters of PC's.

Procs	1	2	4	8
Nodes	1	1	1	2
OpenMP [s]	81.0	49.1	34.1	-
Iterations	582	582	582	-
Speed-up	*1*	*1.65*	*2.38*	-
MPI [s]	81	71.1	36.2	20.1
Iterations	582	999	1210	1283
Speed-up	*1*	*1.14*	*2.24*	*4.03*
Combined [s]	81	-	-	28.9
Iterations	582	-	-	1181
Speed-up	*1*	-	-	*2.8*

Table 2. Solution of linear system: results on Compaq.

Procs	1	2	4	8	16
Nodes	1	1	2	4	8
OpenMP [s]	606.7	338.2	-	-	-
Iterations	582	582	-	-	-
Speed-up	*1*	*1.79*	-	-	-
MPI [s]	606.7	573.3	286.9	116.0	44.7
Iterations	582	995	1179	1258	1252
Speed-up	*1*	*1.06*	*2.11*	*5.23*	*13.57*
Combined [s]	606.7	-	280.5	128.5	52.6
Iterations	582	-	993	1179	1266
Speed-up	*1*	-	*2.16*	*4.72*	*11.53*

Table 3. Solution of linear system: results on cluster of PCs.

As expected the number of CG-iterations remains the same for OpenMP. For MPI, however the number of iterations increases with the number of processors, or equivalently with the number of subdomains. Due to this effect OpenMP is more efficient on one node. The combination of OpenMP on a node and MPI between nodes reduces the number of subdomains that are needed which makes this combination competitive with pure MPI. For example, mixed OpenMP and MPI is faster than pure MPI on two nodes of the cluster of PC's.

11.6 Conclusions

We have discussed the combination of OpenMP and MPI to parallelize an existing vectorized finite element code. Parallelization with OpenMP proved to be a straightforward task. Making an MPI-implementation has been done in combination with a domain decomposition method. This requires major changes to the code, including algorithmic changes. MPI and OpenMP can be combined by exploiting loop parallelism per subdomain.

Experiments with a time integration technique show a satisfactory performance of OpenMP. The speed-ups for MPI are even super-linear due to better use of the cache memory. The pure MPI-implementation also outperforms the combined approach.

The domain decomposition deteriorates the numerical properties of the preconditioner that is used in the the solution of a linear system, our second test case. As a result the CG-algorithm takes more iterations for an increasing number of subdomains. Due to this effect, the speed-ups are less than optimal. The combination of OpenMP and MPI is particularly of interest to reduce the adverse effect of the domain decomposition on the preconditioner. By using OpenMP on the nodes and MPI between the nodes, the number of subdomains is reduced from the number of processors to the number of nodes. We have shown an example where the combined method was for this reason more efficient than pure MPI.

References

1. L. Giraud, Combining Shared and Distributed Memory Programming Models on Clusters of Symmetric Multiprocessors: Some Basic Promising Experiments, *Int. J. High Perf. Comput. Appl.* **16** (2002).

2. OpenMP Architecture review Board. OpenMP Fortran Application Program Interface, Technical Report Version 2.0 (2000).

3. Message Passing Interface Forum. MPI: A message-passing interface standard, *Internat. J. Supercomputer Appl. and High Performance Computing* **8(3/4)** (1994).

4. F.B. Jensen et al. *Computational Ocean Acoustics*, AIP Series in Modern Acoustics and Signal Processing, Section 7.4, American Institute of Physics, New York, 1994.

5. T.J.R. Hughes, R.M. Ferencz, and J.O. Hallquist, Large-scale vectorized implicit calculations in solid mechanics on a Cray X-MP/48 utilizing EBE preconditioned conjugate gradients, *Comput. Meth. Appl. Mech. Engrg.* **61** (1987), 215–248.

6. M.B. van Gijzen, Parallel ocean flow computations on a regular and on an irregular grid, in *Lect. Notes Comp. Sci.*, **1067**, Springer-Verlag, 1996, 207–212.

7. M.R. Hestenes and E. Stiefel, Methods of conjugate gradients for solving linear systems, *J. Res. Natl. Bur. Stand.* **49** (1954), 409–436.

12 Models for the Simulation of Electrostatic Precipitators

Roger Godard, Jen Shi Chang, and Xiaoyi Xu

12.1 Introduction

The prime objective of this work is to present models for the prediction of electrostatic collection (precipitation) of fine particulate from industrial flue gases. This technique includes a coupling between fluid flow, electric parameters, and fine particle transport inside electrostatic precipitators [1,2,3]. The final objective is the optimization of the parameters involved in the reactor.

12.2 The Initial Value Problem for Negative Ions: the Clean Air Conditions

We consider a cylindrical apparatus of radius r_2 having an inner electrode of length L and of radius r_1. We assume that the length of the inner electrode is long enough that we can neglect edge effects. In order to present the model, we first consider clean air conditions with no dust. When the applied potential $V(r_1)$ is negative enough, the air is ionized and we observe the creation of a thin space charge of electrons and positive ions around the inner electrode. Because $V(r_1)$ is deeply negative, the positive ion layer is concentrated at the proximity of the inner electrode. From attachment of electrons to neutral species, we have creation of negative ions, which flow towards the outer electrode. The governing equations for a steady-state problem are as follows, in the domain $D_1 := (r_1 < r < r_2)$:

$$0 = \eta\mu_e||\vec{E}||N_e - \mu_n\nabla \cdot (N_n\vec{E}),$$
$$0 = \alpha\mu_e||\vec{E}||N_e - \mu_p\nabla \cdot (N_p\vec{E}),$$
$$0 = [\alpha - \eta]\mu_e||\vec{E}||N_e - \mu_e\nabla \cdot (N_e\vec{E}),$$
$$\nabla^2 V = -e(N_p - N_e - N_n)/\epsilon_0. \qquad (12.1)$$

The above equations constitute a system of non-linear equations with variable coefficients, which are linked together. Here: $\eta = \eta(||\vec{E}||)$ is the attachment coefficient, $\alpha = \alpha(||\vec{E}||)$ is the ionization coefficient, $\mu_e = 4.01||E||^{-0.501}$, where E in (V/cm), is the electron mobility, μ_n is the negative ion mobility, μ_p is the positive ion mobility, N_e, N_n, N_p are the electron and ion number densities respectively. E is the electric field and $V(r)$ is the electric potential. In the Poisson equation (12.1), e is the elementary charge and ϵ_0 is permittivity in vacuum. The Dirichlet boundary conditions are

$$V(r_1) = V_{dc}, \quad V(r_2) = 0, \quad N_p(r_1) = N_e(r_1), \quad N_n(r_1) = 0.$$

$N_e(r_1)$ is derived from the collected current, which is an input to the model. We need only one boundary condition for each species because the partial differential equations are of the first order. If we write the finite difference continuity equation for the electrons with a constant stepsize, from an upwind scheme we obtain

$$N_e[i] := \frac{r[i-1] \times N_e[i-1] \times E[i-1]}{(r[i] \times E[i]) \times (1 - \Delta r \times (\alpha - \eta))}.$$

A constant stepsize Δr was chosen along the computational domain for the next computations even if for clean air conditions, it is faster to introduce a variable stepsize. To avoid numerical oscillations, we have the following constraint: $\Delta r < 1/|\alpha - \eta|$. The finite difference Poisson equation is represented by a tridiagonal system, which is solved by Gaussian elimination. The major loop on the potential is obtained by underrelaxation. We observed a smooth variation of the negative ion density and a saturation process. With respect to a flat uniform guess field of negative ion density, an approximation which is often used in other models, we found a variation of 8% to 10% depending upon the applied voltage.

In comparison, a typical ion density profile for the wire plate electrode is more complex for a wire-plate geometry. For example, a saddle point in density can be observed near the wire electrode.

12.3   Electrostatic Precipitation

Because the negative ions are dominant, the charged particles consist upon negative ions and charged dust particles. Given $z = 0$, the level at which the dust is introduced, the dust (or aerosols) is not ionized and the ion profile is the initial ion profile from the clean air conditions. The computational domain becomes $D_2 := \{(r,z)|r_1 < r < r_2, \ 0 < z < L\}$. We assume that initially, the dust is uniformly distributed along the radial direction, i.e. $N_d(r_1 < r < r_2, \ z = 0) = 1$, with $N_d(r_1, z) = 0$ and $N_d(r_2, z) = 0$. Therefore, we assume that N_d is initially introduced uniformly over the cross section of the inlet:

The continuity equation for the dust is:

$$\vec{U}_g \cdot \nabla N_d - D_d \nabla^2 N_d = -k N_n N_d. \tag{12.2}$$

The first term corresponds to a convection/advection problem, while the second term is the diffusion term, and $D_d = C_m \frac{kT_g}{6\pi\mu_g r_d}$ is the diffusion coefficient; $k_n = k_n(\|\vec{E}\|)$ is the particle charging rate for the ionized dust [5,6]; U_g is the gas velocity. We neglect the gravity effects for the size of particles below 10 μm. Equation (12.2) means that the dust decreases through the effects of charging and transport, and the sink is proportional to the density of the negative ions and the density of the dust. Negative ions have the same loss term, and the continuity equation becomes

$$\vec{U}_g \cdot \nabla N_n - \mu_n \nabla \cdot (N_n \vec{E}) = -k_n N_n N_d.$$

The continuity equation for the charged dust is

$$\vec{U}_g \cdot \nabla N_{cd} - \mu_{cd} \nabla \cdot (N_{cd} \vec{E}) - D_d \nabla^2 N_{cd} = +k_n N_n N_d.$$

At $z = 0$, we can assume that $N_{cd}(r, 0) = 0$. For the negatively charged dust, we include convection effects, diffusion processes, and the effect of the electric field. Because, we consider diffusion terms, the partial differential equation is elliptic, and we have chosen the Dirichlet boundary conditions as

$$N_{cd}(r_1) = 0; \quad N_{cd}(r_2) = 0,$$

but could have chosen floating boundary conditions as well, that is, Neumann conditions. The above system of equations is linked together through the Poisson equation.

12.4 The 1.5D Model

In the 1.5D model, we include only the effect of the axial velocity for a viscid laminar flow or turbulent flow, and Navier–Stokes or Reynolds equations are decoupled from the transport equations [4].

We have solved the above equations in their finite difference approximations, with a constant stepsize in the radial direction. For the convection contribution, we selected an upwind scheme, so that the tridiagonal matrix is always diagonally dominant. We emphasize that source and sink are proportional to $N_d(i)$. Therefore we always have to solve systems of linear equations. At each z, we solve the 1.5D problem. Because the solution at the level $(k + 1)$ depends upon the solution at the level (k), once the solution $(k+1)$ is reached, we erase the level (k), and we dump the contents of level (k) into an auxiliary memory for display. And the Courant number would be: $\Delta z \leq (\Delta r)^2 R_{ed}/2$, where R_{ed} is the Reynolds number. Here, we have assumed for the numerical tests, that we only have one population of negative ions, one population of dust, one population of negatively charged dust particles.

We now present some preliminary numerical results. Given an averaged collected current, and the applied potential, the program computes automatically the reference negative ion number density and the self-consistent profile. The negative ion density is very stable as a function of the vertical distance z. As a function of z, we observe a decrease of the dust (pollutant), and a variation of the negatively charged dust. Figure 1 shows the vertical profile of the decrease of the dust density in the variable $n_d = N_d/N_{d0}$ at a radial distance $(r_1 + r_2)/2$, while Figure 2 shows the vertical profile of the charged dust density $n_{cd} = N_{cd}/N_{d0}$ at the same radial distance. We observe an increase of the charged dust density from the level $z = 0$, then a decrease, because the density of the dust is also decreasing as a function of z.

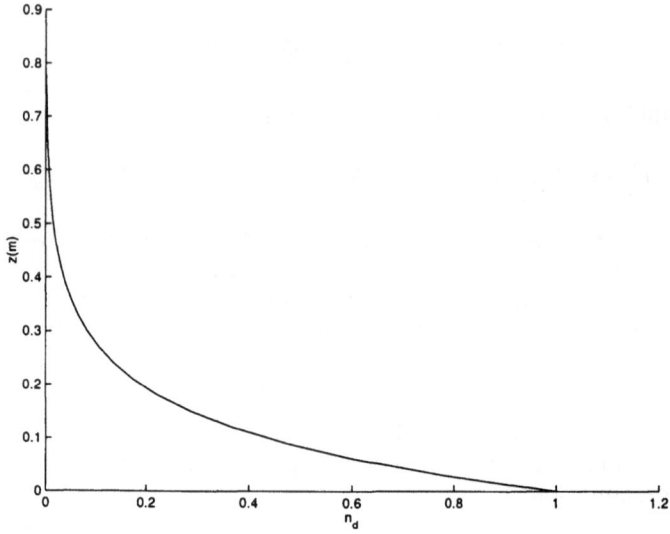

Fig. 1. Variation of the dust density with the vertical distance z.

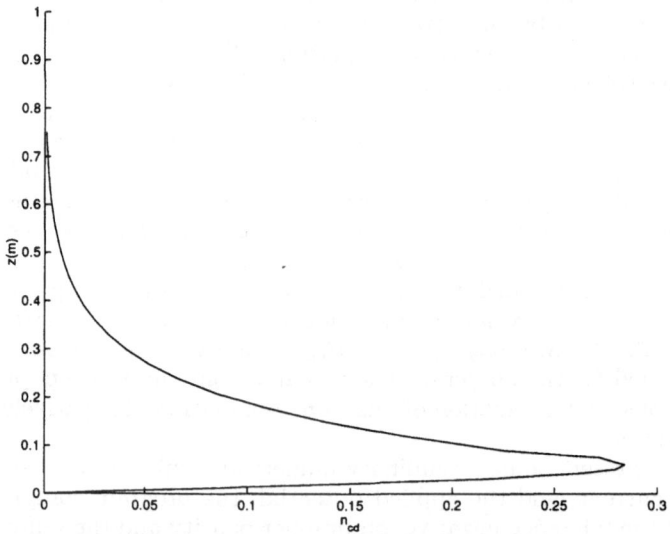

Fig. 2. Variation of the charged dust density with the vertical distance z.

We would like to improve the code and to add a full coupling between the fluid equations and the charge transport equations, to consider gravity effects, and turbulence.

12.5 The Initial Value Problem for the Pulse Corona: the Time-Dependent and Pseudo-Spectral Approach

Our objective is not to present a full numerical code for the electrostatic precipitation for the difficult problem of a periodic pulse boundary condition, but to emphasize new numerical difficulties that we encounter. Firstly, we observe that in a dynamic system, or a filter, an input pulse is translated by a non-linear convolution integral for the output. In order to follow the response of the system, we have basically the choice between two approaches: the spectral or frequency approach, i.e. a Fourier analysis of the system, or to follow the time-space problem for the duration of a period. The period has to be adjusted so that there is no piling-up of ionization waves, and that there is a complete absorption at the surface of the outer electrode and no reflection. An important parameter is the computation of the travel time of the ionization wave and its velocity. Here, we consider only the clean air conditions for the time-dependent problem. The governing equations become:

$$\frac{\partial N_n}{\partial t} = \eta \mu_e ||\vec{E}|| N_e(r, t - \tau) - \mu_n \nabla \cdot (N_n \vec{E}), \tag{12.3}$$

$$\frac{\partial N_p}{\partial t} = \alpha \mu_e ||\vec{E}|| N_e(r, t - \tau) - \mu_p \nabla \cdot (N_p \vec{E}), \tag{12.4}$$

$$\frac{\partial N_e}{\partial t} = [\alpha - \eta] \mu_e ||\vec{E}|| N_e(r, t - \tau) - \mu_e \nabla \cdot (N_e \vec{E}), \tag{12.5}$$

$$\nabla^2 V = -\frac{e}{\epsilon_0} (N_p - N_e - N_n) \tag{12.6}$$

in the domain $D_3 := \{(r, t) | r_1 < r < r_2,\ 0 < t < T\}$, where T is the period of the pulse. τ is an attempt to model the time delay and the convolution effects. The boundary conditions become:
(a) for the electric potential:

$$V(r_1, t) = V_m \frac{t}{t_{\max}} e^{(1 - t/t_{\max})} + V_{dc}, \quad V(r_1, t + T) = V(r_1, t),$$
$$V(r_2, t) = 0; \tag{12.7}$$

here, V_m is the peak voltage of the negative pulse and $t_{\max}$ is the rise time;
(b) for the densities:

$$N_n(r_1, t) = 0, \quad N_p(r_1, t) = N_e(r_1, t). \tag{12.8}$$

The main difficulty comes from the crucial boundary condition for the electron density at the inner electrode, which is a non-local boundary condition. A reasonable Dirichlet boundary condition is to consider (from experimental data) an empirical fit of the type: $N_e(r_1, t) = f(||\vec{E}(r_1, t)||)$.

Because the ionization coefficient increases exponentially with $||\vec{E}||$, electrons respond very fast to the negative potential pulse at the inner boundary condition, while the negative ions respond more slowly and smoothly.

The explanations come from the Dirichlet boundary condition $N_n(r_1) = 0$ which is maintained to zero and the fact that the attachment coefficient has a much slower increase as a function of $\|\vec{E}\|$. Therefore the system of equations (12.3)–(12.8) introduces a process of regularization in the negative ion profile. Also the geometry of the reactor is an important factor. For example the wire-plate geometry shows a more complex negative ion density profile.

Because of the large computer time required for a time dependent simulation, and the Courant–Friedrich–Levy condition, we may consider a spectral approach. A true spectral model is a 3D model. Because the boundary condition for the electric potential has the shape of a pulse, its Fourier representation takes roughly 60 terms instead of the 2000 time steps required for the time computation. We may even go further, and to search a simplified approach for the industry based upon an average negative ion content and an equivalent average applied potential where we can use the 1.5 DC model, with an estimation of the radial propagation velocity of the negative ion wave.

It is promising to solve the following 2.5 pseudo DC model based on the following remarks. Due to the frequency of the applied voltage, we can consider that we get intense RF waves, or more exactly a superposition of RF signals. Each wave is equivalent to a superposition of static potentials, and this algorithm enables us to compute an equivalent DC potential profile:

$$\varphi_n(r) = \left(\frac{q}{4m_i \varpi_n^2} \right) E_n^2(r).$$

References

1. R. Morrow, Theory of negative corona in oxygen, *Phys. Rev.* **32** (1985), 1799–1809.

2. I. Gallimberti, Recent advancements in the physical modelling of electrostatic precipitators, *J. Electrostatics* **43** (1998), 219–247.

3. E.A. Gerteisen, W. Egli, and U. Kogelschatz, Solution methods for ion flows in electrostatic precipitators, in *ECCOMAS, Computational Fluid Dynamics '98, Proceedings Fourth European Comput. Fluid Dynamics Conf.*, Athens, Greece, September 1998.

4. R. Godard, X. Xu, J.S. Chang, and A.A. Berezin, Development of electrostatic precipitator optimization codes for control of fine particulate in industrial flue gas, *Conf. Proceedings, Deuxième Conférence sur L'Électrostatique SFE 2000*, Montpellier, France, July 2000.

5. J.S. Chang, Electrostatic charging of particles, in *Handbook of Electrostatic Processes*, J.S. Chang, A.J. Kelly, and J.M. Crowley editors, Marcel Dekker, New York, 1995.

6. P.A. Lawless, Particle charging bounds, symmetry relations, and an analytical charging rate model for the continuum regime, *J. Aerosol Sci.* **27** (1996), 191–215.

13 A Method For Modeling Nonharmonic Periodic Acoustic Radiation from a Loudspeaker

Paul J. Harris, Hui Wang, Roma Chakrabarti, and David Henwood

13.1 Introduction

In the past most of the work on the problem of determining the acoustic field radiated by a loudspeaker, or similar structure, has concentrated on the harmonic problem [1,2,3,4,5,6,7,8]. This has the advantage that the frequency domain problem is essentially a purely boundary value type problem and methods exist for its efficient numerical solution. However, in many situations the radiation is either periodic but not harmonic, or transient.

This paper presents a method for extending the analysis of the time-harmonic problem to problems where the vibrations of the structure are still periodic but no longer harmonic. For example, an object may undergo a short vibration at regular time-intervals, and it may be necessary to determine how the sound pulse produced by each vibration radiates out into the surrounding medium. The methods used in this paper are based around expanding the time-dependent part of the object's vibrations in a Fourier series and then solving a series of time-harmonic problems at these frequencies. The time-harmonic calculations will be carried out using the boundary integral method and compared to an exact analysis for the vibrations of a sphere.

13.2 Mathematical Model

Let D_- denote the space occupied by a finite object in $\mathbf{R}^3$ and let S denote the surface of the object. The otherwise unbounded region D_+ exterior to S is assumed to be filled with compressible fluid of uniform density ρ and speed of sound c. It is well known that small amplitude pressure waves propagating through D_+ can be described by the linear wave equation [9]

$$\nabla^2 \Phi(x, t) = \frac{1}{c^2} \frac{\partial^2 \Phi(x, t)}{\partial t^2}, \tag{13.1}$$

This work was partly supported by B & W Loudspeakers Ltd., Steyning, West Sussex.

where $\Phi(x,t)$ denotes the excess pressure. The boundary conditions for the problem are that the surface S undergoes small amplitude displacements with corresponding velocities $V(x,t)$, both of which are assumed to be periodic with period T. The surface velocity is related to the normal derivative of the pressure by [9]

$$\frac{\partial \Phi(x,t)}{\partial n_x} = -\rho \frac{\partial V(x,t).n_x}{\partial t}, \quad x \in S, \tag{13.2}$$

where n_x denotes the unit normal to S at x directed outwards towards D_+. The requirements that all radiated waves are outgoing at infinity is expressed as the radiation condition

$$\lim_{r \to \infty} r \left\{ \frac{\partial \Phi}{\partial r} + \frac{1}{c} \frac{\partial \Phi}{\partial t} \right\} = 0, \tag{13.3}$$

which must be satisfied in all directions $\frac{\mathbf{r}}{r}$ where $r = |\mathbf{r}|$.

Since the surface displacements and velocities are periodic with period T, the solution $\Phi(x,t)$ will also be periodic with period T and hence it is possible to express the solution as the real part of the Fourier series

$$\Phi(x,t) = \sum_{j=1}^{\infty} \phi_j(x) e^{-i\omega_j t}, \tag{13.4}$$

where $\omega_j = \frac{2\pi j}{T}$. It is noted that $\phi_0(x) = 0$ since the displacements of the surface are periodic. It can be shown that if $\frac{\partial \Phi(x,t)}{\partial t}$ is piecewise continuous in time for all $x \in S \cup D_+$, then (13.4) is uniformly convergent for all t.

Substituting (13.4) into (13.1) leads to the requirement

$$\nabla^2 \phi_j(x) + k_j^2 \phi_j(x) = 0, \tag{13.5}$$

where $k_j = \frac{\omega_j}{c}$. That is, the Fourier coefficients $\phi_j(x)$ must satisfy an appropriate Helmholtz or reduced wave equation. Further, substituting the series (13.4) into radiation condition (13.3) shows that each Fourier coefficient must also satisfy the usual Sommerfeld radiation condition.

The normal derivative of the Fourier coefficients can be expressed in terms of the boundary condition (13.2) as

$$\frac{\partial \phi_j(x)}{\partial n_x} = -\frac{2\rho}{T} \int_0^T \frac{\partial V(x,t).n_x}{\partial t} e^{i\omega_j t} \ dt. \tag{13.6}$$

Hence the solution to the wave equation is transformed into the problem of solving the sequence of Helmholtz equations (13.5) subject to the boundary conditions (13.6) and the Sommerfeld radiation conditions.

One of the most widely used methods for solving the exterior Helmholtz problem is the boundary integral method. Using Green's second theorem it is possible to rewrite the governing differential equation (13.5) as the integral equation

$$\int_S \left(\phi_j(y) \frac{\partial G_k(x,y)}{\partial n_y} - G_k(x,y) \frac{\partial \phi_j(y)}{\partial n_y} \right) dS_y = \frac{1}{2} \phi_j(x), \qquad (13.7)$$

where $G_k(x,y) = \frac{e^{ik|x-y|}}{4\pi|x-y|}$ is the free-space Green's function or fundamental solution for the Helmholtz equation (note that for clarity the suffix j has been omitted from the wavenumber k in this and subsequent boundary integral equations).

However, it is well known that for certain discrete values of the wavenumber, called characteristic wavenumbers, (13.7) does not have a unique solution. Although it is possible that the wavenumbers corresponding to the angular frequencies $\omega_j = \frac{2j\pi}{T}$ are not equal to any of the characteristic wavenumbers, problems still arise whenever the wavenumber is close to one of the characteristic wavenumbers due to the integral equation becoming ill-conditioned. This problem is more apparent at higher wavenumbers as the density of the characteristic wavenumbers increases [3].

A number of methods for overcoming this problem with the integral equation formulation have been developed, [2,6,7], and it is the method of Burton and Miller [2] that has been used here. It can be shown that a linear combination of (13.7) and its normal derivative at x of the form

$$-\frac{1}{2}\phi_j(x) + \int_S \left(\frac{\partial G_k(x,y)}{\partial n_y} + \alpha \frac{\partial^2 G_k(x,y)}{\partial n_x \partial n_y} \right) \phi_j(y) \, dS_y$$

$$= \frac{\alpha}{2} \frac{\partial \phi_j(x)}{\partial n_x} + \int_S \left(G_k(x,y) + \alpha \frac{\partial G_k(x,y)}{\partial n_x} \right) \frac{\partial \phi_j(y)}{\partial n_y} \, dS_y, \qquad (13.8)$$

where α is a coupling constant, has a unique solution for all real and positive wavenumbers, provided the imaginary part of the coupling parameter α is chosen to be nonzero. Further, it can be shown that the choice $\alpha = \frac{i}{k}$ is the "almost optimal" choice in the sense that it is the value of α which almost minimises the condition number of the resulting integral operator [1,10].

The drawback of the Burton and Miller method is that it introduces the hyper-singular operator with kernel function $\frac{\partial^2 G_k}{\partial n_x \partial n_y}$. In fact, the derivative with respect to n_x in (13.8) should not be taken inside the integral, but this is nearly always done in practise. Meyer et al [5] show that it is possible to write

$$\int_S \phi_j(y) \frac{\partial^2 G_k}{\partial n_x \partial n_y} \, dS_y = \int_S (\phi_j(y) - \phi_j(x)) \frac{\partial^2 G_k}{\partial n_x \partial n_y} \, dS_y$$

$$+ k^2 \int_S G_k(x,y) n_x \cdot n_y \, dS_y. \qquad (13.9)$$

In the work presented here, the boundary element method based on a piecewise constant collocation scheme has been used to solve (13.8), along with

(13.9) to interpret the second derivative of the Green's function, in order to compute the Fourier coefficients ϕ_j. Further details of the collocation method can be found in [11], for example.

Once the Fourier coefficients on the surface of the structure has been determined it is possible to compute the Fourier coefficients in the exterior domain using

$$\int_S \left(\phi_j(y) \frac{\partial G_k(x,y)}{\partial n_y} - G_k(x,y) \frac{\partial \phi_j(y)}{\partial n_y} \right) dS_y = \phi_j(x),$$

which, in turn, can be used in (13.4) to compute the solution to the wave equation.

13.3 Numerical Results

The first problem considered is the determination of the acoustic radiation from a sphere whose surface is vibrating with a uniform nonharmonic periodic wave. This problem was considered as a test problem as it is possible to obtain the exact solution. Fig. 1 presents the results showing the pressure at different distances from the surface of the sphere at time $t = 0.0025$ after the current period has started. Here the speed of sound is $340ms^{-1}$ and the radius of the sphere is $0.1m$. As can be seen, there is excellent agreement between the exact solution and the computed numerical solution.

Fig. 2 compares the computed pressure with the experimentally determined pressure at a point $1.5m$ in front of a typical loudspeaker. Here the surface velocities used as the boundary data for the boundary integral method were found by measuring the velocity of the loudspeaker cone as it was vibrating. The numerical results have been scaled so that the magnitude of the first peak in the pressure is the same as the first peak in the experimental data. This is due to the unknown calibration constant for the equipment measuring the velocity of the surface of the loudspeaker. These results shows that initially there is good agreement with the experimentally determined pressure, but at later times the agreement is not so good. However, this is almost certainly due to experimental error such as the small echoes that can occur, even in an anechoic chamber.

13.4 Conclusion

The results presented in this paper show that the Fourier approach is an accurate method for modelling non-harmonic periodic acoustic radiation, provided the wavenumbers corresponding to the Fourier frequencies are not too large. However, calculating the solutions to the sequence of Helmholtz equations is computationally expensive, especially at higher frequencies where a large number of boundary elements are needed.

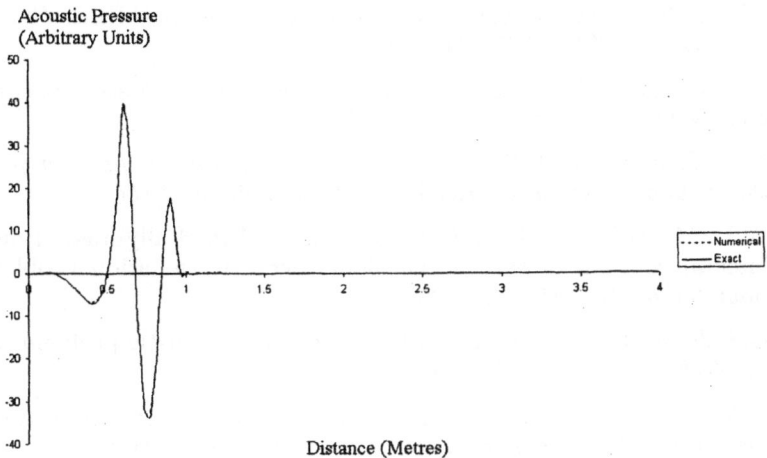

Fig. 1. Comparing the computed acoustic pressure radiated from a sphere at time $t = 0.0025$ with the exact solution.

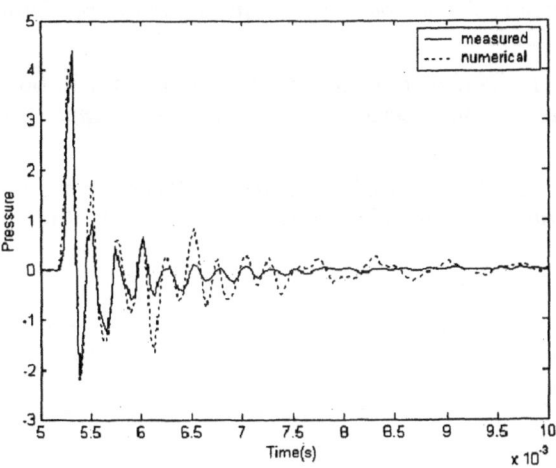

Fig. 2. A comparison of the predicted and experimental acoustic radiation from a typical loudspeaker.

References

1. S. Amini, P.J. Harris, and D.T. Wilton, *Coupled boundary and finite element methods for the solution of the dynamic fluid-structure interaction problem,* Lect. Notes in Engng. Springer-Verlag, Berlin-New York-London, 1992.

2. A.J. Burton and G.F. Miller, The application of integral equation methods to the numerical solution of boundary value problems. *Proc. Roy. Soc. Lond.* **A232** (1971), 201–210.

3. D. Colton and R. Kress, *Integral equation methods in scattering theory*, John Wiley and Sons, New York, 1983.

4. M.A. Jaswon and G.T. Symm, *Integral equation methods in potential theory and elastostatics*, Academic Press, London, 1977.

5. W.L. Meyer, W.A. Bell, B.T. Zinn, and M.P. Stallybrass, Boundary integral solution of three-dimensional acoustic radiation problems, *J. Sound Vib.* **59** (1978), 245–262.

6. D.S. Jones, Integral equations for the exterior acoustic problem, *Quart. J. Mech. Appl. Math.* **27** (1976), 129–142.

7. H.A. Shcenck, Improved integral equation formulations for acoustic radiation problems, *J. Acoust. Soc. Amer.* **44** (1968), 41–58.

8. A.F. Seybert, B. Soenkarko, F.J. Rizzo, and D.J. Shippy, A special integral equation formulation for acoustic radiation and scattering for axisymmetric bodies and boundary conditions, *J. Acoust. Soc. Amer.* **80** (1986), 1241–1247.

9. G.K. Batchelor, *An introduction to fluid dynamics*, Cambridge University Press, 1967.

10. S. Amini, On the choice of the coupling parameter in boundary integral formulations of the exterior acoustic problem, *Appl. Anal.* **35** (1989), 75–92.

11. K.E. Atkinson, *A survey of numerical methods for the solution of Fredholm integral equations of the second kind*, SIAM, Philadelphia, 1976.

14 Implicit Function Theorems and Discontinuous Implicit Differential Equations

Seppo Heikkilä

14.1 Introduction

In this paper we shall first prove an existence result for an implicit functional equation. The proof is based on a fixed point result in a Banach lattice derived in [1]. The so obtained implicit function theorem is then applied to an initial value problem of an implicit functional differential equation. The functions in the considered equations may be discontinuous in all their arguments. Special cases and a concrete example are given to demonstrate the obtained results.

14.2 A Fixed Point Lemma

Assume that $E = (E, \| \cdot \|, \leq)$ is a weakly complete Banach lattice, that J is a compact real interval, and that the space $L^p(J, E)$, $1 \leq p < \infty$, is ordered a.e. pointwise.

Lemma 1. *Assume that a mapping $G : L^p(J, E) \to L^p(J, E)$ is increasing, and that $\|Gv\|_p \leq M + h(\|v\|_p)$ for all $v \in L^p(J, E)$, where $M \geq 0$, $h : R_+ \to R_+$ is increasing and $M + h(r) \leq r$ for some $r \geq 0$. Then G has a fixed point.*

Proof. The weak completeness of any Banach lattice is by [2, Theorem 1.c4] equivalent to strong convergence of its bounded and monotone sequences. It then follows from [3, Proposition 5.8.7] that bounded and monotone sequences of $L^p(J, E)$ converge. By a hypothesis there exists an $r \geq 0$ such that $M + h(r) \leq r$. Since h is increasing, we have $G[P] \subset P$, where $P = \{v \in L^p(J, E) \mid \|v\|_p \leq r\}$. Since P is closed and bounded, its bounded and monotone sequences have limits in P. Because E is a Banach lattice, it is elementary to verify that $\| \sup\{0, v\}\|_p \leq \|v\|_p$, i.e. $\sup\{0, v\} \in P$ for each $v \in P$. Thus it follows from [1, Proposition 4] that G has a fixed point v which can be obtained in the following manner: The union C of those well-ordered subsets A of P (each nonempty subset of A has a minimum) whose elements satisfy $w = \sup\{0, \{Gu \mid u \in A, u < w\}\}$, is well-ordered, $b = \max C$ exists and $Gb \leq b$. The union D of those inversely well-ordered subsets B of P (each nonempty subset of B has a maximum) whose elements are of the form $w = \inf\{Gu \mid u \in B,\, u > w\}$, is inversely well-ordered, $v = \min D$ exists and $Gv = v$.

14.3 An Implicit Function Theorem

We shall prove an existence result for equation

$$v(t) = F(t, \phi(v), v) \quad \text{for a.e. } t \in J. \tag{14.1}$$

The following hypotheses are imposed on $\phi : L^p(J, E) \to L^p(J, E)$ and $F : J \times L^p(J, E) \times L^p(J, E) \to E$.

(ϕ) ϕ is increasing, and $\|\phi \circ v\|_p \leq m + c\|v\|_p$ for all $v \in L^p(J, E)$, where $m,\, c \geq 0$.

(F1) $t \mapsto F(t, u, v)$ is strongly measurable in J whenever $u,\, v \in L^p(J, E)$.

(F2) $(u, v) \mapsto F(\cdot, u, v) + \lambda v$ is increasing for some $\lambda \geq 0$.

(F3) $\|F(\cdot, u, v)\|_p \leq K + a\|u\|_p^\alpha + b\|v\|_p^\beta$ for all $u,\, v \in L^p(J, E)$, where $K, a, b \geq 0$, and either

(i) $0 < \alpha,\, \beta < 1$, or

(ii) $\alpha = \beta = 1$ and $ca + b < 1$, where c, a and b are as in (ϕ) and in (F3).

Theorem 1. *The equation (14.1) has under the hypotheses (F1)–(F3) and (ϕ) a solution v in $L^p(J, E)$, i.e., $v = F(\cdot, \phi(v), v)$.*

Proof. The hypotheses (F1) and (ϕ) imply that for each $v \in L^p(J, E)$ the relation

$$Gv = \frac{F(\cdot, \phi(v), v) + \lambda v}{1 + \lambda} \tag{14.2}$$

defines for each $v \in L^p(J, E)$ a strongly measurable function $Gv : J \to E$. To show that (14.2) defines a mapping $G : L^p(J, E) \to L^p(J, E)$, let $v \in L^p(J, E)$ be given. Applying the growth condition of (F3) and the hypothesis (ϕ) we obtain

$$(1 + \lambda)\|Gv\|_p \leq \|F(\cdot, \phi(v), v)\|_p + \lambda\|v\|_p$$

$$\leq K + a\,\|\phi \circ v\|_p^\alpha + b\,\|v\|_p^\beta + \lambda\|v\|_p$$

$$\leq K + a\,(m + c\|v\|_p)^\alpha + b\,\|v\|_p^\beta + \lambda\|v\|_p.$$

Thus $Gv \in L^p(J, E)$, and

$$\begin{cases} \|Gv\|_p \leq M + h(\|v\|_p), \quad \text{where} \\ M = \frac{K}{1+\lambda}, \quad h(r) := \frac{a\,(m+cr)^\alpha + b\,r^\beta + \lambda\,r}{1+\lambda}. \end{cases} \tag{14.3}$$

If $v_1, v_2 \in L^p(J, E)$, $v_1 \leq v_2$, then $\phi(v_1) \leq \phi(v_2)$ by (ϕ). These inequalities and the hypothesis (F2) imply that

$$(1 + \lambda)Gv_1 = F(\cdot, \phi(v_1), v_1) + \lambda v_1 \leq F(\cdot, \phi(v_2), v_2) + \lambda v_2 = (1 + \lambda)Gv_2.$$

This proves that G is increasing.

a) Assume first that the hypotheses (F3)(i) are satisfied. Since $0 < \alpha, \beta < 1$, then the mapping $h : R_+ \to R_+$ defined in (14.3) is increasing, and $\frac{M+h(r)}{r} \to \frac{\lambda}{1+\lambda} < 1$ as $r \to \infty$. Thus $M + h(r) \le r$ when r is large enough.

b) Assume next that the hypotheses (F3)(ii) hold. Since $\alpha = \beta = 1$, then h given by (14.3) is of the form $h(r) = \frac{a(m+cr)+br+\lambda r}{1+\lambda}$. Since $ca + b < 1$, then $M + h(r) \le r$ when $r \ge \frac{M+am}{1-(ca+b)}$.

The above proof shows that G satisfies the hypotheses of Lemma 1 in the cases a) and b), whence G has a fixed point $v \in L^p(J, E)$. This implies by the definition (14.2) of G that $v = F(\cdot, \phi(v), v)$, i.e., v satisfies (14.1).

14.4 An Existence Result for an Initial Value Problem

As an application of Theorem 1 we prove an existence result for the implicit initial value problem (IVP)

$$Au(t) = F(t, u, Au) \text{ for a.e. } t \in J, \quad u(t_0) = u_0, \tag{14.4}$$

where

$$Au(t) = \frac{d}{dt}(\psi(t)u(t)), \text{ with } \frac{1}{\psi} \in L^1_+(J). \tag{14.5}$$

Theorem 2. *Assume that $F : J \times L^1(J, E) \times L^1(J, E) \to E$ satisfies the hypotheses (F1)–(F3) with $p = 1$ and $c = \|\frac{1}{\psi}\|_1$. Then the IVP (14.4),(14.5) has for each choice of $t_0 \in J$ and $u_0 \in E$ a solution in the set $W = \{u \in L^1(J, E) \mid \psi \cdot u \text{ is absolutely continuous and a.e. differentiable}\}$.*

Proof. The relation

$$\phi(v)(t) := \frac{1}{\psi(t)} \left(\psi(t_0)u_0 + \int_{t_0}^t v(s) \, ds \right), \quad t \in J, \tag{14.6}$$

defines mapping $\phi : L^1(J, E) \to L^1(J, E)$, which is increasing and satisfies

$$\|\phi \circ v\|_1 \le \left\| \frac{1}{\psi} \right\|_1 (\psi(t_0)\|u_0\| + \|v\|_1) \text{ for all } v \in L^1(J, E).$$

Thus the hypothesis (ϕ) holds with $p = 1$ and $c = \|1/\psi\|_1$. It then follows from Theorem 1 that the equation $v = F(\cdot, \phi(v), v)$ has a solution v in $L^1(J, E)$. Denoting $u = \phi(v)$ we obtain by (14.5) and (14.6)

$$Au(t) = \frac{d}{dt}(\psi(t)u(t)) = v(t) \text{ for a.e. } t \in J.$$

Thus

$$Au = v = F(\cdot, \phi(v), v) = F(\cdot, u, Au),$$

whence the first equation of (14.4) holds. Since $u = \phi(v)$ satisfies by (14.6) also the initial condition $u(t_0) = u_0$, then u is a solution of the IVP (14.4), (14.5) in W.

14.5 Special Cases

As an application of Theorem 1 we obtain an existence result for equation

$$v(t) = f(t, \phi(v)(t), v(t)) \quad \text{for a.e. } t \in J, \tag{14.7}$$

when the function $f : J \times E \times E \to E$ satisfies the following hypotheses.
(f1) f is sup-measurable, i.e., $t \mapsto f(t, u(t), v(t))$ is strongly measurable in J whenever $u, v : J \to E$ are strongly measurable.
(f2) $(y, z) \mapsto f(t, y, z) + \lambda z$ is increasing for a.e. $t \in J$ and for some $\lambda \geq 0$.
(f3) $\|f(t, y, z)\| \leq k(t) + c_1(t)\|y\|^{\alpha} + c_2(t)\|z\|^{\beta}$ for a.e. $t \in J$ and for all $y, z \in E$, where $k \in L^p(J)$, $c_1 \in L^{\frac{p}{1-\alpha}}(J)$ and $c_2 \in L^{\frac{p}{1-\beta}}(J)$, and either
(i) $0 < \alpha, \beta < 1$, or
(ii) $\alpha = \beta = 1$ and $c\|c_1\|_{\infty} + \|c_2\|_{\infty} < 1$.

Theorem 3. *The equation (14.7) has under the hypotheses (f1)–(f3) and (ϕ) a solution v in $L^p(J, E)$.*

Proof. The given hypotheses imply that a function $F : J \times L^p(J, E) \times L^p(J, E) \to E$ given by

$$F(t, u, v) = f(t, u(t), v(t)), \quad t \in J, \ u, v \in L^p(J, E). \tag{14.8}$$

F satisfies the hypotheses (F1)–(F3) with $K = \|k\|_p$, $a = \|c_1\|_{\frac{p}{1-\alpha}}$ and $b = \|c_2\|_{\frac{p}{1-\beta}}$. Thus the equation $v = F(\cdot, \phi(v), v)$ has by Theorem 1 a solution v in $L^p(J, E)$. This and (14.8) imply that v is also a solution of (14.7).

As an application of Theorem 3 one can prove an existence result for the implicit initial value problem

$$\frac{d}{dt}(\psi(t)u(t)) = f(t, u(t), \frac{d}{dt}(\psi(t)u(t))) \quad \text{for a.e. } t \in J, \quad u(t_0) = u_0. \tag{14.9}$$

Theorem 4. *Assume that $1/\psi \in L^1_+(J)$, and that $f : J \times E \times E \to E$ satisfies the hypotheses (f1)–(f3) with $p = 1$ and $\kappa = \|1/\psi\|_1$. Then the IVP (14.9) has for each choice of $t_0 \in J$ and $u_0 \in E$ a solution in W.*

Remarks. The given hypotheses allow the functions F, f, ϕ and ψ to be discontinuous in all their arguments.
The following spaces are examples of weakly complete Banach lattices:
– A reflexive (e.g., a uniformly convex) Banach lattice.
– A uniformly monotone Banach lattice in the sense defined in [4].

– A finite-dimensional normed space ordered by a cone generated by a basis.

– A separable Hilbert space whose order cone is generated by an orthonormal basis.

– l^q, $1 \leq q < \infty$, normed by q-norm and ordered componentwise.

– $L^q(\Omega)$, $1 \leq q < \infty$, normed by q-norm and ordered a.e. pointwise.

Thus all the results derived above hold when E is one of the spaces listed above. These results hold also when E is a reflexive lattice-ordered Banach space whose lattice operation $x \mapsto x^+ = \sup\{0, x\}$ is continuous and also nonexpansive, i.e., $\|x^+\| \leq \|x\|$ for all $x \in E$. Thus E can be, e.g., one of the Sobolev spaces $W^{1,q}(\Omega)$ or $W_0^{1,q}(\Omega)$, $1 < q < \infty$, ordered a.e. pointwise, where $\Omega \subset R^N$ is a bounded domain with Lipschitz boundary (cf. [5, C4]).

The results of Theorems 1 and 3 hold also when the compact interval J is replaced by any measure space Ω. This allows applications to partial differential equations. (cf. [6]).

14.6 An Example

Choose $J = [0, 3]$, and consider the following problem.

$$u'(t) = \frac{[u(3-t) + 2t - 1]}{8} + \frac{[\int_0^3 [u'(s)]ds]}{8} + \frac{1}{10} \quad \text{a.e. in } J, \ u(0) = 0,$$

$$(14.10)$$

where $[z]$ denotes the greatest integer $\leq z$. Problem (14.9) is of the form (14.4), where $\psi(t) \equiv 1$ and

$$F(t, u, v) = \frac{[u(3-t) + 2t - 1]}{8} + \frac{[\int_0^3 [v(s)]ds]}{8} + \frac{1}{10}, \quad t \in J, \ u, \ v \in L^1(J).$$

Obviously, the hypotheses (F1), (F2) and (F3)(ii) are valid when $\lambda = 0$ $c = \|1/\psi\|_1 = 3$, $a = \frac{1}{8}$ and $b = \frac{3}{8}$. It then follows from Theorem 2 that the IVP (14.10) has a solution. By the proofs of Theorems 1 and 2 a solution of (14.10) is of the form

$$u(t) = \phi(v)(t) = \int_0^t v(s)\, ds,$$

where $v \in L^1(J)$ is a fixed point of the operator $G : L^1(J) \rightarrow L^1(J)$ defined by

$$Gv = F(\cdot, \phi(v), v).$$

It turns out that for this mapping G the sets C and D constructed in the proof of Lemma 1 are finite. Thus a fixed point of G is the last member of the finite sequence $C \cup D$, which can be determined by the following

Algorithm. $v_0 \equiv 0$: For n from 0 while $v_n \neq Gv_n$ do: If $v_n < \sup\{0, Gv_n\}$, then $v_{n+1} = \sup\{0, Gv_n\}$, else $v_{n+1} = Gv_n$ od;

Inserting $Gv_n = F(\cdot, \phi(v_n), v_n))$ in the above algorithm and approximating the integrals $\phi(v_n)$ by Simpson rule one can calculate an estimate for a solution of (14.10):

$$
u(t) \approx
\begin{cases}
-.151\,t, & 0 < t \le .324, \\
-.041 - .026\,t, & .324 < t \le .941, \\
-.162 + .1\,t, & .941 < t \le 1.505, \\
-.351 + .224\,t, & 1.505 < t \le 2.033, \\
-.605 + .349\,t, & 2.033 < t \le 2.527, \\
-.92 + .474\,t, & 2.527 < t \le 3.
\end{cases}
$$

With the help of (14.10) one can infer an exact formula for u:

$$
u(t) =
\begin{cases}
-\frac{3}{20}t, & 0 \le t \le \frac{1626572}{5061499}, \\[2mm]
-\frac{406643}{10122998} - \frac{1}{40}t, & \frac{1626572}{5061499} \le t \le \frac{4752672}{5062499}, \\[2mm]
-\frac{1594811}{10122998} + \frac{1}{10}t, & \frac{4752672}{5062499} \le t \le \frac{144634582}{96168481}, \\[2mm]
-\frac{66460041}{192336962} + \frac{9}{40}t, & \frac{144634582}{96168481} \le t \le \frac{195249518}{96168481}, \\[2mm]
-\frac{12133939}{20245996} + \frac{7}{20}t, & \frac{195249518}{96168481} \le t \le \frac{12785397}{5062499}, \\[2mm]
-\frac{37053275}{40491992} + \frac{19}{40}t, & \frac{12785397}{5062499} \le t \le 3.
\end{cases}
$$

References

1. S. Heikkilä, A method to solve discontinuous boundary value problems, in *Proc. Third World Congress on Nonlinear Analysts*, Nonlinear Anal. **47** (2001), 2387–2394.

2. J. Lindenstraus and L. Tzafriri, *Classical Banach Spaces. II. Function Spaces*, Springer-Verlag, Berlin, 1979.

3. S. Heikkilä and V. Lakshmikantham, *Monotone Iterative Techniques for Discontinuous Nonlinear Differential Equations*, Marcel Dekker, New York-Basel-Hong Kong, 1994.

4. G. Birkhoff, *Lattice Theory*, Amer. Math. Soc. Publ. **25**, Providence, Rhode Island, 1973.

5. S. Carl and S. Heikkilä, *Nonlinear Differential Equations in Ordered Spaces*, Chapman & Hall/CRC, London, 2000.

6. S. Carl and S. Heikkilä, Existence of solutions for discontinuous functional equations and elliptic boundary value problems, *Electronic J. Differential Equations* **61** (2002), 1–10.

15 Numerical Method for Solving Differential Algebraic Equations by Power Series

Hiroshi Hirayama

15.1 Introduction

The arithmetic operations and functions of a power series can be easily defined by the Fortran 90, C++ [1], and C# languages. The functions represented by these languages, which consist of arithmetic operations, predefined functions and conditional statements, can be expanded in power series.

We consider the differential algebraic equation (DAE) and initial conditions

$$F(y, y', x) = 0, \quad y(x_0) = y_0, \quad y'(x_0) = y_1, \tag{15.1}$$

where y is a sufficiently differentiable vector function of x. The solutions of (15.1) can be assumed to have the form

$$y = y_0 + y_1 x + e x^2, \tag{15.2}$$

where e is a vector of the same dimension as y.

Substituting (15.2) into (15.1) and neglecting higher order terms, we obtain for e a linear equation of the form

$$Ae = B, \tag{15.3}$$

where A and B are a constant matrix and a constant vector, respectively. Solving equation (15.3), we determine the coefficient of x^2 in (15.2). Repeating this procedure for higher order terms, we can construct a power series of arbitrary order for the solutions of (15.1). The arithmetic and functions of this type of power series can be defined easily in same manner as those for general power series (see [2]–[4]).

The power series given by the above procedure can be transformed into a Padé series [5]. According to E. Hairer and G. Wanner ([6], p. 60), if $M + 2 \geq L \geq M$, where M and L are the degrees of the numerator and denominator in the Padé series, respectively, then this Padé series gives an A-stable formula of arbitrary order for differential algebraic equations.

15.2 Power Series Solution for DAEs

In order to substitute (15.2) into (15.1), we define a different type of power series of the form

$$f(x) = f_0 + f_1 x + f_2 x^2 + \cdots + (f_n + p_1 e_1 + \cdots + p_m e_m)x^n, \qquad (15.4)$$

where $p_1, p_2, \ldots, p_m$ are constants, m is the dimension of the vector e in (15.3), and $e_1, e_2, \ldots, e_m$ are the vectors of a basis for m-dimensional space. The arithmetic and functions of this type of power series can be defined easily for $n \geq 0$ in the same manner as for general power series. The vector y in (15.2) has m components, each of which can be represented by a power series of the form (15.4), that is,

$$y_i = y_{i,0} + y_{i,1}x + y_{i,2}x^2 + \cdots + e_i x^n, \qquad (15.5)$$

where y_i is the ith component of y. Substituting (15.5) into (15.1), we arrive at the equality

$$f_i = (f_{i,n} + p_{i,1}e_1 + \cdots + p_{i,m}e_m)x^{n-j} + O(x^{n-j+1}), \qquad (16.6)$$

where f_i is the ith component of $f(y, y', x)$ in (15.1) and j is 0 if $f(y, y', x)$ contains y' and 1 if does not. From (15.6), we can determine the linear equation in (15.3) as

$$\begin{aligned} A_{i,j} &= p_{i,j}, \\ B_i &= -f_{i,n}. \end{aligned} \qquad (15.7)$$

Solving this linear equation, we find the coefficients e_i $(i = 1, \ldots, m)$, which, substituted in (15.5), yield the y_i $(i = 1, \ldots, m)$ as polynomials of degree n.

Repeating the procedure (15.5)–(15.7), we can construct a power series of arbitrary order for the solution of the DAE in (15.1). Let the step size for x be h. Using it in the power series for y and y', we determine y and y' at $x = x_0 + h$. This leads us to a numerical solution of (15.1).

15.3 Numerical Examples

Two simple examples are shown below. The first example (15.8) cannot be solved by ATOMFT [7], while the second one, (15.14), can be solved by both ATOMF and our method.

15.3.1 Simple Example

We consider the differential algebraic system [8]

$$v_1' + v_3 v_2' - (v_2 + 1)v_3' = -v_1 + 1 + \sin x,$$
$$(v_3 + 1)v_1' + v_1 v_2' = -e^{-x}, \qquad (15.8)$$
$$0 = v_1 v_2 v_3 - 0.5e^{-x}\sin(2x)$$

and initial values

$$v(0) = \begin{pmatrix} 1 \\ 0 \\ 1 \end{pmatrix}, \quad v'(0) = \begin{pmatrix} -1 \\ 1 \\ 0 \end{pmatrix}.$$

The exact solution of this initial value problem is

$$v_1(x) = e^{-x}, \quad v_2(x) = \sin x, \quad v_3(x) = \cos x.$$

Taking into account the initial values, we may assume that the solution of (15.8) is of the form

$$v_1(x) = 1 - x + e_1 x^2, \quad v_2(x) = x + e_2 x^2, \quad v_3(x) = 1 + e_3 x^2. \quad (15.9)$$

Substituting (15.9) in (15.8) and neglecting higher order terms, we find that

$$e_2 x^2 + O(x^3) = 0,$$
$$(-2 + 4e_1 + 2e_2)x + O(x^2) = 0,$$
$$(-2 + 2e_1 + 2e_2 - 2e_3)x + O(x^2) = 0.$$

These formulas corresponds to (15.6). The linear equation that corresponds to (15.3) can be written as

$$\begin{pmatrix} 0 & 1 & 0 \\ 4 & 2 & 0 \\ 2 & 3 & -2 \end{pmatrix} e = \begin{pmatrix} 0 \\ 2 \\ 2 \end{pmatrix}. \qquad (15.10)$$

From equation (15.10), we have

$$v_1(x) = 1 - x + 0.5x^2, \quad v_2(x) = x, \quad v_3(x) = 1 - 0.5x^2. \qquad (15.11)$$

Equalities (15.11) suggest, as above, that the solution of (15.8) is of the form

$$v_1(x) = 1 - x + 0.5x^2 + e_1 x^3,$$
$$v_2(x) = x + e_2 x^3, \qquad (15.12)$$
$$v_3(x) = 1 - 0.5x^2 + e_3 x^3.$$

Substituting (15.12) in (15.8) and neglecting higher order terms, we arrive at

$$(0.166667 + e_2)x^3 + O(x^4) = 0,$$
$$(-1.5 + 6e_1 + 3e_2)x^2 + O(x^3) = 0, \qquad (15.13)$$
$$(1 + 3e_1 + 3e_2 - 3e_3)x^2 + O(x^3) = 0.$$

System (15.13) yields the linear matrix equation

$$\begin{pmatrix} 0 & 1 & 0 \\ 6 & 3 & 0 \\ 3 & 3 & -3 \end{pmatrix} \begin{pmatrix} e_1 \\ e_2 \\ e_3 \end{pmatrix} = \begin{pmatrix} -0.166667 \\ -1.5 \\ -1 \end{pmatrix},$$

with solution

$$v_1(x) = 1 - x + 0.5x^2 - 0.166667x^3,$$
$$v_2(x) = x - 0.166667x^3,$$
$$v_3(x) = 1 - 0.5x^2.$$

Repeating the above procedure, we find that

$$
\begin{aligned}
v_1 = {}& 1 - x + 0.5x^2 - 0.166667x^3 + 0.0416667x^4 - 0.00833333x^5 \\
& + 0.00138889x^6 - 0.000198413x^7 + 2.48016 \times 10^{-5}x^8 \\
& - 2.75573 \times 10^{-6}x^9,
\end{aligned}
$$

$$
\begin{aligned}
v_2 = {}& x - 0.166667x^3 + 5.55112 \times 10^{-17}x^4 + 0.00833333x^5 \\
& + 1.00614 \times 10^{-16}x^6 - 0.000198413x^7 + 7.71952 \times 10^{-17}x^8 \\
& + 2.75573 \times 10^{-6}x^9,
\end{aligned}
$$

$$
\begin{aligned}
v_3 = {}& 1 - 0.5x^2 + 1.38778 \times 10^{-17}x^3 + 0.0416667x^4 + 4.59702 \times 10^{-17}x^5 \\
& - 0.00138889 * x^6 + 4.07931 \times 10^{-18}x^7 + 2.48016 \times 10^{-5}x^8 \\
& - 2.849 \times 10^{-18}x^9.
\end{aligned}
$$

This is the actual computer output. The power series for v_1 can be transformed into a Padé series with numerator and denominator

$$pp = 1 - 0.5x + 0.107143x^2 - 0.0119048x^3 + 0.000595238x^4,$$
$$qq = 1 + 0.5x + 0.107143x^2 + 0.0119048x^3 + 0.000595238x^4.$$

We can also obtain Padé series for v_2 and v_3. Substituting the step size h into v_1, v_2, and v_3, we construct $v_1(h)$, $v_2(h)$, and $v_3(h)$, from which we also compute $v_1'(h)$, $v_2'(h)$, and $v_3'(h)$.

If the DAE in (15.1) is stiff, we can use this Padé series to compute $v(h)$ and $v'(h)$. This gives an A-Stable method for the DAE.

If the power series for v_1, v_2, and v_3 converge very fast, then the transformation of v_1, v_2, and v_3 into Padé series becomes unnecessary.

Considering $v_1(h)$, $v_2(h)$, $v_3(h)$, $v_1'(h)$, $v_2'(h)$, and $v_3'(h)$ as initial values at $x = h$ for the differential algebraic equation (15.8), we can calculate

$v_1(x)$, $v_2(x)$, $v_3(x)$, $v_1'(x)$, $v_2'(x)$, and $v_3'(x)$ at $x = 2h$ by the above procedure.

In Table 1 we show the solution of (15.8) computed by the above numerical method. Problem (15.8) is very difficult to solve by the numerical method proposed in [8], but very easy to solve by ours. The numerical values in Table 1 coincide with those of the exact solution of (15.8).

x	v_1	v_2	v_3
0.1	0.904837	0.0998334	0.995004
0.2	0.818731	0.1986690	0.980067
0.3	0.740818	0.2955200	0.955336
0.4	0.670320	0.3894180	0.921061
0.5	0.606531	0.4794260	0.877583
0.6	0.548812	0.5646420	0.825336
0.7	0.496585	0.6442180	0.764842
0.8	0.449329	0.7173560	0.696707
0.9	0.406570	0.7833270	0.621610
1.0	0.367879	0.8414710	0.540302

Table 1. Numerical solution of (15.8).

15.3.2 The Simple Pendulum

We consider the simple DAE [7]

$$\frac{d^2x}{dt^2} = -x\lambda(t), \quad x(0) = \sin(1.2), \ x'(0) = 0,$$

$$\frac{d^2y}{dt^2} = -y\lambda(t), \quad y(0) = -\cos(1.2), \ y'(0) = 0, \tag{15.14}$$

$$x^2 + y^2 - 1 = 0.$$

To solve this system, the following codes must be written in C++:

```
s[0] = diff(x,2))+x*lambda ;
s[1] = diff(y,2))+y*lambda-9.8 ;
s[2] = x*x+y*y-1 ;
```

We do not need the Jacobian of these equations; therefore, it is very easy to use this program to get accurate numerical results.

The power series for $x(t)$, $y(t)$, and $\lambda(t)$ at $t = 0$ are

$$x(t) = 0.932039 + 1.65488t^2 - 9.23024t^4 - 12.5981t^6 + 22.9718t^8,$$

$$y(t) = -0.362358 + 4.25661t^2 + 5.03856t^4 - 15.3707t^6 - 26.4184t^8,$$

$$\lambda(t) = -3.55111 + 125.144t^2 + 148.134t^4 - 451.899t^6.$$

15.4 Conclusion

An arbitrary A-stable formula for differential algebraic equations is given. The algorithm is very simple and effective for most DAEs. It is easy to convert the above formula to an adaptive one, but we do not show this here. Our algorithm can be easily programmed in C++, C#, or Fortran 90.

Although the algorithm can also be programmed in older languages that do not have overload functions, such as C, FORTRAN 77, or Java, it is not recommended to do so because these programs become very complex. We can also extend our method to the expansion of implicit functions in power series and Padé series.

References

1. M.A. Ellis and B. Stroustrup, *The Annotated C++ Reference Manual*, Addison-Wesley, New York, 1990.

2. P. Henrici, *Applied Computational Complex Analysis, Vol. 1*, John Wiley & Sons, New York, 1974.

3. H. Hirayama, Numerical method for solving ordinary differential equation by Picard's method, in *Integral Methods in Science and Engineering*, Birkhäuser, Boston-Basel, 2002, 111–116.

4. L.B. Rall, *Automatic Differentiation-Technique and Applications*, Lect. Notes Comp. Sci. **120**, Springer-Verlag, Berlin-Heidelberg-New York, 1981.

5. W.H. Press, B.P. Flannery, S.A. Teukolsky, and W.T. Vetterling, *Numerical Recipes*, Cambridge University Press, Cambridge, 1988.

6. E. Hairer and G. Wanner, *Solving Ordinary Differential Equations. II*, Springer-Verlag, Berlin, 1991.

7. Y.F. Chang and G. Corliss, ATOMFT: solving ODEs and DAEs by Taylor series, *Computer Math. Appl.* **28** (1994), 209–233.

8. T. Watanabe, HIDMAS - a computer program for initial value problems for ordinary differential equations, *JSIAM* **1** (1991), 135–163 (Japanese).

16 On Hemivariational Inequalities in Coupled Thermoelasticity

Jiri V. Horák

16.1 Introduction

The mathematical formulation of one class of problems in mechanics leads to an inclusion type problem. One of these is a thermoelastic plate strip or beam bending problem. For the sake of brevity, we restricted our contribution only to the linearized theory of *coupled thermoelasticity* [1], [2], and only to the most illustrative combination of classical, non-classical and inclusion type boundary conditions. The goal is to show how a model problem can be formulated and solved if one takes into account conditions including non-monotone behaviour of support. Then, the resulting formulation of the problem has a form of an inclusion type problem. Such a problem generally reads as follows: find a function u such that

$$u \in \mathcal{V}: \quad \mathcal{F} - \mathcal{A}u \in \mathcal{B}(u) \quad \text{in } \mathcal{V}^*,$$

where $\mathcal{F} \in \mathcal{V}^*$ is given, $\mathcal{A} : \mathcal{V} \to \mathcal{V}^*$ is a single-valued mapping, while $\mathcal{B} : \mathcal{V} \to \mathcal{V}^*$ is generally a multivalued mapping. We discuss only typical boundary conditions; thus, we assume that $\mathcal{B}(u) \equiv \bar{\partial}j(u)$, where $\bar{\partial}j$ is the generalized Clarke gradient of a given superpotential j. Then the inclusion problem can be equivalently expressed (see, for example, [3]) by

$$\{u, \Xi\} \in \mathcal{V} \times \mathcal{X}: \quad \mathcal{A}u + \Xi = \mathcal{F} \quad \text{in } \mathcal{V}^*,$$
$$\Xi \in \bar{\partial}j(u) \quad \text{in } \mathcal{V}^*,$$

where $\mathcal{X} \subset \mathcal{V}^*$ is an appropriate function space. Superpotential $j : \mathcal{V} \to \mathbf{R}$ has different forms, including nonconvex and nonsmooth parts. Special cases of quadratic or convex potentials imply reduction of hemivariational inequality to the variational inequality of the 1st (2nd) kind (see [3], [4]) or to the linear (non-linear) variational equation.

Weak formulation of the problem in the form of evolutional constrained hemivariational inequality is introduced and its solvability is briefly discussed for special types of potentials and semi-coercive cases. Due to nonconvexity of the superpotential and nonmonoticity of its subgradient, we can not generally get uniqueness of the solution even in the coercive case.

This work was supported by a grant from the Council of Czech Government, Project J14/98: 1531 000 11.

16.2 Model Problem

The linear space of vector functions $V = \{v_1, v_2, \eta_1, \eta_2\}$ with finite energy has the form $H = H^1(\Omega) \times H^2(\Omega) \times H^1(\Omega) \times H^1(\Omega)$, while the linear space of virtual displacement and temperatures is defined by $\mathcal{H} = \prod_{i=1}^{4} \mathcal{V}_i$; its components can be given by $\mathcal{V}_1 = H^1(\Omega)$, $\mathcal{V}_2 = \{v \in H^2(\Omega) \mid \gamma_2^{(0)}(v)(0) = \gamma_2^{(1)}(v)(0) = 0\}$ or only $\mathcal{V}_2 = \{v \in H^2(\Omega) \mid \gamma_1(v)(0) = 0\}$ in the semi-coercive case, $\mathcal{V}_3 = H^1(\Omega)$, $\mathcal{V}_4 = H^1(\Omega)$, where $H^k(\Omega)$ are Sobolev spaces and $\Omega = (0, L)$.

The convex set $\mathcal{K}$ of kinematically admissible "displacements" has the form $\mathcal{K} = \prod_{i=1}^{4} \mathcal{K}_i$, where $\mathcal{K}_1 = \{V \in \mathcal{H} \mid \gamma_1(v_1)(0) \geq 0\}$, $\mathcal{K}_2 \equiv \mathcal{V}_2$, $\mathcal{K}_3 = \{V \in \mathcal{H} \mid \gamma_3(v_3)(0) \geq 0\}$, $\mathcal{K}_4 = \{V \in \mathcal{H} \mid \gamma_4(v_4)(0) \leq 0\}$. Notation for the Dirichlet trace operator $\gamma : H \to \mathbf{R}^5 \times \mathbf{R}^5$ has been used in the form $\boldsymbol{\gamma} = \{\gamma_1, \gamma_2, \gamma_3, \gamma_4\}$, $\gamma_2 = \{\gamma_2^{(0)}, \gamma_2^{(1)}\}$.

Auxiliary bilinear forms $\mathbf{a}^{(i)}$, $\mathbf{b}$ are given by

$$\mathbf{a}^{(i)}(u, v) = (D^i u, D^i v)_{L_2(\Omega)} \quad u, v \in H^i(\Omega), \; i = 0, 1, 2,$$

$$\mathbf{b}(\vartheta, \eta) = (D\vartheta, \eta)_{L_2(\Omega)}, \quad \vartheta, \eta \in H^1(\Omega),$$

and bilinear forms representing the potential energy of stretching, bending, and coupling are expressed by

$$\mathcal{A}_S(\boldsymbol{U}, \boldsymbol{V}) = -\mathbf{a}^{(1)}(u_1, v_1) + \mathbf{a}^{(1)}(\vartheta_1, \eta_1) + a_{1,1}\mathbf{a}^{(0)}(\vartheta_1, \eta_1),$$

$$\mathcal{A}_B(\boldsymbol{U}, \boldsymbol{V}) = \mathbf{a}^{(2)}(u_2, v_2) + \mathbf{a}^{(1)}(\vartheta_2, \eta_2) + a_{2,2}\mathbf{a}^{(0)}(\vartheta_2, \eta_2),$$

$$\mathcal{B}_S(\boldsymbol{U}, \boldsymbol{V}) = a_1(\vartheta_1, \eta_1)_{L_2(\Omega)} - a_2\mathbf{a}^{(0)}(u_1, D\eta_1),$$

$$\mathcal{B}_B(\boldsymbol{U}, \boldsymbol{V}) = a_1(\vartheta_2, \eta_2)_{L_2(\Omega)} + a_2\mathbf{a}^{(1)}(u_2, \eta_2),$$

$$\mathcal{B}_C(\boldsymbol{U}, \boldsymbol{V}) = -a_{1,2}\mathbf{a}^{(0)}(\vartheta_2, \eta_1) + a_{2,1}\mathbf{a}^{(0)}(\vartheta_1, \eta_2),$$

$$\mathcal{C}_S(\boldsymbol{U}, \boldsymbol{V}) = \alpha\mathbf{b}(\vartheta_1, v_1), \quad \mathcal{C}_B(\boldsymbol{U}, \boldsymbol{V}) = \alpha\mathbf{a}^{(1)}(\vartheta_2, v_2) \quad \boldsymbol{U}, \boldsymbol{V} \in \mathcal{H},$$

where the following notation was used (see [2] and [4]): $a_1 = \frac{c}{k}$, $a_{2,1} = \frac{\alpha_u + \alpha_d}{kH}$, $a_{2,2} = \frac{12}{H^2} + 3a_{2,1}$, $a_{4,1} = \frac{\alpha_u - \alpha_d}{2k}$, $a_{4,2} = \frac{12}{H^2}a_{4,1}$, $a_3 = \theta_0\frac{E\alpha}{k}$, H is height, E is Young's modulus, and k and α are coefficients of thermal conductivity and linear expansion of the material, respectively.

The resulting forms for the model problem are then as follows:

$$\mathcal{A}(\boldsymbol{U}, \boldsymbol{V}) = \mathcal{A}_S(\boldsymbol{U}, \boldsymbol{V}) + \mathcal{A}_B(\boldsymbol{U}, \boldsymbol{V}),$$

$$\mathcal{B}(\boldsymbol{U}, \boldsymbol{V}) = \mathcal{B}_S(\boldsymbol{U}, \boldsymbol{V}) + \mathcal{B}_B(\boldsymbol{U}, \boldsymbol{V}) + \mathcal{B}_C(\boldsymbol{U}, \boldsymbol{V}),$$

$$\mathcal{C}(\boldsymbol{U}, \boldsymbol{V}) = \mathcal{C}_S(\boldsymbol{U}, \boldsymbol{V}) + \mathcal{C}_B(\boldsymbol{U}, \boldsymbol{V}) \quad \text{for } \boldsymbol{U}, \boldsymbol{V} \in \mathcal{H}.$$

Next, we define linear forms corresponding to the potential energy of the load q_i and sources r_i: $\mathcal{F}(\boldsymbol{V}) = \mathcal{F}_S(\boldsymbol{V}) + \mathcal{F}_B(\boldsymbol{V})$, $\mathcal{F}_S(\boldsymbol{V}) = (q_1, v_1)_{(1)} + \langle r_1, \vartheta_1 \rangle_{(3)}$, $\mathcal{F}_B(\boldsymbol{V}) = (q_2, v_2)_{(2)} + \langle r_2, \vartheta_2 \rangle_{(4)} + \hat{M}_L Dv_2(L)$, where $(.,.)_{(i)}$ and $\langle .,. \rangle_{(i+2)}$ denote the duality on $\mathcal{V}_{(i)}^* \times \mathcal{V}_{(i)}$, $i = 1, 2$.

Finally, we introduce boundary forms $h^{(i)}(u_i, v_i) = [(k_i f(u_i) v_i)(x)]_0^L$, $f(u_i) = u_i$ or u_i^+, corresponding potentials $j_N(f; v_i) = 1/2(k_i f(v_i)^2)(x)$, $x = 0, L$; then $\mathcal{H} = \sum_i \mathcal{H}^{(i)}$, $\mathcal{H}^{(i)}(\boldsymbol{U}, \boldsymbol{V}) = h^{(i)}(u_i, v_i)$. Superpotentials representing non-monotone behaviour and possible destruction of support (see [3], for example) are defined ($\hat{u}_i$ given) by

$$
j_{N_D}(v_i)(x) = \begin{cases} 1/2 k_i(x) v_i^2(x) & \text{if } v_i(x) \in (-\infty, \hat{u}_i), \\ 1/2 k_i(x) \hat{u}_i^2 & \text{if } v_i(x) \geq \hat{u}_i, \end{cases}
$$

$$
j_{N_D^+}(v_i)(x) = \begin{cases} 0 & \text{if } v_i(x) \leq 0, \\ 1/2 k_i(x) v_i^2(x) & \text{if } v_i(x) \in (0, \hat{u}_i), \\ 1/2 k_i(x) \hat{u}_i^2 & \text{if } v_i(x) \geq \hat{u}_i. \end{cases}
$$

The resulting forms of the potential and superpotentials read as follows:

$$
\mathcal{J}_N^{(i)}(\boldsymbol{V}) \equiv j_N(f; v_i), \ \mathcal{J}_{N_D}^{(i)}(\boldsymbol{V}) \equiv j_{N_D}(v_i), \ \mathcal{J}_{N_D^+}^{(i)}(\boldsymbol{V}) \equiv j_{N_D^+}(v_i), \ \boldsymbol{V} \in \mathcal{H}.
$$

Definition. Let $I = (0, T)$, $\mathcal{J} = \sum_i \mathcal{J}_{N_D}^{(i)} + \sum_i \mathcal{J}_{N_D^+}^{(i)}$, and a quadruple of abstract and a couple of real functions

$$
\{q_1, q_2, r_1, r_2\} \in L_2(I; \mathcal{V}_1^* \times \mathcal{V}_2^* \times \mathcal{V}_3^* \times \mathcal{V}_4^*), \ \{\vartheta_{1,0}, \vartheta_{2,0}\} \in [L_2(\Omega)]^2
$$

be given. Then a couple $\{\boldsymbol{U}, \boldsymbol{\Xi}\}$ of abstract and vector functions such that

$$
\{\boldsymbol{U}, \boldsymbol{\Xi}\} \in L_2(I; \mathcal{K}) \cap AC(I; \mathcal{V}_1 \times \mathcal{V}_2 \times [L_2(\Omega)]^2) \times L_2(I; \boldsymbol{R}^4),
$$
$$
D_t \boldsymbol{U} \in L_2(I; \mathcal{V}_1 \times \mathcal{V}_2 \times [L_2(\Omega)]^2), \quad \boldsymbol{\Xi}(t) \in \bar{\partial}\mathcal{J}(\boldsymbol{U}(t)) \text{ for a.a. } t \in I,
$$
$$
\{\vartheta_1, \vartheta_2\}(0) = \{\vartheta_{1,0}, \vartheta_{2,0}\} \quad (\text{in } C(I; [L_2(\Omega)]^2)),
$$

and

$$
\int_I \mathcal{A}(\boldsymbol{U}(t), \boldsymbol{V}(t) - \boldsymbol{U}(t)) \, dt + \int_I \mathcal{C}(\boldsymbol{U}(t), \boldsymbol{V}(t) - \boldsymbol{U}(t)) \, dt
$$
$$
+ \int_I \mathcal{H}(\boldsymbol{U}(t), \boldsymbol{V}(t) - \boldsymbol{U}(t)) \, dt + \int_I \boldsymbol{\Xi}(t)(\boldsymbol{V}(t) - \boldsymbol{U}(t)) \, dt
$$
$$
+ \int_I \mathcal{B}(D_t \boldsymbol{U}(t), \boldsymbol{V}(t) - \boldsymbol{U}(t)) \, dt \geq \int_I \mathcal{F}(\boldsymbol{V}(t) - \boldsymbol{U}(t)) \, dt
$$

holds $\forall \boldsymbol{V} \in L_2(I; \mathcal{K})$, is said to be a weak solution of the model problem $(\mathcal{P}_\mathcal{H})$ of constrained hemivariational inequality.

For the sake of simplicity, let us suppose only a part of superpotential $\mathcal{J} = \mathcal{J}_{N_D}^{(1)} + \mathcal{J}_{N_D^+}^{(2)}$ is given. Then, we can interpret the model problem $(\mathcal{P}_\mathcal{H})$ solution and find that components $\{\{u_1, u_2\}, \{\vartheta_1, \vartheta_2\}\}$ of $\boldsymbol{U}$ fulfill the following set of four coupled equations:

$$\frac{\partial^2 \vartheta_1}{\partial x^2} - \frac{\alpha_h + \alpha_d}{kH} \vartheta_1 + \frac{\alpha_h - \alpha_d}{2k} \vartheta_2 - \theta_0 \frac{E\alpha}{k} \frac{\partial^2 u_1}{\partial x \partial t} + r_1 = a \frac{\partial \vartheta_1}{\partial t},$$

$$\frac{\partial^2 \vartheta_2}{\partial x^2} - \left(\frac{12}{H^2} + \frac{3(\alpha_h + \alpha_d)}{kH} \right) \vartheta_2 - \frac{6(\alpha_h - \alpha_d)}{kH^2} \vartheta_1$$

$$+ \theta_0 \frac{E\alpha}{k} \frac{\partial^2}{\partial x^2} \left(\frac{\partial u_2}{\partial t} \right) + r_2 = a \frac{\partial \vartheta_2}{\partial t},$$

$$\frac{\partial}{\partial x} \left(EH \frac{\partial}{\partial x} u_1 \right) - \frac{\partial}{\partial x} (\alpha EH \vartheta_1) = q_1,$$

$$\frac{\partial^2}{\partial x^2} \left(EJ \frac{\partial^2}{\partial x^2} u_2 \right) + \frac{\partial^2}{\partial x^2} (\alpha EJ \vartheta_2) = q_2$$

in Q, boundary conditions and the inclusion type relation

$$\gamma_1(u_1) \geq 0, \ Du_1 \geq 0, \ \gamma_1(u_1)Du_1 = 0 \quad \text{on } \{0\} \times I,$$

$$\gamma_2^{(0)}(u_2) = 0, \ \gamma_2^{(1)}(u_2) = 0 \text{ or } \gamma_N^{(2)}(u_2) = \hat{M}_0 \quad \text{on } \{0\} \times I,$$

$$\gamma_N^{(1)}(u_1) \times \gamma_N^{(3)}(u_2) \in \bar{\partial} j_{N_D}(u_1) \times \bar{\partial} j_{N_D^+}(u_2) \quad \text{on } \{L\} \times I,$$

$$\gamma_N^{(2)}(u_2) = \hat{M}_L \quad \text{on } \{L\} \times I,$$

$$\gamma_3(\vartheta_1) \geq 0, \ D\vartheta_1 \geq 0, \ \gamma_3(\vartheta_1)D\vartheta_1 = 0 \quad \text{on } \Gamma,$$

$$\gamma_4(\vartheta_2) \leq 0, \ D\vartheta_2 \geq 0, \ \gamma_4(\vartheta_2)D\vartheta_2 = 0 \quad \text{on } \Gamma,$$

and initial conditions $\vartheta_1 = \vartheta_{0,1}$, $\vartheta_2 = \vartheta_{0,2}$ in Ω_0.

Remark. Other types of relations in the interpretation of the problem $(\mathcal{P}_\mathcal{H})$ can be obtained through the changes of function sets $\mathcal{K}_i$ and $\mathcal{V}_i$ as well as forms and types of potentials and superpotentials $\mathcal{J}_{(.)}$.

Boundary conditions of the inclusion type can be used for temperature and its gradient: $\gamma_3(\vartheta_1) \times \gamma_4(\vartheta_2) \in \bar{\partial} j_{(.)}(\vartheta_1) \times \bar{\partial} j_{(.)}(\vartheta_2)$ on $\Gamma \equiv \partial \Omega \times I$.

16.3 Solvability of the Model Problem

The solvability of the model problem can be studied for different combinations of boundary conditions and superpotentials. We introduce here only three model combinations of conditions and superpotentials leading to coercive and semi-coercive bilateral and unilateral cases. The results are as follows.

Theorem 1. *Let* $\mathcal{V}_2 = \{v \in H^2(\Omega) \mid \gamma_2^{(0)}(v)(0) = \gamma_2^{(1)}(v)(0) = 0\}$ *and assume that* $\alpha_h = \alpha_d$ *holds. Then, for given forms of superpotentials* $\mathcal{J}_N$, $\mathcal{J}_{N_D}$ *and* $\mathcal{J}_{N_D^+}$, *the problem* $(\mathcal{P}_\mathcal{H})$ *has at least one solution.*

Proof. Due to the properties of the material and the assumption that $\alpha_h = \alpha_d$, the identity $\mathcal{B}_C \equiv 0$ holds and the model problem $(\mathcal{P}_\mathcal{H})$ can be split into two mutually independent problems representing only the "stretching" and "bending" effects: $(\mathcal{P}_{\mathcal{H}}(\boldsymbol{U})) \equiv \left((\mathcal{P}_{\mathcal{H}}^{Str}(\boldsymbol{U}_1)) \bowtie (\mathcal{P}_{\mathcal{H}}^{Ben}(\boldsymbol{U}_2))\right)$. Here we have used the following reordering and splitting of the vector function $\boldsymbol{U}$ into two couples of functions:

$$\boldsymbol{U} = \{\{u_1, u_2\}, \{\vartheta_1, \vartheta_2\}\} \equiv (\boldsymbol{U}_1 = \{u_1, \vartheta_1\} \bowtie \boldsymbol{U}_2 = \{u_2, \vartheta_2\}),$$

and analogous redefinitions of function sets $\mathcal{H} = \mathcal{H}^{Str} \times \mathcal{H}^{Ben}$, where $\mathcal{H}^{Str} = \mathcal{H}_1 \times \mathcal{H}_3, \mathcal{H}^{Ben} = \mathcal{H}_2 \times \mathcal{H}_4$, and $\mathcal{K} = \mathcal{K}^{Str} \times \mathcal{K}^{Ben}, \mathcal{K}^{Str} = \mathcal{K}_1 \times \mathcal{K}_3, \mathcal{K}^{Ben} = \mathcal{K}_2 \times \mathcal{K}_4$. Thus, the new problems $(\mathcal{P}_{\mathcal{H}}^{Str})$ and $(\mathcal{P}_{\mathcal{H}}^{Ben})$ are obtained from $(\mathcal{P}_{\mathcal{H}})$ by transformation of linear and bilinear forms $\mathcal{F}$ and $\mathcal{A}, \mathcal{B}, \mathcal{C}$ (functions of $\boldsymbol{U}, \boldsymbol{V}$) into $\mathcal{F}_S, \mathcal{A}_S, \mathcal{B}_S, \mathcal{C}_S$ (functions of $\boldsymbol{U}_1, \boldsymbol{V}_1$) and $\mathcal{F}_\mathcal{B}, \mathcal{A}_\mathcal{B}, \mathcal{B}_\mathcal{B}, \mathcal{C}_\mathcal{B}$ (functions of $\boldsymbol{U}_2, \boldsymbol{V}_2$).

The general form of hemivariational inequality in the definition of the weak solution of the problem $(\mathcal{P}_{\mathcal{H}}^{\beta})$ then reads as follows:

$$\int_I \mathcal{A}_\alpha(\boldsymbol{U}_i(t), \boldsymbol{V}_i(t) - \boldsymbol{U}_i(t))\,dt + \int_I \mathcal{C}_\alpha(\boldsymbol{U}_i(t), \boldsymbol{V}_i(t) - \boldsymbol{U}_i(t))\,dt$$

$$+ \int_I \mathcal{H}_\alpha(\boldsymbol{U}_i(t), \boldsymbol{V}_i(t) - \boldsymbol{U}_i(t))\,dt + \int_I \boldsymbol{\Xi}_i(t)(\boldsymbol{V}_i(t) - \boldsymbol{U}_i(t))\,dt$$

$$+ \int_I \mathcal{B}_\alpha(D_t\boldsymbol{U}_i(t), \boldsymbol{V}_i(t) - \boldsymbol{U}_i(t))\,dt \geq \int_I \mathcal{F}_\alpha(\boldsymbol{V}_i(t) - \boldsymbol{U}_i(t))\,dt$$

for $\forall \boldsymbol{V}_i \in L_2(I; \mathcal{K}^\beta)$, $\alpha = \mathcal{S}, \mathcal{B}$, $\beta = Str, Ben$, $i = 1, 2$; the character of the terms $\mathcal{H}_\alpha, \boldsymbol{\Xi}_i(t)$ and sets $\mathcal{K}^\beta$ depends on the choice of the concrete type of boundary condition and superpotential.

Both problems $(\mathcal{P}_{\mathcal{H}}^{Str})$ and $(\mathcal{P}_{\mathcal{H}}^{Ben})$ remain still coupled (in components $\{u_i, \vartheta_i\}$) and coercive but they can be studied independently [4].

Then, the statement of the theorem can be proved for the "standard" second-order problem $(\mathcal{P}_{\mathcal{H}}^{Str})$ by the methods introduced in [3], while the method of proof for the combined fourth-order and second-order coupled problem $(\mathcal{P}_{\mathcal{H}}^{Ben})$ depends on the choice of potential/superpotential and character of the load. In case of "active" reaction in support, the existence of the solution of $(\mathcal{P}_{\mathcal{H}}^{Ben})$ can be obtained by using the Rothe method of discretization in time and a-priori estimations (see [2], for example), while in other cases the method of factorization [2], [4] can be used to get another decomposition and simplification of the problem. Finally, the coupled thermoelastic problem $(\mathcal{P}_{\mathcal{H}}^{Ben})$ can be split into only temperature effects (second-order parabolic problem) and bending effects (fourth-order elasticity problem—hemivariational inequality) and solved by classical methods. Some auxiliary results and useful methods of solution can be found in [2], [3] and [4].

In semicoercive cases, the existence of the model problem solution can be guaranteed only through some additional conditions. For bilateral and unilateral boundary conditions with "possible break" given by Newton superpotentials $\mathcal{J}_{N_D}$ and $\mathcal{J}_{N_D}^+$ (representing only limited load-reaction transfer [3]: elastic support may sustain only force/deflection not exceeding the given stress/displacement value $\hat{\boldsymbol{T}}/\hat{u}_2$), we have the following assertions.

Theorem 2. *Let us assume* $V_2 = \{v \in H^2(\Omega) \mid \gamma_2^{(0)}(v)(0) = 0\}$, $\alpha_h = \alpha_d$
holds and superpotential $\mathcal{J} \equiv \mathcal{J}_{N_D}$ $(\equiv \mathcal{J}_{N_D}^{(2)})$ *is given. If the problem* $(\mathcal{P}_{\mathcal{H}})$
has a solution, then the necessary condition

$$(q_2(t)(x), x/L)_{(2)} \leq \hat{T}(\hat{u}_2), \quad q_2(t) \in L_2(\Omega) \text{ for a.a. } t \in I$$

holds (for given "reaction limit" $\hat{T}(\hat{u}_2) > 0$). *A sufficient condition for the
existence of a solution has the form*

$$(q_2(t)(x), x/L)_{(2)} < \hat{T}(\hat{u}_2), \quad q_2(t) \in L_2(\Omega) \text{ for a.a. } t \in I.$$

Proof. A sketch of the proof can be written in steps analogous to the previous one: we only need to use a sufficient condition to force the existence of the active reaction in support and then use the method of factorization for decomposition and simplification of that transformed problem.

To get a solution of the model problem with unilateral Newton boundary condition we again have to formulate conditions guaranteeing solvability of the problems. A statement can then be proved as in Theorems 1 and 2.

Theorem 3. *Let* $V_2 = \{v \in H^2(\Omega) \mid \gamma_2^{(0)}(v)(0) = 0\}$, *and suppose that*
$\alpha_h = \alpha_d$ *holds and that the superpotential* $\mathcal{J} \equiv \mathcal{J}_{N_D^+}$ $(\equiv \mathcal{J}_{N_D^+}^{(2)})$ *is given. If
the problem* $(\mathcal{P}_{\mathcal{H}})$ *has a solution, then the condition*

$$0 \leq (q_2(t)(x), x/L)_{(2)} \leq \hat{T}(\hat{u}_2), \quad q_2(t) \in L_2(\Omega) \text{ for a.a. } t \in I$$

holds (for given $\hat{T}(\hat{u}_2) > 0$). *A sufficient condition for the existence of a
solution has the form*

$$0 \leq (q_2(t)(x), x/L)_{(2)} < \hat{T}(\hat{u}_2), \quad q_2(t) \in L_2(\Omega) \text{ for a.a. } t \in I.$$

References

1. D.E. Carlson, Linear thermoelasticity, in *Encyclopedia of Physics*, ed. S. Flüge, Vol. VIa/2, *Mechanics of Solids II*, Springer-Verlag, Berlin, 1972.

2. J.V. Horák, On solvability of one special problem of coupled thermoelasticity, Part I., *Acta Univ. Palackianae Olomucensis, Facultas Rer. Nat., Math.* **34** (1995), 39–58.

3. J. Haslinger, M. Miettinen, and P.D. Panagiotopoulos, *Finite Element Method for Hemivariational Inequalities. Theory, Methods and Applications*, Kluwer, Dordrecht, 1999.

4. J.V. Horák, Applied variational inequalities, *VUT Brno, Scientific Reports of Technical Univ. Brno* **86** (2002) (Czech).

17 Asymptotic Stability in Functional Differential Equations with Delay

Alexander O. Ignatyev

17.1 Introduction

Let $x = (x_1, \ldots, x_n) \in R^n, t \in R, |x| = \sqrt{x_1^2 + \ldots + x_n^2}$. For a given $h > 0, C$ denotes the space of continuous functions mapping $[-h, 0]$ into R^n and for $\phi \in C, \|\phi\| = \sup_{-h \leq \theta \leq 0} |\phi(\theta)|$. According to [1] we denote

$$C_H = \{\phi \in C : \|\phi\| \leq H\}.$$

If x is a continuous function of u defined on $-h \leq u < A, A > 0$, and if t is a fixed number satisfying $0 \leq t < A$, then x_t denotes the restriction of x to the segment $[t-h, t]$ so that x_t is an element of C defined by $x_t(\theta) = x(t+\theta)$ for $-h \leq \theta \leq 0$. Consider a system of functional differential equations

$$\frac{dx}{dt} = f(t, x_t) \tag{17.1}$$

and obtain conditions on a Lyapunov functional to insure that the zero solution is asymptotically stable. In this system dx/dt denotes the right-hand derivative of x at t, t is time, and $f(t, \phi) \in R^n$ is defined on $[0, \infty) \times C_H$; $f(t, 0) \equiv 0$. According to [1,2,3] we denote by $x(t_0, \phi)$ a solution of (17.1) with initial condition $\phi \in C_H$ where $x_{t_0}(t_0, \phi) = \phi$ and we denote by $x(t, t_0, \phi)$ the value of $x(t_0, \phi)$ at t and $x_t(t_0, \phi) = x(t+\theta, t_0, \phi), -h \leq \theta \leq 0$. It is assumed that the vector-valued functional $f(t, \phi)$ is continuous on $[0, \infty) \times C_H$ so that a solution will exist for each continuous initial condition. For continuation of solutions, we suppose that f takes closed bounded sets of $[0, \infty) \times C_H$ into closed bounded sets of R^n. Let $V(t, \phi)$ be a continuous functional defined for $t \geq 0, \phi \in C_H$. The upper right-hand derivative of V along solutions of (17.1) is defined to be [4]

$$\frac{dV(t, x_t(t_0, \phi))}{dt} = \varlimsup_{\Delta t \to +0} \{V(t + \Delta t, x_{t+\Delta t}(t_0, \phi)) - V(t, x_t(t_0, \phi))\} \frac{1}{\Delta t}.$$

If V satisfies a Lipschitz condition in the second argument, then this limit is uniquely determined. The classical criterion of asymptotic stability of zero solution of equations (17.1), which was obtained by N.N. Krasovskii [4], assumes the existence of positive definite functional V and negative definite functional dV/dt. In applications one can construct a positive definite functional V, which derivative is not negative definite but is less than or equal to zero. Exactly for such cases J.Hale [5] created the effective asymptotic stability criterion if the functional f in equations (17.1) is autonomous (f does not depend on t) and N.N.Krasovskii [4] created such criterion for the case where operator f is periodic in t. For the general case of a nonautonomous operator f, V.M. Matrosov [6] proved that this criterion is not right even for ordinary differential equations. The goal of this paper is to prove this criterion for the case when f is almost periodic in t. This case is a particular one of the class of nonautonomous operators.

17.2 Definitions and Preliminary Results

Definition 1. [7–10] A continuous function $F(t) : R \to R^n$ is called almost periodic if for every $\epsilon > 0$ there exists $l = l(\epsilon) > 0$ such that any segment $[\alpha, \alpha + l], \alpha \in R$, contains at least one number τ such that $|F(t + \tau) - F(t)| < \epsilon$ for every $t \in R$. A number τ is called an ϵ-almost period of F.

Let us introduce the following definition which is analogous to [10].

Definition 2. A continuous functional $F(t, \phi) : R \times C_r \to R^n (0 < r < \infty)$ is called uniformly almost periodic in t if for every $\epsilon > 0$ there exists $l = l(\epsilon, r) > 0$ such that any segment $[\alpha, \alpha + l], \alpha \in R$, contains at least one number τ such that $|F(t + \tau, \phi) - F(t, \phi)| < \epsilon$ for every $t \in R, \phi \in C_r$.

Lemma 1. *[9] Let $F_1(t), ..., F_N(t) : R \to R^n$ be almost periodic functions. Then for every $\epsilon > 0$ there exists $l = l(\epsilon) > 0$ such that any segment $[\alpha, \alpha + l], \alpha \in R$, contains a number τ such that*

$$|F_i(t + \tau) - F_i(t)| < \epsilon, \qquad i = 1, 2, ..., N; \ t \in R.$$

We denote

$$C_{H(L)} = \{\phi \in C_H : |\phi(x_1) - \phi(x_2)| \le L|x_1 - x_2| \text{ for each } x_1, x_2 \in [-h, 0]\}.$$

Lemma 2. *If the functional $F(t, \phi) : R \times C_{H(L)} \to R^n$ is Lipschitzian in ϕ and almost periodic in t for every fixed $\phi \in C_{H(L)}$, then it is uniformly almost periodic in t.*

Proof. Since the functional $F(t, \phi)$ satisfies Lipschitz conditions in ϕ, then

$$|F(t, \phi) - F(t, \psi)| \le L_1 \|\phi - \psi\|, \tag{17.2}$$

where L_1 is Lipschitz constant. Let $\epsilon > 0$ be any real number. $C_{H(L)}$ is the set of uniformly bounded equicontinuous functions, therefore $C_{H(L)}$ is a compact set. Hence there is a finite set of functions $\phi_1, ..., \phi_N$ such that $\phi_j \in C_{H(L)}$ ($j = 1, ..., N$) and for each $\phi \in C_{H(L)}$ there exists such number i ($1 \le i \le N$) that

$$\|\phi - \phi_i\| < \frac{\epsilon}{3L_1}. \tag{17.3}$$

From Lemma 1 it follows that there exists $l > 0$ such, that in any segment $[\alpha, \alpha + l]$ there exists such number τ, that

$$|F(t, \phi_i) - F(t + \tau, \phi_i)| < \frac{\epsilon}{3} \qquad (17.4)$$

for each $t \in R, i = 1, ..., N$. Now we will show that for every $\phi \in C_{H(L)}$, each number τ, which satisfies inequality (17.4), is an ϵ-almost period of the functional $F(t, \phi)$. Let ϕ_k be the same element of the set $\phi_1, ..., \phi_N$ for which $\|\phi - \phi_k\| < \epsilon/(3L_1)$. Then from (17.2)–(17.4) we obtain

$$|F(t + \tau, \phi) - F(t, \phi)| \leq |F(t + \tau, \phi) - F(t + \tau, \phi_k)|$$
$$+ |F(t + \tau, \phi_k) - F(t, \phi_k)| + |F(t, \phi_k) - F(t, \phi)|$$
$$< \frac{\epsilon}{3} + 2L_1 \cdot \frac{\epsilon}{3L_1} = \epsilon. \qquad (17.5)$$

Inequality (17.5) proves Lemma 2.

17.3 Main Results

In this section we consider the system of functional differential equations (17.1) under assumptions above. Besides we assume that the functional $f(t, \phi)$ is Lipschitzian in ϕ and almost periodic in t for every fixed $\phi \in C_H$.

Lemma 3. *Consider the solution $x(t_0, \phi_0)$ of system (17.1). We suppose that $x_t(t_0, \phi_0)$ belongs to C_r, $0 < r < H$, for $t \geq 0$. Let $\{\epsilon_k\}$ be monotonically approaching zero sequence of positive numbers and $\{\tau_k\}$ be some sequence of ϵ_k-almost periods of $f(t, \phi)$ (for every ϵ_k there corresponds ϵ_k-almost period τ_k). Then the limit relation*

$$\lim_{k \to \infty} \|x_{t^*}(t_0, \phi_k) - x_{t^* + \tau_k}(t_0, \phi_0)\| = 0 \qquad (17.6)$$

holds, where $\phi_k = x_{t_0 + \tau_k}(t_0, \phi_0)$ and t^* is some fixed moment of time which is more than t_0 $(t^* > t_0)$.

Proof. Consider the solutions of the system (17.1)

$$x(t_0, \phi_k) \qquad (17.7)$$

and

$$x(t_0 + \tau_k, \phi_k). \qquad (17.8)$$

For the time $\Delta t = t^* - t_0$ the function ϕ_k moves to the function $x_{t^*}(t_0, \phi_k)$ along the trajectory (17.7) and ϕ_k moves to the function

$$x_{t^* + \tau_k}(t_0 + \tau_k, \phi_k) = x_{t^* + \tau_k}(t_0, \phi_0)$$

along the solution (17.8). The restriction of the solution x of the system (17.1) $x_t(t_0 + \tau_k, \phi_k)$ with initial boundary value problem $\phi_k = x_{t_0 + \tau_k}$ may be interpreted as the system

$$\frac{dx}{dt} = f(t + \tau_k, x_t) \tag{17.9}$$

with initial function ϕ_k and initial moment of time t_0. If t is large enough, then $x_t \in C_{H(L)}$. But according to Lemma 2, the right-hand side of the system (17.1) is uniformly almost periodic in t on the set $R \times C_{H(L)}$, therefore, the right-hand sides of the systems (17.1) and (17.9) differ from each other by however little if k is a large enough natural number. Hence, the limit relation (17.6) follows.

Theorem. *Let functional differential equations (17.1) satisfy the above conditions and suppose that there is a continuous functional $V(t, \phi) : R \times C_H \to R$ which is locally Lipschitz in ϕ and such that the following conditions are fulfilled on the set $R \times C_H$:*
(i) $a(|\phi(0)|) \leq V(t, \phi) \leq b(\|\phi\|)$, where $a, b \in K$; K is the class of Hahn's functions [11, 12];
(ii) $V(t, \phi)$ is almost periodic in t for each fixed $\phi \in C_H$;
(iii) $dV/dt \leq 0$, $dV/dt \not\equiv 0$ on each nonzero solution of the system (17.1).
Then the solution

$$x = 0 \tag{17.10}$$

of the functional differential equations (17.1) is asymptotically stable.

Proof. From conditions (i),(iii) it follows that the solution (17.10) is uniformly stable [4,5]. Let $\epsilon \in (0, H)$ be any positive number. Denote by $t_0 \in R$ the initial moment of time. By the stability of the zero solution there exists $\delta > 0$ such that if $\phi \in C_\delta$, then $x_t(t_0, \phi) \in C_\epsilon$ for every $t \geq t_0$. Choose such $\delta > 0$ and show that any solution $x(t_0, \phi)$ with $\phi \in C_\delta$ tends to zero as $t \to \infty$. Suppose that this is not true, i.e. there exist $\eta > 0$ and $\phi_0 \in C_\delta$ such that $|x(t, t_0, \phi_0)| > \eta > 0$ as $t \geq t_0$. The function $V(t) = V(t, x_t(t_0, \phi_0))$ is monotonically nonincreasing because $dV/dt \leq 0$. Hence there exists

$$\lim_{t \to \infty} V(t) = \lim_{t \to \infty} V(t, x_t(t_0, \phi_0)) = V_0 \geq a(\eta) > 0,$$

and it is easy to see that $V(t, x_t(t_0, \phi_0)) \geq V_0$ for $t \in [t_0, \infty)$. Consider some sequence $\{\epsilon_k\}$ of positive numbers monotonically convergent to zero, where ϵ_1 is sufficiently small. By Lemma 2, for every ϵ_i there is a sequence of ϵ_i-almost periods $\tau_{i,1}, \tau_{i,2}, \ldots, \tau_{i,n}, \ldots \to \infty$ for the functionals $f(t, \phi)$ and $V(t, \phi)$ such that the inequalities

$$|V(t + \tau_{i,n}, \phi) - V(t, \phi)| < \epsilon_i,$$

$$|f(t + \tau_{i,n}, \phi) - f(t, \phi)| < \epsilon_i$$

hold for each $t \in R, \phi \in C_\epsilon$. Without loss of generality one can suppose $\tau_{i,n} < \tau_{i+1,n}$ for every i, n. Designate $\tau_k = \tau_{k,k}$. Consider the sequence of functions $\phi_k = x_{t_0+\tau_k}(t_0, \phi_0)$ $(k = 1, 2, ...)$. It is a bounded sequence of equicontinuous functions because $\phi_k \in C_\epsilon$, therefore there is a limit function ϕ^* of this sequence. Without loss of generality one can assume the sequence ϕ_k itself converges to ϕ^*. Because of continuity and almost periodicity of the functional $V(t, \phi)$ we obtain

$$
\begin{aligned}
V(t_0, \phi^*) &= \lim_{n\to\infty} V(t_0, \phi_n) \\
&= \lim_{k\to\infty} \lim_{n\to\infty} V(t_0 + \tau_k, \phi_n) = \lim_{n\to\infty} V(t_0 + \tau_n, \phi_n) \\
&= \lim_{n\to\infty} V(t_0 + \tau_n, x_{t_0+\tau_n}(t_0, \phi_0)) = V_0.
\end{aligned}
$$

Now consider the solution $x(t_0, \phi^*)$. From condition (iii) of the theorem follows the existence of a moment of time t^*, $t^* > t_0$, when the inequality

$$
V(t^*, x_{t^*}(t_0, \phi^*)) = V_1 < V_0
$$

holds. Solutions of the functional differential equations (17.1) are continuous in the initial data, so one can write

$$
\lim_{k\to\infty} \| x_{t^*}(t_0, \phi_k) - x_{t^*}(t_0, \phi^*) \| = 0
$$

because $\lim_{k\to\infty} \| \phi_k - \phi^* \| = 0$. Hence,

$$
\lim_{k\to\infty} V(t^*, x_{t^*}(t_0, \phi_k)) = V_1. \tag{17.11}
$$

Using the uniform almost periodicity property of $f(t, \phi)$ and the limit relation (17.6), we obtain the inequality

$$
\| x_{t^*}(t_0, \phi_k) - x_{t^*+\tau_k}(t_0, \phi_0) \| \le \gamma_k \tag{17.12},
$$

where $\gamma_k \to 0$ as $k \to \infty$. Because of the uniform almost periodicity of $V(t, \phi)$ we have

$$
|V(t^*, \phi) - V(t^* + \tau_k, \phi)| < \epsilon_k \tag{17.13}
$$

for every $\phi \in C_H$, and from conditions (17.11) and (17.12) it follows that

$$
|V(t^*, x_{t^*+\tau_k}(t_0, \phi_0)) - V_1| < \eta_k, \tag{17.14}
$$

where $\eta_k \to 0$ as $k \to \infty$. From (17.13) we obtain

$$
|V(t^*, x_{t^*+\tau_k}(t_0, \phi_0)) - V(t^* + \tau_k, x_{t^*+\tau_k}(t_0, \phi_0))| < \epsilon_k. \tag{17.15}
$$

From (17.14) and (17.15) we have

$$
|V(t^* + \tau_k, x_{t^*+\tau_k}(t_0, \phi_0)) - V_1| < \eta_k + \epsilon_k, \tag{17.16}
$$

where $\eta_k + \epsilon_k \to 0$ as $k \to \infty$. On the other hand,

$$\lim_{k \to \infty} V(t^* + \tau_k, x_{t^* + \tau_k}(t_0, \phi_0)) = V_0. \tag{17.17}$$

Relations (17.16) and (17.17) contradict the inequality $V_1 < V_0$, and the Theorem is proved.

References

1. T.A. Burton, Uniform asymptotic stability in functional differential equations, *Proc. Amer. Math. Soc.* **68** (1978), 195–199.

2. A.O. Ignatyev, On the partial equiasymptotic stability in functional differential equations, *J. Math. Anal. Appl.* **268** (2002), 615–628.

3. L. Hatvani, On the asymptotic stability for nonautonomous functional differential equations by Lyapunov functionals, *Trans. Amer. Math. Soc.* **354** (2002), 3555–3571.

4. N.N. Krasovskii, *Stability of Motion*, Stanford University Press, Stanford, California, 1963.

5. J. Hale, *Theory of Functional Differential Equations*, Springer-Verlag, New York-Heidelberg-Berlin, 1977.

6. V.M. Matrosov, On the theory of stability, *J. Appl. Math. Mech.* **26** (1962), 1506–1522.

7. A.S. Besicovich, *Almost Periodic Functions*, Dover, New York, 1954.

8. H. Bohr, *Almost Periodic Functions*, Chelsea, New York, 1947.

9. C. Corduneanu, *Almost Periodic Functions*, 2nd ed., Chelsea Publ. Co., New York, 1989.

10. A.O. Ignatyev, On the stability of equilibrium for almost periodic systems, *Nonlinear Anal.* **29** (1997), 957–962.

11. W. Hahn, *Stability of Motion*, Springer-Verlag, New York-Berlin-Heidelberg, 1967.

12. N. Rouche, P. Habets, and M. Laloy, *Stability Theory by Liapunov's Direct Method*, Springer-Verlag, New York, 1977.

18 On Optimal Stabilization of Nonautonomous Systems

Alexey A. Ignatyev

18.1 Introduction

Consider a controlled system of differential equations of perturbed motion

$$\dot{x} = X(t, x; u), \qquad (18.1)$$

where $x = (x_1, ..., x_n)$, $X = (X_1, ..., X_n)$, $u = (u_1, ..., u_r)$. Suppose that functions $X(t, x; u)$ are defined, continuous, and satisfying a Lipschitz condition in x in the domain

$$t \in R, \quad \|x\| < H \quad (H = const). \qquad (18.2)$$

It is known [1] that in the optimal stabilization problem the performance criterion of motion $x(t)$ is given as

$$\int_{t_0}^{\infty} \omega(t, x_1(t), ..., x_n(t); u_1(t), ..., u_r(t)) \, dt, \qquad (18.3)$$

where $\omega(t, x; u)$ is a nonnegative function defined in the domain (18.2). Let us denote by $u_j(t) = u_j(t, x_1(t), ..., x_n(t))$ the control actions (as a function of time only), which are realized in the system (18.1) when $u_j = u_j(t, x)$. The symbols $x_i(t)$ designate the solution of (18.1) which is generated by control function $u_j(t) = u_j(t, x_1(t), ..., x_n(t))$. To emphasize that motion $x_i(t)$ is generated by some fixed control function $u_j = u_j^*(t, x)$ we shall write $x_i^*(t)$ and $u_j^*(t)$. The optimal stabilization problem consists in finding control functions $u_1^0(t, x), ..., u_r^0(t, x)$ that resolve the stabilization problem, and also for each such control function $u^*(t, x)$ to establish that

$$\int_{t_0}^{\infty} \omega(t, x^0(t); u^0(t)) \, dt \leq \int_{t_0}^{\infty} \omega(t, x^*(t); u^*(t)) \, dt \qquad (18.4)$$

holds. N.N. Krasovskii proved a theorem [1] in which sufficient conditions for the functions $u_j^0(t, x)$ are given. One of these conditions is that the function $w(t, x) = \omega(t, x; u^0(t, x))$ is positive definite in the domain (18.2).

Let us consider the situation when $X(t, x; u(t, x))$ are almost periodic functions in t. In this paper we show that in this case it is sufficient for the function $w(t, x)$ to be nonnegative only, provided some additional conditions are fulfilled.

18.2 Main Results

We shall use the following definitions [2].

Definition 1. A continuous function $f(t) : R \to R^n$ is almost-periodic if for any $\varepsilon > 0$ there exists $L = L(\varepsilon)$ such that there exists at least one number τ in any interval $[\alpha, \alpha + L(\varepsilon)]$, $\alpha \in (-\infty; +\infty)$ such that

$$\|f(t) - f(t + \tau)\| < \varepsilon \quad (-\infty < t < +\infty)$$

holds.

Definition 2. A continuous function $f(x, t) : R^m \times R \to R^n$ is uniformly almost periodic in t if for any $\varepsilon > 0$ and for any $r > 0$ there exists such $L = L(\varepsilon, r)$ that there exists at least one number τ in any interval $[\alpha, \alpha + L(\varepsilon, r)]$, $\alpha \in (-\infty; +\infty)$ such that

$$\|f(x, t) - f(x, t + \tau)\| < \varepsilon \quad (-\infty < t < +\infty, \ \|x\| < r)$$

holds.

Lemma 1. *Functions $X(x, t)$ and $V(x, t)$ are uniformly almost periodic.*

The proof of Lemma 1 is shown in [2].

Lemma 2. *For any $\varepsilon > 0$ there is an infinitely increasing sequence of ε-almost-periods $\{\tau_i\}$ and these ε almost-periods are the same for the functions $X(x, t)$ and $V(x, t)$ such that inequalities*

$$\|X(x, t) - X(x, t + \tau_i)\| < \varepsilon, \ |V(x, t) - V(x, t + \tau_i)| < \varepsilon$$

hold.

Kronecker's theorem [2] implies the proof of Lemma 2.

Lemma 3. *Let $x(x_0, t_0, t)$ $(t_0 < t < +\infty)$ be a semitrajectory of the system (18.1) which satisfies condition $x(x_0, t_0, t_0) = x_0$ and is situated on the set (18.2). If $\{\varepsilon_k\}$ is a sequence of the positive numbers monotonically tending to zero and $\{\tau_k\}$ is some sequence of ε_k- almost-periods of vector-function $X(x, t)$ (where for each ε_k there corresponds its ε_k- almost-period τ_k) and the sequence $\{\tau_k\}$ is monotonically increasing ($\tau_k \to \infty$ as $k \to \infty$), then*

$$\lim_{k \to \infty} \|x(x_k, t_0, t^*) - x(x_0, t_0, t^* + \tau_k)\| = 0, \qquad (18.5)$$

holds, where $x_k = x(x_0, t_0, t_0 + \tau_k)$, t^* is some moment of time which is more than t_0.

The proof of Lemma 3 can be found in [3].

Theorem. *If there exists an almost periodic in t, positive definite, continuously differentiable function $V^0(t, x)$, and functions $u_j^0(t, x)$ satisfying in the domain (18.2) the following conditions:*

1) $w(t, x) = \omega(t, x; u^0(t, x))$ is nonnegative and $w(t, x)$ may equal zero only at the points of a set which does not include any semitrajectory of the system (18.1) $x(x_0, t_0, t)$, $t_0 < t < +\infty$, entirely (except the trivial solution);

2) the equality

$$B[V^0; t, x; u^0(t, x)] = 0 \tag{18.6}$$

holds, where

$$B[V; t, x; u] = \frac{\partial V}{\partial t} + \sum_{i=1}^{n} \frac{\partial V}{\partial x_i} X_i(t, x; u) + \omega(t, x; u) = \frac{dV}{dt} + \omega(t, x; u); \tag{18.7}$$

3) the equality

$$B[V^0; t, x; u] \geq 0 \tag{18.8}$$

holds for each control function u_j,
then the functions $u_j^0(t, x)$ solve the problem of optimal stabilization and the equality

$$\int_{t_0}^{\infty} \omega(t, x^0[t]; u^0[t]) \, dt = \min \int_{t_0}^{\infty} \omega(t, x[t]; u[t]) \, dt = V^0(t_0, x(t_0)) \tag{18.9}$$

holds.

Proof. Let us show first that the control functions $u_j^0(t, x)$ ensure the asymptotic stability of the solution $x = 0$. When $u = u^0(t, x)$, the function V^0 satisfies all the conditions of Lyapunov's theorem [1]. Its derivative dV^0/dt with respect to the system (18.1) (when $u = u^0(t, x)$) is defined as

$$\frac{dV^0}{dt} = -\omega(t, x; u^0) \leq 0. \tag{18.10}$$

Therefore, $x = 0$ is stable. We shall prove that the solution $x = 0$ is an attractor, i.e., $\|x(x_0, t_0, t)\| \to 0$ as $t \to \infty$. Suppose the opposite. The function $V(x(x_0, t_0, t), t)$ is monotonically nonincreasing because $dV/dt \leq 0$. Hence, there exists

$$\lim_{t \to \infty} V^0(x(x_0, t_0, t), t) = V_0,$$

and $V^0(x(x_0, t_0, t), t) \geq V_0$. Our assumption implies $V_0 \neq 0$. Let $\{\varepsilon_i\}$ be a sequence of positive numbers monotonically tending to zero. For each ε_i

there exists a sequence of almost-periods $\tau_{i1}, \tau_{i2}, ..., \tau_{in} \to \infty$ corresponding to number ε_i for the functions $V^0(x,t)$ $X(x,t; u(t,x))$. One can write

$$|V^0(x,t) - V^0(x, t + \tau_{in})| < \varepsilon_i,$$
$$\|X(x,t; u(t,x)) - X(x, t + \tau_{in}; u(t + \tau_{in}, x))\| < \varepsilon_i,$$
$$\|x\| \le \varepsilon, \ (-\infty < t < +\infty).$$

We shall assume that $\tau_{in} < \tau_{i+1,n}$ and denote $\tau_{kk} = \tau_k$. Let us consider the sequence of points $x_k = x^0(x_0, t_0, t_0 + \tau_k)$, $(k = 1, 2, ...)$. This sequence is bounded because solution $x = 0$ is stable. One can select a convergent subsequence from the bounded sequence. For simplicity we assume that $\{x_k\}$ itself is convergent. Let x^* be a limit point of the sequence $\{x_k\}_{k=1}^{\infty}$. Our assumption implies that $x^* \ne 0$. Since $V^0(x,t)$ is continuous and almost periodic in t, one can write

$$V^0(x^*, t_0) = \lim_{k \to \infty} V^0(x_k, t_0 + \tau_k) = \lim_{k \to \infty} V^0(x^0(x_0, t_0, t_0 + \tau_k), t_0 + \tau_k) = V_0.$$

Let us consider the semitrajectory $x^0(x^*, t_0, t)$, $t_0 < t < \infty$. The hypothesis of the theorem implies the existence of points on this semi-trajectory where $dV^0(x^0(x^*, t_0, t), t)/dt < 0$, i.e., it is possible to indicate $t^* > t_0$ such that $V^0(x^0(x^*, t_0, t^*), t^*) = V_1 < V_0$. Because of the continuous dependence of the solution on the initial data, the equality $x^0(x^*, t_0, t^*) = \lim_{k \to \infty} x^0(x_k, t_0, t^*)$ holds; hence,

$$\lim_{k \to \infty} V^0(x^0(x_k, t_0, t^*), t^*) = V_1. \tag{18.11}$$

Since $X(x,t; u(t,x))$ is an almost-periodic function and (18.5) holds, we obtain

$$\|x^0(x_k, t_0, t^*) - x^0(x_0, t_0, t^* + \tau_k)\| \le \gamma_k, \tag{18.12}$$

where $\lim_{k \to \infty} \gamma_k = 0$. Because $V^0(x,t)$ is uniformly almost-periodic function, the inequality

$$|V^0(x^0, t^*) - V^0(x^0, t^* + \tau_k)| < \varepsilon_k \tag{18.13}$$

holds. Relations (18.11) and (18.12) imply that

$$|V^0(x^0(x_0, t_0, t^* + \tau_k), t^*) - V_1| < \eta_k, \tag{18.14}$$

where $\lim_{k \to \infty} \eta_k = 0$. Using (18.13), one can write

$$|V^0(x^0(x_0, t_0, t^* + \tau_k), t^*) - V^0(x^0(x_0, t_0, t^* + \tau_k), t^* + \tau_k)| < \varepsilon_k. \tag{18.15}$$

Adding inequalities (18.14) and (18.15), we obtain

$$|V^0(x^0(x_0, t_0, t^* + \tau_k), t^* + \tau_k) - V_1| < \eta_k + \varepsilon_k, \tag{18.16}$$

where $\eta_k + \varepsilon_k \to 0$ as $k \to \infty$. But

$$\lim_{k \to \infty} V^0(x^0(x_0, t_0, t^* + \tau_k), t^* + \tau_k) = V_0. \qquad (18.17)$$

Relations (18.16) and (18.17) contradict the inequality $V_1 < V_0$. Therefore, the solution $x = 0$ is an attractor and is asymptotically stable.

Let us prove equality (18.12). The motion $x^0(t)$ satisfies the condition $\|x^0(t)\| < H$. Hence, along this motion for all $t \geq t_0$ equality (18.7) holds; therefore, inequality (18.10) holds. The asymptotic stability of $x = 0$ implies that

$$\lim_{t \to \infty} V^0(t, x^0(t)) = 0. \qquad (18.18)$$

Integrating (18.10) along the motion $x^0(t)$ from $t = t_0$ to $t = \infty$ and taking (18.18) into account, we get

$$V^0(t, x(t_0)) = \int_{t_0}^{\infty} \omega(t, x^0(t); u^0(t)) \, dt. \qquad (18.19)$$

Let $u_1^*(t, x), ..., u_r^*(t, x)$ be other control functions that solve the stabilization problem for the motion $x = 0$. Then along the motion $x^*(t)$ the inequality (18.8) is fulfilled, i.e., the inequality

$$\frac{dV^0}{dt} \geq -\omega(t, x^*(t)) \qquad (18.20)$$

holds, where dV^0/dt is the derivative of V^0 along the solution $x^*(t)$. Integrating (18.20) from $t = t_0$ to $t = \infty$ and taking into account the equality

$$\lim_{t \to \infty} V^0(t, x^*(t)) = 0,$$

we obtain

$$V^0(t_0, x(t_0)) \leq \int_{t_0}^{\infty} \omega(t, x^*(t); u^*(t)), \, dt.$$

Taking (18.19) into account, we arrive at (18.9).

References

1. I.G. Malkin, *The Stability of Motion Theory*, Moscow, Nauka, 1966.

2. B.M. Levitan and V.V. Zhikov, *Almost Periodic Functions and Differential Equations*, Cambridge University Press, Cambridge, 1982.

3. A.Ya. Savchenko and A.O. Ignatyev, *Some Problems of Stability of Nonautonomous Dynamical Systems*, Naukova Dumka, Kiev, 1989.

19 A New Superconvergent Projection Method

Rekha P. Kulkarni

19.1 Introduction

Let X be a complex Banach space and T be a bounded linear operator defined on X. We are interested in the eigenvalue problem

$$T\phi = \lambda\phi, \ 0 \neq \lambda \in C, \ 0 \neq \phi \in X.$$

As the above problem, in general, can not be solved exactly, it is approximated by

$$T_n\phi_n = \lambda_n\phi_n,$$

where T_n is a sequence of finite rank operators converging to T. The eigenvalue problem associated with the operator T_n is equivalent to a matrix eigenvalue problem and λ_n and ϕ_n provide approximations to the exact eigenelements λ and ϕ, respectively.

If π_n is a sequence of bounded projections on X converging to the identity operator I pointwise, then in the classical Galerkin method, T is approximated by $T_n^G = \pi_n T \pi_n$. In the iterated version of the Galerkin method, proposed by Sloan, T is approximated by $T_n^S = T\pi_n$.

We here propose a new approximating operator

$$T_n^M = \pi_n T \pi_n + \pi_n T (I - \pi_n) + (I - \pi_n) T \pi_n.$$

We show that this new method has better performance as compared to the Galerkin or the Sloan method. More specifically, let T be an integral operator with a smooth kernel. Let π_n be either the orthogonal projection or the interpolatory projection at r Gauss points with the range as the space of piecewise polynomials of degree $\leq r-1$ with respect to a partition with norm h. Then the error in the eigenvalue approximation using the new method is of the order of h^{4r}, whereas in the Galerkin/Sloan method it is of the order of h^{2r}. For the spectral subspace approximation the error in the new method is of the order of h^{3r}, in the Galerkin method it is of the order of h^r and in the Sloan method it is of the order of h^{2r}.

We set the following notation. Let $\sigma(T)$ and $\rho(T)$ denote the spectrum of T and the resolvent set of T, respectively. Let λ be a nonzero isolated eigenvalue of T with finite algebraic multiplicity m. Let ϵ be such that

$0 < \epsilon < \text{dist}(\lambda, \sigma(T) \ \{\lambda\})$ and Γ_ϵ the positively oriented circle with center λ and radius ϵ. Let

$$P = -\frac{1}{2\pi i} \int_{\Gamma_\epsilon} (T - zI)^{-1} dz$$

denote the spectral projection associated with T and λ. Then rank $P = m$. For nonzero subspaces Y and Z of X, let

$$\delta(Y, Z) = \sup\{\text{dist}(y, Z) : y \in Y, \ \|y\| = 1\}.$$

Then

$$\hat{\delta}(Y, Z) = \max\{\delta(Y, Z), \delta(Z, Y)\}$$

is known as the gap between Y and Z. For a bounded linear operator T on X, we denote the range space of T by $R(T)$.

19.2 Main Results

Let T be a compact linear operator. Then

$$\|T - T_n^M\| = \|(I - \pi_n)T(I - \pi_n)\| \to 0,$$

whereas T_n^G and T_n^S converge to T in collectively compact fashion. Then for all large enough n, there are m eigenvalues of T_n^G, T_n^S and T_n^M inside Γ_ϵ, counted according to their algebraic multiplicities. Let $\hat{\lambda}_n^G$, $\hat{\lambda}_n^S$ and $\hat{\lambda}_n^M$ denote respectively their arithmetic means and let P_n^G, P_n^S and P_n^M be the associated spectral projections. We prove the following result using a slight modification of results of Osborn [1].

Theorem 1. *For all large n,*

$$\hat{\delta}(R(P), R(P_n^M)) \le C_1 \|(I - \pi_n)T(I - \pi_n)\|,$$

$$|\lambda - \hat{\lambda}_n^M| \le C_1 \| T(I - \pi_n)T(I - \pi_n)T\|,$$

where C_1 is a generic constant independent of n.

Remark. Since $\|(I - \pi_n)T\| \to 0$ as $n \to \infty$, we have

$$\hat{\delta}(R(P), R(P_n^M)) = O(\|(I - \pi_n)T\|)^2,$$

$$|\lambda - \hat{\lambda}_n^M| = O(\|(I - \pi_n)T\|)^2.$$

On the other hand,

$$\hat{\delta}(R(P), R(P_n^G)) = O(\|(I - \pi_n)T\|),$$

$$\hat{\delta}(R(P), R(P_n^S)) = O(\|(I - \pi_n)T\|),$$

$$|\lambda - \hat{\lambda}_n^G| = O(\|(I - \pi_n)T\|).$$

If the dimension of $R(\pi_n)$ is N, then the rank of T_n^G and T_n^S is N, whereas the rank of T_n^M is $2N$. Thus, in the new method we need to solve a matrix eigenvalue problem of double the size as compared to the size of the eigenvalue problem in the Galerkin/Sloan method. However, the improvement in the order of convergence makes the new method economical as compared to Galerkin/Sloan methods.

Note that if T is a compact integral operator with a continuous kernel, then the above results are applicable.

19.2.1 Orthogonal Projection

We now specialise to the case when T is an integral operator with a smooth kernel. Let $X = L^2[0,1]$ and

$$(Tx)(s) = \int_0^1 k(s,t)x(t)dm(t), \quad s \in [0,1],$$

where the kernel $k(.,.) \in C([0,1] \times [0,1])$. Consider a quasiuniform partition

$$0 = t_0 < t_1 < \cdots < t_n = 1$$

of $[0,1]$ and let $h = \max\{t_i - t_{i-1} : i = 1, \ldots, n\}$ denote the norm of the partition. Choose $r \geq 1$ and $-1 \leq \nu \leq r - 2$. Let $X_n = S_{r,n}^\nu$, the space of all piecewise polynomials of degree $\leq r - 1$ with breakpoints at $t_1, \ldots, t_{n-1}$ and with ν continuous derivatives. Here $\nu = 0$ corresponds to the case of continuous piecewise polynomials. If $\nu = -1$, there is no continuity requirements at the breakpoints. Let $\pi_n : X \to X_n$ denote the orthogonal projection. Then $\pi_n u \to u$ as $n \to \infty$ for each $u \in X$. We quote the following result from Chatelin [2]. For $u \in C^r[0,1]$,

$$\|(I - \pi_n)u\|_\infty \leq C_2 \|u^{(r)}\|_\infty h^r,$$

where $u^{(r)}$ denotes the r-th derivative of u and C_2 is a constant independent of n. Using the above estimate and Theorem 1 we prove the following result.

Theorem 2. If $k(.,.) \in C^r([0,1] \times [0,1])$, then

$$\hat{\delta}(R(P), R(P_n^M)) = O(h^{3r}),$$

$$|\lambda - \hat{\lambda}_n^M| = O(h^{4r}).$$

We see that using a piecewise polynomial space of degree $\leq r - 1$, we obtain the order of convergence h^{3r} for spectral subspace approximation and h^{4r} for eigenvalue approximation. Thus the new method exhibits superconvergence.

We quote the following results from Chatelin [2] for comparison.

$$\hat{\delta}(R(P), R(P_n^G)) = O(h^r),$$
$$\hat{\delta}(R(P), R(P_n^S)) = O(h^{2r}), \qquad (19.1)$$
$$|\lambda - \hat{\lambda}_n^G| = O(h^{2r}).$$

19.2.2 Collocation at Gauss Points

Now let $X = C[0, 1]$ with the supremum norm, $X_n = S_{r,n}^{-1}$, the space of all discontinuous piecewise polynomials of degree $\leq r - 1$ and let $\{\tau_1, \ldots, \tau_r\}$ be the set of r Gauss points in $[-1, 1]$. Let

$$\left\{ \eta_{ij} = \frac{1 - \tau_j}{2} t_{i-1} + \frac{1 + \tau_j}{2} t_i, \ \ i = 1, \ldots, n, \ \ j = 1, \ldots, r \right\}$$

be the set of nr collocation points. We define

$$\pi_n : C[0, 1] \to X_n$$

to be the interpolating operator such that for $u \in C[0, 1]$,

$$\pi_n u \in X_n, \ (\pi_n u)(\eta_{ij}) = u(\eta_{ij}), \ \ 1 \leq i \leq n, \ 1 \leq j \leq r.$$

Then $\pi_n u \to u$ as $n \to \infty$ for each $u \in C[0, 1]$ and for $u \in C^r[0, 1]$,

$$\|(I - \pi_n)u\|_\infty \leq C_2 \|u^{(r)}\|_\infty h^r$$

(see Chatelin [2]).

The following assertion holds.

Theorem 3. *If $k(., .) \in C^{2r}([0, 1] \times [0, 1])$, then*

$$\hat{\delta}(R(P), R(P_n^M)) = O(h^{3r}),$$
$$|\lambda - \hat{\lambda}_n^M| = O(h^{4r}).$$

Note that we need a higher order of smoothness of the kernel as compared to the case of the orthogonal projection.

The orders of convergence for the Galerkin and the Sloan method are the same as given in (19.1).

19.2.3 Iterative Refinement

We now consider the case when $m = 1$, that is, λ is a simple eigenvalue of T. Even though the eigenelements of T_n provide approximations for the eigenelements of T, in order to achieve the desired accuracy, we may have to choose n large resulting in a large matrix eigenvalue problem, which is costly in terms of computer memory and time. We describe below an iterative refinement scheme in which an eigenvalue problem of relatively small size is solved. The iteration scheme has these eigenelements as the starting step of the iteration and at each step of the iteration, we need to solve a system of linear equations.(See Ahues et al [3].)

Let

$$T_n \phi_n = \lambda_n \phi_n, \quad \|\phi_n\| = 1,$$

$$T_n^* \phi_n^* = \overline{\lambda}_n \phi_n^*, \quad \langle \phi_n, \phi_n^* \rangle = 1,$$

where T_n^* denotes the adjoint of T_n. Let S_n denote the reduced resolvent associated with T_n and λ_n. Then

$$S_n(T_n - \lambda_n I) = (T_n - \lambda_n I)S_n = I - P_n, \quad S_n P_n = P_n S_n = 0.$$

Fixed Point Iteration

$$\lambda_n^{(0)} = \lambda_n, \quad \phi_n^{(0)} = \phi_n,$$

and for $k = 1, 2, \ldots$

$$\lambda_n^{(k)} = \langle T\phi_n^{(k-1)}, \phi_n^* \rangle,$$

$$\phi_n^{(k)} = \phi_n^{(k-1)} + S_n[\lambda_n^{(k)}\phi_n^{(k-1)} - T\phi_n^{(k-1)}].$$

We define

$$\phi_{(n)} = \frac{\phi}{\langle \phi, \phi_n^* \rangle}.$$

Theorem 3. *Let T be an integral operator with a kernel r times continuously differentiable and let π_n be the orthogonal projection defined in Section (2.1). Then for $k = 0, 1, \ldots,$*

$$|\lambda_n^{(k)} - \lambda| = O(h^{4r}(h^{2r})^k),$$

$$\|\phi_n^{(k)} - \phi_{(n)}\| = O(h^{3r}(h^{2r})^k).$$

In the case of the Galerkin method, the error in both the eigenvalue and eigenvector iterates is of the order of $h^r(h^r)^k$, while in the Sloan method, the error is of the order of $h^{2r}(h^r)^k$.

In the case of the interpolatory projection, the new method once again has higher orders of convergence than in the Galerkin/Sloan methods.

19.2.4 Extensions

Instead of the collocation at Gauss points described above, it is possible to consider the new method with the collocation at Lobatto points.

As in the case of the Sloan method, the new approximating operator has asymptotic series expansion. Using the Richardson extrapolation we can improve the order of convergence h^{4r} for eigenvalue approximation to h^{4r+2}.

The new operator T_n^M can be used for solution of operator equations.

The technique of accelerated spectral approximation can be used for the new method.

Finally the discrete version of the new method can be analyzed.

Some of the results described above are reported in [4] and [5].

References

1. J.E. Osborn, Spectral Approximation for Compact operators, *Math. Comp.* **29** (1975), 712–725.

2. F. Chatelin, *Spectral Approximation of Linear Operators*, Academic Press, New York, 1983.

3. M. Ahues, A. Largillier, and B.V. Limaye, *Spectral Computations for Bounded Operators*, Chapman and Hall/CRC, New York, 2001.

4. R.P. Kulkarni, A new superconvergent projection method for approximate solutions of eigenvalue problems (communicated).

5. R.P. Kulkarni and N. Gnaneshwar, Spectral refinement using a new projection method (communicated).

20 Spectral Approximation for Compact Integral Operators

Balmohan V. Limaye

20.1 Compact Integral Operators, the Spectral Problem

Several problems in science and engineering can be modeled and represented with the help of function spaces and integral operators on them.

Let X be a Banach space over $\mathbb{C}$ and T be a linear operator on X which is *compact*, that is, if (x_n) is any bounded sequence in X, then the sequence (Tx_n) has a convergent subsequence. This requirement is stronger than the continuity of T; in fact, if T is compact and (x_n) converges weakly to x in X, then (Tx_n) converges to Tx in X. Hence compact operators are also known as completely continuous operators.

We introduce integral operators as follows. Let μ be a measure on a measurable space E, and let X denote a Banach space over $\mathbb{C}$ of complex-valued measurable functions on E. Let $k(.,.) : E \times E \to \mathbb{C}$ be a function such that for every $x \in X$, the function $y : E \to \mathbb{C}$ defined by

$$y(s) := \int_E k(s,t)x(t)d\mu(t), \quad s \in E,$$

is well-defined and belongs to X. Define $T : X \to X$ by letting $Tx := y$ for $x \in X$, that is,

$$Tx(s) := \int_E k(s,t)x(t)d\mu(t), \quad x \in X, \ s \in E.$$

Then T is known as an *integral operator* on X, and $k(.,.)$ is called the *kernel* of T. Clearly, T is linear. We now give several examples wherein T is compact.

Examples.

(i) Let $n \in \mathbb{N}$, the set of natural numbers, $E := \{1, \ldots, n\}$, μ the counting measure, $X := \mathbb{C}^n$ with any norm and $k(i,j) \in \mathbb{C}$ for $i, j = 1, \ldots, n$. Then the $n \times n$ matrix $[k(i,j)]$ defines a compact integral operator $T : \mathbb{C}^n \to \mathbb{C}^n$ given by

$$(Tx)(i) := \sum_{j=1}^{n} k(i,j)x(j), \quad x \in \mathbb{C}^n, \ i = 1, \ldots, n.$$

(ii) Let $E := \mathbb{N}$, μ the counting measure and $X := \ell^p$. For $1 \le p < \infty$,

$$\ell^p := \{(x(1), x(2), \ldots) : x(j) \in \mathbb{C}, \sum_{j=1}^{\infty} |x(j)|^p < \infty\}$$

and for $x := (x(1), x(2), \ldots) \in \ell^p$,

$$\|x\|_p = \Big(\sum_{j=1}^{\infty} |x(j)|^p\Big)^{1/p}.$$

Also, for $p = \infty$,

$$\ell^\infty := \{(x(1), x(2), \ldots) : x(j) \in \mathbb{C}, (x(j)) \text{ is bounded}\}$$

and for $x := (x(1), x(2), \ldots) \in \ell^\infty$,

$$\|x\|_\infty = \sup\{|x(j)| : j = 1, 2, \ldots\}.$$

Let $k(i,j) \in \mathbb{C}$ for $i, j = 1, 2 \ldots$. Suppose that the infinite matrix $[k(i,j)]$ defines an operator $T : \ell^p \to \ell^p$ given by

$$Tx(i) := \sum_{j=1}^{\infty} k(i,j)x(j), \quad x \in \ell^p, \ i = 1, 2, \ldots.$$

Under the following conditions, T is in fact a compact integral operator on $\ell^p([1], 17.4(a))$.

(a) $p = 1$: For every $j = 1, 2, \ldots$, the jth column sum

$$\gamma(j) := \sum_{i=1}^{\infty} |k(i,j)|$$

is finite and $\gamma(j) \to 0$ as $j \to \infty$.

(b) $p = \infty$: For every $i =, 1, 2, \ldots$, the ith row sum

$$\delta(i) := \sum_{j=1}^{\infty} |k(i,j)|$$

is finite and $\delta(i) \to 0$ as $i \to \infty$.

(c) $1 < p < \infty$: One of the sequences $(\gamma(j))$ and $(\delta(j))$ defined above is bounded and the other converges to 0.

(d) $1 < p < \infty$: $\sum_{i=1}^{\infty} \Big(\sum_{j=1}^{\infty} |k(i,j)|^q\Big)^{p/q} < \infty$, where $\dfrac{1}{p} + \dfrac{1}{q} = 1$.

(iii) Let $E := [a, b] \subseteq \mathbb{R}$, μ the Lebesgue measure, and $X := L^p([a, b])$, the space of equivalence classes of p-integrable complex-valued functions on $[a, b]$ (if $1 \leq p < \infty$) or the essentially bounded complex-valued functions on $[a, b]$ (if $p = \infty$) with the usual $\| \ \|_p$ norm, or let $X := C([a, b])$, the set of all continuous complex-valued functions on $[a, b]$ with the supremum norm. Let $k(s, t) \in \mathbb{C}$ for $s, t \in [a, b]$. If $k(., .) : [a, b] \times [a, b] \to \mathbb{C}$ is a continuous function, then the integral operator $T : X \to X$ defined by

$$Tx(s) := \int_a^b k(s, t)x(t)dt, \quad x \in X, \ s \in [a, b],$$

is compact ([1], 17.4(b)).

Note that Example (iii) is the continuous analogue of the discrete case given in Example (ii).

We now turn to the spectral problem for an integral operator T on a Banach space X. The *spectrum* of T is defined by

$$\text{sp}(T) := \{\lambda \in \mathbb{C} : T - \lambda I \text{ is either not one-one or not onto}\}.$$

If $\lambda \in \text{sp}(T)$, we say that λ is a *spectral value* of T. For $\lambda \in \mathbb{C}$, $T - \lambda I$ is not one-one if and only if there is a nonzero $x \in X$ such that $Tx = \lambda x$, that is, λ is an *eigenvalue* of T (with x as a corresponding eigenvector).

If T is a compact operator and $\lambda \in \mathbb{C}$ with $\lambda \neq 0$, then it is well known that λ is a spectral value of T if and only if λ is an eigenvalue of T, and then it is an isolated point of $\text{sp}(T)$. Thus finding nonzero spectral values of T is equivalent to finding nonzero eigenvalues of T. However, even when X is finite dimensional, it is not easy to find eigenvalues of T, especially if they cluster together.

Let Λ be a finite subset of $\text{sp}(T)$ and assume that $0 \notin \Lambda$. The *spectral subspace* associated with T and Λ is the largest closed subspace Y_Λ of X such that Y_Λ is invariant under T and the spectrum of the restriction of T to Y_Λ equals Λ. Further, the *spectral projection* associated with T and Λ is the (continuous) projection P_Λ of X onto Y_Λ such that $P_\Lambda T = T P_\Lambda$.

If T is a compact operator, then so is P_Λ. This implies that the range Y_Λ of the projection P_Λ is finite dimensional. Another way of stating this result is as follows: The algebraic multiplicity of every eigenvalue belonging to Λ is finite ([2], 1.34).

The *spectral problem* for a compact operator T consists of finding all nonzero eigenvalues of T and if Λ is any finite set of such values, then of finding a basis for the (finite dimensional) spectral subspace Y_Λ.

If $x_1, \ldots, x_m$ be a basis for Y_Λ, then there are $\theta_{i,j} \in \mathbb{C}$, $i, j = 1, \ldots, m$, such that

$$Tx_1 = \theta_{1,1}x_1 + \cdots + \theta_{m,1}x_m,$$

$$\vdots \qquad \vdots \qquad \vdots \qquad \vdots$$

$$Tx_m = \theta_{1,m}x_1 + \cdots + \theta_{m,m}x_m.$$

The above m equations can be written in a matrix notation as $\underline{T}\underline{x} = \underline{x}\Theta$, where $\underline{T}\underline{x} = [Tx_1, \ldots, Tx_m]$, $\underline{x} = [x_1, \ldots, x_m]$ and the (i,j)th element of the $m \times m$ matrix Θ is $\theta_{i,j}, i, j = 1, \ldots, m$. Notice that if $m = 1$, this equation reduces to $Tx = x\theta$, which is usually written as $Tx = \theta x$; it says that θ is an eigenvalue of T with a corresponding eigenvector x. Thus in this case, the spectral problem for T reduces to the eigenvalue problem.

20.2 Approximation by Finite Rank Operators

Since it is difficult to solve the spectral problem $\underline{T}\underline{x} = \underline{x}\Theta$ for a compact operator exactly, one attempts to approximate a compact operator T : $X \to X$ by a finite rank continuous operator $T_n : X \to X$, and then look for a solution of the spectral problem $\underline{T}_n \underline{x}_n = \underline{x}_n \Theta_n$, $n = 1, 2, \ldots$. Here the sequence (T_n) should "converge" to T in such a way that $\underline{x}_n$ and Θ_n provide approximations of $\underline{x}$ and Θ respectively.

Two classical modes of such convergence are as follows.

1. **Norm convergence**: $\|T_n - T\| \to 0$. This mode is well-studied in Kato's book [3].

2. **Collectively compact convergence**: $T_n x \to Tx$ for every $x \in X$ and the set $\{T_n x : x \in X, \|x\| \leq 1, n = 1, 2, \ldots\}$ is relatively compact in X. This mode was introduced by Atkinson [4] and Anselone [5]. It is well-studied in the book of Chatelin [6].

A new mode of convergence which retains the essential features of the above-mentioned two modes has evolved from the works of Ahues [7] and Nair [8]. It is well-studied in the recent book of Ahues, Largillier and Limaye [2]. For lack of a better name, it is called the ν-convergence:

(i) $(\|T_n\|)$ is a bounded sequence,

(ii) $\|(T_n - T)T\| \to 0$,

(iii) $\|(T_n - T)T_n\| \to 0$.

The three conditions which comprise ν-convergence are general enough to encompass a wide variety of approximation methods, and at the same time, they are simple enough to be verified in practice. Notice that in this case, $\|(T_n - T)^2\| \to 0$.

The following result shows how ν-convergence implies convergence of spectral values.

Theorem. *Let T be a compact operator, $\lambda \in \mathbb{C}$ with $\lambda \neq 0$ and assume that (T_n) is ν-convergent to T.*

(i) *(Upper semicontinuity) If $\lambda_n \in \mathrm{sp}(T_n)$ for each $n \in \mathbb{N}$ and $\lambda_n \to \lambda$, then $\lambda \in \mathrm{sp}(T)$.*

(ii) *(Lower semicontinuity). If $\lambda \in \mathrm{sp}(T)$, then there is $\lambda_n \in \mathrm{sp}(T_n)$ for each $n \in \mathbb{N}$ such that $\lambda_n \to \lambda$.*

The proof follows from Corollary 2.7 and Corollary 2.13 of [2].

We shall now consider error estimates when (T_n) is ν-convergent to T, $\lambda_n \in \text{sp}(T_n)$ and $\lambda_n \to \lambda$. First we consider the case when λ is a *simple* eigenvalue of T, that is, the corresponding spectral projection is of rank 1.

Theorem. *Let T be a compact operator and λ be a nonzero simple eigenvalue of T. Then for each large $n \in \mathbb{N}$, T_n has a simple eigenvalue λ_n such that $\lambda_n \to \lambda$.*

Let ϕ and ϕ_n be the corresponding eigenvectors of T and T_n respectively, and let P denote the spectral projection associated with T and λ. Then

(i) $|\lambda_n - \lambda| = O(\|(T_n - T)T\|), \quad |\lambda_n - \lambda| = O(\|(T_n - T)T_n\|),$

(ii) $\dfrac{\|\phi_n - P\phi_n\|}{\|\phi_n\|} = O(\|(T_n - T)T_n\|),$ *and*

(iii) there is $c_n \in \mathbb{C}$, $c_n \neq 0$ such that $\dfrac{\|c_n\phi_n - \phi\|}{\|\phi\|} = O(\|(T_n - T)T\|).$

The proof follows from Theorem 2.17 of [2].

Let us now consider a cluster $\Lambda := \{\lambda(1), \ldots, \lambda(m)\}$ of nonzero eigenvalues of T. Let $\widehat{\lambda}$ denote the arithmetic mean of the eigenvalues in Λ counted according to their algebraic multiplicities. If Γ is a simple closed contour separating Λ from the rest of $\text{sp}(T)$ and from 0, then for each large $n \in \mathbb{N}$, the set

$$\Lambda_n := \text{sp}(T_n) \cap \text{Interior of } \Gamma$$

consists of m eigenvalues of T_n counted according to their algebraic multiplicities, and if $\widehat{\lambda}_n$ denotes the arithmetic mean of the eigenvalues of T_n inside Γ counted according to their algebraic multiplicities, then we have

$$|\widehat{\lambda}_n - \widehat{\lambda}| = O(\|(T_n - T)T\|) \quad \text{and} \quad |\widehat{\lambda}_n - \widehat{\lambda}| = O(\|(T_n - T)T_n\|).$$

Also, if $\underline{\phi} = [\phi(1), \ldots, \phi(m)]$ is a basis for the spectral subspace associated with T and Λ, and $\underline{\phi}_n = [\phi_n(1), \ldots, \phi_n(m)]$ is a basis for the spectral subspace associated with T_n and Λ_n, then we have

$$\frac{\|\underline{\phi}_n - P\underline{\phi}_n\|_\infty}{\|\underline{\phi}_n\|_\infty} = O(\|(T_n - T)T_n\|),$$

where P is the spectral projection associated with T and Λ, and there is a nonsingular $m \times m$ matrix C_n of complex numbers such that

$$\frac{\|\underline{\phi}_n - \underline{\phi}\, C_n^{-1}\|_\infty}{\|\underline{\phi}\, C_n^{-1}\|_\infty} = O(\|(T_n - T))T\|).$$

For proofs of these error estimates, see Theorem 2.18 of [2].

Thus the approximations $\lambda_n, \widehat{\lambda}_n, \phi_n, \underline{\phi}_n$ of $\lambda, \widehat{\lambda}, \phi, \underline{\phi}$ are at least as accurate as $\|(T_n - T)T\|$ and/or $\|(T_n - T)\overline{T}_n\|$.

20.3 Reduction of the Spectral Problem for a Finite Rank Operator to the Spectral Problem for a Finite Matrix

We began in Section 1 by considering the spectral problem for an operator T defined on a possibly infinite dimensional domain X, whose range is also a possibly infinite dimensional subspace of X. In Section 2, we saw how such an operator can be approximated by operators defined on X whose ranges are finite dimensional and how this approximation can be used to set up an approximate spectral problem. In this section, we shall give a canonical method of reducing this approximate spectral problem to a spectral problem for an operator whose domain as well as range are finite dimensional, that is, to a spectral problem for a finite matrix.

Let $n \in \mathbb{N}$ and $T_n : X \to X$ be a continuous operator of finite rank. Then there are $x_{1,n}, \ldots, x_{n,m_n}$ in X and $f_{n,1}, \ldots, f_{n,m_n}$ in the dual X' of X such that

$$T_n x = f_{n,1}(x) x_{n,1} + \cdots + f_{n,m_n}(x) x_{n,m_n} \quad \text{for all } x \in X.$$

We remark that neither the elements $x_{n,1}, \ldots, x_{n,m_n}$ of X nor the elements $f_{n,1}, \ldots, f_{n,m}$ of X' are assumed to be linearly independent. Thus the above representation of T_n is not unique in any sense.

For ease in notation, let us assume that $m_n = n$ for all $n \in \mathbb{N}$. Consider the $n \times n$ matrix

$$A_n := [f_{n,j}(x_{n,i})], \quad i, j = 1, \ldots, n.$$

The following result shows that solving the eigenvalue problem for the finite rank operator T_n is the same as solving the eigenvalue problem for the associated matrix A_n (except possibly for the eigenvalue 0).

Lemma. *Let $\lambda_n \in \mathbb{C}$ and $\lambda_n \neq 0$. Then for any $x \in X$ and $u \in \mathbb{C}^n$,*

$$T_n x = \lambda_n x \text{ and } f_{n,j}(x) = u(j), \ j = 1, \ldots, n$$
$$\Longleftrightarrow A_n u = \lambda_n u \text{ and } \sum_{j=1}^{n} u(j) x_{n,j} = \lambda_n x.$$

For proof, see Lemma 5.2 of [2] with $p = 1$ and $y = 0$.

In particular, in the notation of the above lemma, $x \neq 0$ if and only if $u \neq 0$, so that λ_n is an eigenvalue of T_n if and only if it is an eigenvalue of A_n, and then the eigenvector x can be constructed from u as follows:

$$x = \frac{1}{\lambda_n} \sum_{j=1}^{n} u(j) x_{n,j}.$$

Turning now to the (more general) spectral problem for T_n, we give the following generalization of the above lemma.

Let Θ_n be an $m \times m$ nonsingular matrix of complex numbers. Then for $\underline{x} = [x_1, \ldots, x_m] \in X^{1 \times m}$ and $\underline{u} = [u_1, \ldots, u_m] \in \mathbb{C}^{n \times m}$, we have

$$T_n \underline{x} = \underline{x}\, \Theta_n \text{ and } f_{n,j}(x_i) = u_i(j), \; i = 1, \ldots, m, j = 1, \ldots, n$$

$$\Longleftrightarrow A_n \underline{u} = \underline{u}\, \Theta_n \text{ and } \left[\sum_{j=1}^n u_1(j)x_{n,j}, \ldots, \sum_{j=1}^n u_m(j)x_{n,j} \right] = \underline{x}\, \Theta_n.$$

Also, $\{x_1, \ldots, x_m\}$ is a linearly independent subset of X if and only if $\{u_1, \ldots, u_m\}$ is a linearly independent subset of $\mathbb{C}^n$, and then the basis $\underline{x}$ can be constructed from $\underline{u}$ as follows:

$$\underline{x} = \left[\sum_{j=1}^n u_1(j)x_{n,j}, \ldots, \sum_{j=1}^n u_m(j)x_{n,j} \right] \Theta_n^{-1}.$$

We now describe some well-known examples of finite rank approximations of a compact operator T where ν-convergence holds. We shall also give the associated finite matrices.

(A) For $n \in \mathbb{N}$, let $\pi_n : X \to X$ be a finite rank (continuous) projection such that $\pi_n x \to x$ for every $x \in X$.

   (i) *Projection Approximation* $T_n^P := \pi_n T$,

   (ii) *Sloan Approximation* $T_n^S := T \pi_n$,

   (iii) *Galerkin Approximation* $T_n^G := \pi_n T \pi_n$.

If $X = \ell^2$, $\sum_{i,j=1}^{\infty} |k(i,j)|^2 < \infty$, $Tx(i) := \sum_{j=1}^{\infty} k(i,j)x(j)$ for $x \in X$, $i = 1, 2, \ldots$, and $\pi_n(x(1), x(2), \ldots) = (x(1), \ldots, x(n), 0, 0, \ldots)$ for $x \in X$, then

   (i) $T_n^P x(i) = \sum_{j=1}^{\infty} k(i,j)x(j)$ if $1 \le i \le n$, and $T_n^P x(i) = 0$ if $i > n$,

   (ii) $T_n^S x(i) = \sum_{j=1}^{n} k(i,j)x(j), i = 1, 2, \ldots$,

   (iii) $T_n^G x(i) = \sum_{j=1}^{n} k(i,j)x(j)$ if $1 \le i \le n$ and $T_n^G x(i) = 0$ if $i > n$.

Here T may be represented by the infinite matrix A whose (i,j)th element is $k(i,j)$, $i, j = 1, 2, \ldots$. The infinite matrix representing T_n^P is obtained by truncating each column of A at the nth entry and putting all zeros thereafter, while the infinite matrix representing T_n^S is obtained by replacing every column after the nth column of A by a column of all zeros. The infinite matrix representing T_n^G is obtained by carrying out both these operations. In all the three cases, the (i,j)th element of the associated finite matrix A_n is $k(i,j)$, $i, j = 1, \ldots, n$.

(B) Let $X = C([a, b])$, $k(., .)$ be a continuous complex-valued function on $[a, b] \times [a, b]$, and define $T : X \to X$ by

$$Tx(s) := \int_a^b k(s, t)x(t)dt, \quad x \in X, \ s \in [a, b].$$

Consider a sequence (Q_n) of quadrature formulæ such that $Q_n x \to x$ for every $x \in X$. Let Q_n for $n \in \mathbb{N}$ be given by

$$Q_n x := \sum_{j=1}^n w_{n,j} x(t_{n,j}), \quad x \in X.$$

Then the *Nyström approximation* $T_n^N : X \to X$ is given by

$$T_n^N x(s) := \sum_{j=1}^n w_{n,j} k(s, t_{n,j})x(t_{n,j}) \quad x \in X, \ s \in [a, b].$$

In this case, the (i, j)th element of the associated finite matrix A_n is $w_{n,j} k(t_{n,i}, t_{n,j})$, $i, j = 1, \ldots, n$.

The above examples illustrate the step-by-step passage from the infinite to the finite! For some numerical experiments, see Section 5.4 of [2].

References

1. B.V. Limaye, *Functional Analysis*, 2nd ed., New Age International, New Delhi, 1996.

2. M. Ahues, A. Largillier, and B.V. Limaye, *Spectral Computations for Bounded Operators*, Chapman and Hall/CRC, Boca Raton, FL, 2001.

3. T. Kato, *Perturbation Theory for Linear Operators*, 2nd ed., Springer-Verlag, Berlin, 1976.

4. K.E. Atkinson, The numerical solution of eigenvalue problem for compact integral operators, *Trans. Amer. Math. Soc.* **129** (1967), 458–465.

5. P.M. Anselone, *Collectively Compact Operator Approximation Theory and Applications to Integral Equations*, Prentice Hall, Englewood Cliffs, N.J., 1971.

6. F. Chatelin, *Spectral Approximation for Linear Operators*, Academic Press, New York, 1983.

7. M. Ahues, A class of strongly stable operator approximations, *J. Austral. Math. Soc. Ser. B* **28** (1987), 435–442.

8. M.T. Nair, On strongly stable approximations, *J. Austral. Math. Soc. Ser. A* **52** (1992), 251–260.

21 Fundamental Solutions and Functionally Graded Materials

Paul A. Martin

21.1 Introduction

A fundamental solution (or Green's function) is a singular solution of a governing partial differential equation (PDE). Such solutions can be constructed easily when the PDE has constant coefficients. They are useful for reducing boundary-value problems to boundary integral equations (BIEs). We begin by describing simple properties of fundamental solutions, and then comment on the use and construction of half-space Green's functions.

We then move on to consider functionally graded materials (FGMs). These are inhomogeneous materials: their properties vary with position. Modelling FGMs leads to PDEs with variable coefficients, and this makes the construction of fundamental solutions more difficult.

In this paper, we consider FGMs where the properties vary exponentially in one prescribed direction; such 'exponentially graded' materials provide a reasonable model of certain real situations. We discuss the construction of fundamental solutions for steady-state heat conduction and for three-dimensional elasticity. These solutions should be useful in the development of boundary integral methods for FGMs.

21.2 What is a Fundamental Solution?

As a prototypical example, consider Laplace's equation in three dimensions, $\nabla^2 u = 0$. A fundamental solution for this PDE is

$$G(P, P') = G(\mathbf{x}, \mathbf{x}') = R^{-1},$$

where the points P and P' have position vectors $\mathbf{x}$ and $\mathbf{x}'$, respectively, with respect to an origin O, and $R = |\mathbf{x} - \mathbf{x}'|$ is the distance between P and P'. Notice that $\nabla_P^2 G(P, P') = 0$ and $\nabla_{P'}^2 G(P, P') = 0$ for $P \neq P'$.

We can use G in order to reduce boundary-value problems to boundary integral equations. For example, suppose that one wants to solve $\nabla^2 u = 0$ inside a bounded region V with a Dirichlet condition, $u = f$, on the boundary of V, S. A careful application of Green's theorem in V to $u(P)$ and $G(P, P')$, with $P' \in V$, gives the integral representation

$$u(P') = \frac{1}{4\pi} \int_S \left\{ G(p, P') \frac{\partial u}{\partial n} - f(p) \frac{\partial G}{\partial n_p} \right\} ds_p, \quad P' \in V, \qquad (21.1)$$

where the unit normal to S points *out* of V. To obtain this well-known formula, one has to excise a small sphere from V (of radius ε and centred at P') prior to using Green's theorem, and then let the radius $\varepsilon \to 0$.

The unknown boundary values of $\partial u/\partial n$ in (21.1) can then be found by solving a BIE; such equations can be obtained by, for example, considering the limit $P' \to p' \in S$ in (21.1), or by calculating the normal derivative of (21.1) at p'.

So far, we have used the simplest choice for G, namely $G = R^{-1}$. In fact, G could be modified in various ways. Thus, we could use

$$AR^{-1} + H(P, P'),$$

where A is any constant or any function of P', and H is any non-singular solution of $\nabla_P^2 H = 0$ (at least in the neighbourhood of P'). These modifications may sometimes be exploited to good effect.

As a second example, suppose that one wants to consider the radiation of acoustic waves in the unbounded region outside S, with a Neumann condition, $\partial u/\partial n = g$, on S. The governing PDE is the three-dimensional Helmholtz equation, $(\nabla^2 + k^2)u = 0$. In order to have a unique solution, we impose the Sommerfeld radiation condition at infinity; this implies that

$$u \sim r^{-1}\, e^{ikr}\, f(\theta, \phi) \quad \text{as} \quad r \to \infty,$$

where r, θ and ϕ are spherical polar coordinates and f is the (unknown) far-field pattern.

For the three-dimensional Helmholtz equation, a fundamental solution is

$$\frac{\cos kR}{R} \sim \frac{1}{R} \quad \text{as} \quad R \to 0.$$

Another is

$$A\,\frac{\cos kR}{R} + B\,\frac{\sin kR}{R}.$$

Usually, we want a fundamental solution that also satisfies the radiation condition, so we can take $A = 1$ and $B = i$, giving

$$G(P, P') = \frac{e^{ikR}}{R}.$$

We can use G to obtain a BIE for u on S; the standard equation is

$$2\pi u(p) - \int_S u(q)\, \frac{\partial G}{\partial n_q}\, ds_q = -\int_S g(q)\, G(q, p)\, ds_q, \quad p \in S.$$

For more information on BIEs for the Helmholtz equation, see [1].

21.3 Half-Space Green's Functions

It is common to construct (and use) fundamental solutions that also satisfy an additional boundary condition (just as we selected a fundamental
solution that satisfied a radiation condition). To give a flavour of these,
we discuss briefly a few examples of *half-space Green's functions*. These
are singular solutions of a PDE in a half-space $y > 0$, say, that also satisfy a boundary condition on $y = 0$ (together with a condition at infinity).
They are used to derive BIEs when the half-space contains an obstacle with
boundary S; the result is a BIE over S.

The simplest examples of half-space Green's functions are for Laplace's
equation or the Helmholtz equation with the boundary condition $u = 0$ or
$\partial u/\partial y = 0$ on $y = 0$: such fundamental solutions are easily constructed by
the method of images.

If the half-space is filled with water, the governing PDE is Laplace's
equation and the appropriate boundary condition on the free surface is the
Robin condition $Ku + \partial u/\partial y = 0$, where K is a given positive constant.
Appropriate fundamental solutions are known [2]. Fundamental solutions
are also known when the PDE is the Helmholtz equation [3].

If the half-space is filled with a homogeneous isotropic elastic solid, with
a traction-free boundary, corresponding fundamental solutions are known:
the static solutions were obtained by Melan (two dimensions) and Mindlin
(three dimensions) in the 1930s. Time-dependent solutions were obtained
by Lamb in 1904, and are discussed in books on elastic waves [4].

Finally, we mention a recent construction for a bi-material half-plane,
where two solid quarter-planes (made from different materials) are welded
together, and a point force acts inside one of them. This problem can be
solved using Mellin-transform techniques [5]. The solution can be used to
analyse cracks near the intersection of the interface and the traction-free
surface.

All these half-space Green's functions are more complicated than the
corresponding 'full-space' Green's functions. Thus, an issue arises: should
one use a simple full-space Green's function, leading to a BIE over both
S and the half-space boundary; or should one use a half-space Green's
function, leading to a BIE over S only? There is a trade-off here, which
can have computational repercussions. Little has been done by way of
comparison, but see reference [6] for some comparisons in time-harmonic
elastodynamics.

21.4 Steady-State Heat Conduction

Let us now consider inhomogeneous media. We begin with the problem
of steady-state heat conduction in an anisotropic inhomogeneous material.
This is a scalar problem. The governing PDE can be written as

$$\frac{\partial}{\partial x_i}\left(k_{ij}(\mathbf{x})\,\frac{\partial u}{\partial x_j}\right) = 0, \tag{21.2}$$

where the usual summation convention is employed and the conductivity matrix $k(\mathbf{x})$ with entries $k_{ij}(\mathbf{x})$ is symmetric. Little can be done for 'arbitrary' $k(\mathbf{x})$. To make progress, we assume that $k(\mathbf{x})$ has a specific functional form,

$$k_{ij}(\mathbf{x}) = K_{ij} \exp(2\mathbf{b} \cdot \mathbf{x}), \qquad (21.3)$$

where $K_{ij} = K_{ji}$ are constants and $\mathbf{b}$ is a given constant vector. We say that the material is *exponentially graded*, with $\mathbf{b}$ giving the *grading direction*. This choice for $k(\mathbf{x})$ is convenient mathematically, of course, but it also gives a reasonable model for certain thermal barrier coatings; it is also a good prototype for analogous elasticity problems.

Substitution of (21.3) in (21.2) gives

$$K_{ij} \frac{\partial^2 u}{\partial x_i \partial x_j} + 2b_i K_{ij} \frac{\partial u}{\partial x_j} = 0. \qquad (21.4)$$

We are going to transform this equation into a Helmholtz equation. First, we remove the first-derivative terms by changing the dependent variable: putting

$$u = v \exp(-\mathbf{b} \cdot \mathbf{x})$$

gives

$$K_{ij} \frac{\partial^2 v}{\partial x_i \partial x_j} - b_i b_j K_{ij} v = 0.$$

This is beginning to resemble a Helmholtz equation. To go further, we change the independent variables to

$$y_i = \Omega_{ij} x_j \quad \text{with} \quad \Omega K \Omega^T = I;$$

here, $K = (K_{ij})$ and $\Omega = (\Omega_{ij})$. This gives

$$(\nabla_y^2 - \kappa^2)v = 0 \quad \text{with} \quad \kappa^2 = \mathbf{b}^T K \mathbf{b}.$$

The PDE for v is known as the *modified Helmholtz equation*. A typical fundamental solution is

$$A \frac{e^{-\kappa R}}{R},$$

where

$$R^2 = (\mathbf{y} - \mathbf{y}')^T (\mathbf{y} - \mathbf{y}') = (\mathbf{x} - \mathbf{x}')^T K^{-1} (\mathbf{x} - \mathbf{x}').$$

Reverting to the original variables, we find that a fundamental solution for (21.4) is

$$A \exp(-\mathbf{b} \cdot \mathbf{x}) \frac{e^{-\kappa R}}{R},$$

or, with symmetry,

$$B \exp\{-\mathbf{b} \cdot (\mathbf{x} + \mathbf{x}')\} \frac{e^{-\kappa R}}{R}.$$

More details and information on thermal applications of FGMs can be found in [7]. This paper also contains an alternative method, based on the use of Fourier transforms. We will use this method for exponentially graded elastic solids, because the 'transformation method' described above for the scalar equation (21.4) does not extend to vector problems.

21.5 Exponentially Graded Elastic Solids

Consider an anisotropic inhomogeneous elastic solid: the stiffnesses $c_{ijk\ell}$ satisfy $c_{ijk\ell} = c_{jik\ell} = c_{k\ell ij}$. The Green's function $\mathbf{G}(\mathbf{x}; \mathbf{x}')$ is a 3×3 matrix with entries G_{ij} that satisfy

$$\frac{\partial}{\partial x_j} \left\{ c_{ijk\ell}(\mathbf{x}) \frac{\partial G_{\ell m}}{\partial x_k} \right\} = -\delta_{im} \, \delta(\mathbf{x} - \mathbf{x}'), \quad i = 1, 2, 3, \tag{21.5}$$

where δ_{ij} is the Kronecker delta and $\delta(\mathbf{x})$ is the three-dimensional Dirac delta. As usual, $G_{ij}(\mathbf{x}; \mathbf{x}')$ gives the i-th component of the displacement at $\mathbf{x}$ due to a point force acting in the j-th direction at $\mathbf{x}'$. A standard argument shows that $\mathbf{G}$ is symmetric,

$$G_{ij}(\mathbf{x}; \mathbf{x}') = G_{ji}(\mathbf{x}'; \mathbf{x}). \tag{21.6}$$

Evaluating the left-hand side of (21.5) gives

$$c_{ijk\ell}(\mathbf{x}) \frac{\partial^2 G_{\ell m}}{\partial x_j \partial x_k} + \left(\frac{\partial}{\partial x_j} c_{ijk\ell}(\mathbf{x}) \right) \frac{\partial G_{\ell m}}{\partial x_k} = -\delta_{im} \, \delta(\mathbf{x} - \mathbf{x}'). \tag{21.7}$$

We consider a particular inhomogeneous material in which the stiffnesses vary exponentially, so that

$$c_{ijk\ell}(\mathbf{x}) = C_{ijk\ell} \, \exp(2\mathbf{b} \cdot \mathbf{x}),$$

where $\mathbf{b} = (b_1, b_2, b_3)$ and $C_{ijk\ell}$ and b_i are given constants. Hence

$$(\partial / \partial x_j) c_{ijk\ell}(\mathbf{x}) = 2 C_{ijk\ell} \, b_j \, \exp(2\mathbf{b} \cdot \mathbf{x}) = 2 b_j \, c_{ijk\ell}(\mathbf{x}). \tag{21.8}$$

Using (21.8), (21.7) becomes

$$C_{ijk\ell} \frac{\partial^2 G_{\ell m}}{\partial x_j \partial x_k} + 2 b_j \, C_{ijk\ell} \frac{\partial G_{\ell m}}{\partial x_k} = -\delta_{im} \, \exp(-2\mathbf{b} \cdot \mathbf{x}) \, \delta(\mathbf{x} - \mathbf{x}')$$
$$= -\delta_{im} \, \exp(-2\mathbf{b} \cdot \mathbf{x}') \, \delta(\mathbf{x} - \mathbf{x}') \tag{21.9}$$

for $i = 1, 2, 3$. Note that we can replace the right-hand side of (21.9) by

$$-\delta_{im} \, \exp(-\mathbf{b} \cdot [p\mathbf{x} + p'\mathbf{x}']) \, \delta(\mathbf{x} - \mathbf{x}'), \tag{21.10}$$

where p and p' are any constants that satisfy the constraint $p + p' = 2$; this flexibility will be exploited soon.

Let us introduce $\mathbf{G}^0$, the Green's function for a *homogeneous* solid with constant stiffnesses $C_{ijk\ell}$. It is defined by

$$C_{ijk\ell} \frac{\partial^2 G_{\ell m}^0}{\partial x_j \partial x_k} = -\delta_{im}\, \delta(\mathbf{x} - \mathbf{x}'), \quad i = 1, 2, 3. \tag{21.11}$$

Comparing these equations with (21.9) suggests writing

$$\mathbf{G}(\mathbf{x}; \mathbf{x}') = \exp(-2\mathbf{b} \cdot \mathbf{x}') \left\{ \mathbf{G}^0(\mathbf{x}; \mathbf{x}') + \mathbf{G}^1(\mathbf{x}; \mathbf{x}') \right\}, \tag{21.12}$$

whence $\mathbf{G}^1$ is found to satisfy

$$C_{ijk\ell} \frac{\partial^2 G_{\ell m}^1}{\partial x_j \partial x_k} + 2b_j\, C_{ijk\ell} \frac{\partial G_{\ell m}^1}{\partial x_k} = -2b_j\, C_{ijk\ell} \frac{\partial G_{\ell m}^0}{\partial x_k} \tag{21.13}$$

for $i = 1, 2, 3$. Equation (21.13) is a system of three coupled second-order PDEs, with constant coefficients. However, the decomposition (21.12) has a disadvantage: the symmetry property (21.6) is not inherited by $\mathbf{G}^1$. Thus, we change the right-hand side of (21.9), using (21.10) with $p = p' = 1$, giving

$$C_{ijk\ell} \frac{\partial^2 G_{\ell m}}{\partial x_j \partial x_k} + 2b_j\, C_{ijk\ell} \frac{\partial G_{\ell m}}{\partial x_k} = -\delta_{im}\, \exp\{-\mathbf{b} \cdot (\mathbf{x} + \mathbf{x}')\}\, \delta(\mathbf{x} - \mathbf{x}'), \tag{21.14}$$

and we replace (21.12) by

$$\mathbf{G}(\mathbf{x}; \mathbf{x}') = \exp\{-\mathbf{b} \cdot (\mathbf{x} + \mathbf{x}')\} \left\{ \mathbf{G}^0(\mathbf{x}; \mathbf{x}') + \mathbf{G}^g(\mathbf{x}; \mathbf{x}') \right\}, \tag{21.15}$$

so that

$$G_{ij}^g(\mathbf{x}; \mathbf{x}') = G_{ji}^g(\mathbf{x}'; \mathbf{x}).$$

To find an equation for the *grading term* $\mathbf{G}^g$, we simply substitute (21.15) in (21.14), making use of (21.11); the result is

$$C_{ijk\ell} \frac{\partial^2 G_{\ell m}^g}{\partial x_j \partial x_k} + L_{i\ell}\, G_{\ell m}^g(\mathbf{x}; \mathbf{x}') = -L_{i\ell}\, G_{\ell m}^0(\mathbf{x}; \mathbf{x}') \tag{21.16}$$

for $i = 1, 2, 3$, where the first-order differential operator $L_{i\ell}$ is defined by

$$L_{i\ell} = (C_{ijk\ell} - C_{ikj\ell})b_j(\partial/\partial x_k) - C_{ijk\ell}b_j b_k.$$

It remains to solve (21.16); we can do this using three-dimensional Fourier transforms. Before doing that, it is instructive to review the known results for $\mathbf{G}^0$, the so-called *anisotropic Green's function*.

21.6 The Anisotropic Green's Function

Consider solving (21.11) by Fourier transforms, which we define by

$$\mathcal{F}\{u\} = \hat{u}(\mathbf{k}) = \int u(\mathbf{x}) \exp\left(i\mathbf{k} \cdot \mathbf{x}\right) d\mathbf{x},$$

where $\mathbf{k}$ is the vector of transform variables. When inverting the Fourier transform to obtain $\mathbf{G}^0$, we have to integrate over $\mathbf{k}$. The relevant integral turns out to involve the vector $\mathbf{r} = \mathbf{x} - \mathbf{x}'$, and it simplifies by choosing spherical polar coordinates with $\mathbf{r}$ along the polar axis. Moreover, the integrand contains $[\mathbf{Q}(\mathbf{k})]^{-1}$, where

$$Q_{im}(\mathbf{k}) = C_{ij\ell m} k_j k_\ell \tag{21.17}$$

and $C_{ijk\ell}$ are the constant stiffnesses; thus, $\mathbf{Q}$ is homogeneous,

$$\mathbf{Q}(t\mathbf{k}) = t^2 \mathbf{Q}(\mathbf{k}) \quad \text{for any } t \neq 0,$$

and this fact simplifies the calculation. Specifically, we have

$$\mathbf{G}^0 = (2\pi)^{-3} \int [\mathbf{Q}(\mathbf{k})]^{-1} \exp(-i\mathbf{k} \cdot \mathbf{r}) \, d\mathbf{k}$$

$$= (2\pi)^{-3} \iint k^{-2} [\mathbf{Q}(\hat{\mathbf{k}})]^{-1} \cos(kr \cos \varphi) \, k^2 \, dk \, d\hat{\mathbf{k}}$$

where $r = |\mathbf{r}|$, $k = |\mathbf{k}|$, $\mathbf{k} = k\hat{\mathbf{k}}$ and we have observed that both $\mathbf{G}^0$ and $\mathbf{Q}$ are real. Using spherical polar coordinates (k, φ, χ), where $\varphi = 0$ is the polar axis, we have $d\hat{\mathbf{k}} = \sin \varphi \, d\varphi \, d\chi$ whence

$$\mathbf{G}^0 = (2\pi)^{-3} \lim_{X \to \infty} \int_0^\pi \mathbf{S}(\varphi) \int_0^X \cos(kr \cos \varphi) \, dk \, \sin \varphi \, d\varphi$$

$$= \frac{1}{8\pi^3 r} \lim_{X \to \infty} \int_{-1}^1 \mathbf{S}(\cos^{-1} \mu) \, \frac{\sin\left(Xr\mu\right)}{\mu} \, d\mu,$$

where

$$\mathbf{S}(\varphi) = \int_0^{2\pi} [\mathbf{Q}(\hat{\mathbf{k}})]^{-1} \, d\chi.$$

Note that we have evaluated the integral over k and then put $\mu = \cos \varphi$. The integral over μ is known as a *Dirichlet integral*; its limiting value as $X \to \infty$ is $\pi \mathbf{S}(0)$ (see, for example, p. 365 of reference [8]), whence

$$\mathbf{G} = \frac{1}{8\pi^2 r} \oint [\mathbf{Q}(\hat{\mathbf{k}})]^{-1} \, d\chi,$$

where the integral is taken around the unit circle, centred at the origin and lying in the plane perpendicular to $\mathbf{r}$. The remaining one-dimensional integral must be evaluated numerically, in general.

The derivation given above can be found on p. 412 of the book by Synge [9]; other derivations (involving divergent integrals and generalized functions) are available.

21.7 Calculating the Graded Term

Recall that we have to solve (21.11) for $\mathbf{G}^g$, using Fourier transforms. We find that

$$\mathbf{G}^g(\mathbf{x}; \mathbf{x}') = (2\pi)^{-3} \int \mathbf{E}(\mathbf{b}, \mathbf{k}) \, \exp\left(-i\,\mathbf{k} \cdot \mathbf{r}\right) dk, \qquad (21.18)$$

where $\mathbf{r} = \mathbf{x} - \mathbf{x}'$,

$$\mathbf{E}(\mathbf{b}, \mathbf{k}) = -\{\mathbf{Q}(\mathbf{k}) + \mathbf{B}(\mathbf{b}, \mathbf{k})\}^{-1}\mathbf{B}(\mathbf{b}, \mathbf{k}) \left[\mathbf{Q}(\mathbf{k})\right]^{-1},$$
$$B_{im}(\mathbf{b}, \mathbf{k}) = i(C_{ij\ell m} - C_{i\ell jm})b_j k_\ell + C_{ij\ell m}b_j b_\ell$$

and $\mathbf{Q}(\mathbf{k})$ is defined by (21.17). Note that, unlike $\mathbf{Q}(\mathbf{k})$, $\mathbf{E}(\mathbf{b}, \mathbf{k})$ is not a homogeneous function of $\mathbf{k}$.

How should we evaluate (21.18)? The integrand involves *three* vectors, namely $\mathbf{r}$, $\mathbf{b}$ and $\mathbf{k}$, where $\mathbf{r}$ and $\mathbf{b}$ may be regarded as fixed. Compare this with the integral for $\mathbf{G}^0$, which involves *two* vectors, $\mathbf{r}$ and $\mathbf{k}$: there, we evaluated the integral by using spherical polar coordinates for $\mathbf{k}$ with $\mathbf{r}$ along the polar axis. For $\mathbf{G}^g$, it turns out to be better to choose spherical polar coordinates for $\mathbf{k}$ *with* $\mathbf{b}$ *along the polar axis*.

We have not done these calculations in general, but only when the underlying material is isotropic [10]. Thus, we suppose that the solid has Lamé moduli given by

$$\lambda(\mathbf{x}) = \lambda_0 \exp(2\mathbf{b} \cdot \mathbf{x}) \quad \text{and} \quad \mu(\mathbf{x}) = \mu_0 \exp(2\mathbf{b} \cdot \mathbf{x}),$$

where λ_0 and μ_0 are constants. (Evidently, Poisson's ratio is constant for such a solid.) Then, $\mathbf{G}^0$ is known explicitly (it is the *Kelvin solution*) and $\mathbf{E}$ can be calculated explicitly. The details are complicated. The result is that the triple Fourier integral defining $\mathbf{G}^g$, (21.18), can be reduced to the sum of an explicit term, some finite single integrals of modified Bessel functions I_n and some finite double integrals of elementary functions. As $\mathbf{G}^g$ is bounded as $|\mathbf{x} - \mathbf{x}'| \to 0$ (the singularity is contained within the Kelvin solution), having it available only as a computable quantity is not an impediment for a boundary integral implementation.

References

1. D. Colton and R. Kress, *Integral Equation Methods in Scattering Theory*, Wiley, New York, 1983.

2. N. Kuznetsov, V. Maz'ya, and B. Vainberg, *Linear Water Waves*, Cambridge University Press, Cambridge, 2002.

3. S.N. Chandler-Wilde and D.C. Hothersall, Efficient calculation of the Green's function for acoustic propagation above a homogeneous impedance plane, *J. Sound Vib.* **180** (1995), 705–724.

4. J.A. Hudson, *The Excitation and Propagation of Elastic Waves*, Cambridge University Press, Cambridge, 1980.

5. P.A. Martin, On Green's function for a bimaterial elastic half-plane, *Int. J. Solids Structures* **40** (2003), 2101–2119.

6. L. Pan, F. Rizzo, and P.A. Martin, Some efficient boundary integral strategies for time-harmonic wave problems in an elastic halfspace, *Computer Methods Appl. Mech. Engng.* **164** (1998), 207–221.

7. J.R. Berger, P.A. Martin, V. Mantič, and L.J. Gray, Fundamental solutions for steady-state heat transfer in an exponentially graded anisotropic material, *J. Appl. Math. Phys. (ZAMP)* (in press).

8. K. Knopp, *Theory and Application of Infinite Series*, 2nd ed., Blackie, London, 1951.

9. J.L. Synge, *The Hypercircle in Mathematical Physics*, Cambridge University Press, Cambridge, 1957.

10. P.A. Martin, J.D. Richardson, L.J. Gray, and J.R. Berger, On Green's function for a three-dimensional exponentially graded elastic solid, *Proc. Roy. Soc. A* **458** (2002), 1931–1947.

22 Applications of Fixed Point Theorems to a Chemical Reactor Problem

Desmond F. McGhee, Naglaa M. Madbouly, and Gary F. Roach

22.1 Introduction

We consider the mathematical model for an adiabatic tubular chemical reactor which processes an irreversible exothermic chemical reaction. For steady-state solutions, the model can be reduced to the ordinary differential equation

$$u'' - \lambda u' + F(\lambda, \mu, \beta, u) = 0 \qquad (22.1)$$

with boundary conditions

$$u'(0) = \lambda u(0), \quad u'(1) = 0 \qquad (22.2)$$

where

$$F(\lambda, \mu, \beta, u) = \lambda \mu (\beta - u) \exp(u)$$

(see [1]). The unknown u represents the steady-state temperature of the reaction, and the parameters λ, μ and β represent the Peclet number, the Damkohler number and the dimensionless adiabatic temperature rise respectively. This problem has been studied by various authors (e.g. [2,3,4]) who have demonstrated numerically the existence of solutions (sometimes multiple solutions) for particular parameter ranges.

The problem (22.1)–(22.2) can be converted by Green's function techniques into a Hammerstein integral equation

$$u(x) = \mu \int_0^1 k(x, y) f(y, u(y)) dy, \quad 0 \le x \le 1, \qquad (22.3)$$

where

$$k(x, y) = \begin{cases} e^{\lambda(x-y)} & \text{if } 0 \le x < y, \\ 1 & \text{if } y \le x \le 1 \end{cases} \qquad (22.4)$$

and

$$f(y, u) = (\beta - u) \exp(u), \qquad (22.5)$$

which we consider in the space $C[0, 1]$ of continuous functions on the closed interval $[0, 1]$ with the usual sup norm. Throughout, we assume λ and μ are positive, and β is nonnegative.

22.2 Preliminaries

For any positive constant M, we consider the closed ball

$$B(M) := \{u \in C[0,1] : \|u\| \le M\},$$

and define a nonlinear operator N on $C[0,1]$ by

$$(Nu)(y) := f(y, u(y)), \qquad u \in C[0,1], \qquad 0 \le y \le 1.$$

An integral operator K on $C[0,1]$ is defined by

$$(Ku)(x) := \int_0^1 k(x,y)u(y)dy, \qquad 0 \le x \le 1,$$

where the kernel k is given by (22.4) and so

$$\|K\| = \sup_{x \in [0,1]} \int_0^1 |k(x,y)|dy = \sup_{x \in [0,1]} \left[\int_0^x 1 dy + \int_x^1 e^{\lambda(x-y)} dy \right] = 1.$$

Let $T : C[0,1] \longrightarrow C[0,1]$ be defined by

$$(Tu)(x) := \mu(KNu)(x) = \mu \int_0^1 k(x,y)f(y,u(y))dy. \qquad (22.6)$$

Then equation (22.3) takes the form $u = Tu$, $u \in C[0,1]$, so that we are concerned with finding fixed points of the mapping T. We make use of the following well-known fixed point theorems, (see [5]).

Theorem 1. (Schauder's Fixed Point Theorem) *If S is a convex closed subset of a normed linear space and R is a relatively compact subset of S, then every continuous mapping of S into R has a fixed point.*

Theorem 2. (Contraction Mapping Theorem) *Let (X, d) be a complete metric space and let $T : X \longrightarrow X$ be a contraction on X, i.e for any u_1, $u_2 \in X$,*

$$d(Tu_1, Tu_2) \le \gamma d(u_1, u_2) \quad \text{where} \quad 0 < \gamma < 1.$$

Then T has precisely one fixed point in X.

Finally, the following lemma is used in our proofs of existence and uniqueness results; the proof is straightforward.

Lemma 3. *Let T be defined on $C[0,1]$ by (22.6). If N satisfies a Lipschitz condition on $B(M)$ with Lipschitz constant L, i.e. $\|Nf - Ng\| \le L\|f - g\|$ for all $f, g \in B(M)$ where $L < \infty$, then T satisfies a Lipschitz condition on $B(M)$ with Lipschitz constant γ given by $\gamma = |\mu|L = \mu L$ (since $\mu > 0$), and hence T is contractive on $B(M)$ provided $\gamma < 1$, i.e. $\mu < \frac{1}{L}$.*

22.3 Application of Schauder's Fixed Point Theorem

Firstly, we prove the following general result using the Schauder fixed point theorem.

Theorem 4. *Let $k(x,y)$ be continuous for $a \leq x,\ y \leq b$, and $f(y,z)$ be continuous and bounded for $a \leq y \leq b$ for each fixed $z \in \mathbf{R}$, and locally bounded with respect to z. Given $M > 0$, the equation*

$$u(x) = \mu \int_a^b k(x,y) f(y, u(y)) dy,$$

has a solution $u \in B(M) \subset C[a,b]$ provided $\mu \leq M[A C_M (b-a)]^{-1}$, where $A := \sup\limits_{[a,b] \times [a,b]} |k(x,y)|$ and $C_M := \sup\limits_{[a,b] \times [-M, M]} |f(y,z)|$.

Proof. The proof follows that of [5, Theorem 6.45] with appropriate adjustment to cater for local boundedness (as opposed to boundedness) of f with respect to z.

We apply this result to our particular case when k and f are given by (22.4) and (22.5) respectively.

Theorem 5. *Let $M > 0$. For all $\lambda, \mu > 0$, the Hammerstein integral equation (22.3) with k and f given by (22.4) and (22.5) has a solution $u \in B(M) \subset C[0,1]$, provided*

$$\mu \leq M e^{-M} (\beta + M)^{-1}. \tag{22.7}$$

Proof. Consider the operator T defined by (22.6) but acting on the ball $B(M) = \{ u \in C[0,1] : \|u\| \leq M \}$. For any $u \in B(M)$,

$$|f(y, u(y))| = |(\beta - u(y)) \exp(u(y))| \leq (\beta + M) e^M.$$

Since k in (22.3) is defined by (22.4), k is continuous and $|k(x,y)| \leq 1$. Thus, in the notation of Theorem 4, we have $A = 1$, $C_M = (\beta + M)e^M$ and the interval $[a,b]$ is $[0,1]$ and we deduce existence of a solution of (22.3)–(22.5) in $B(M)$ provided the stated condition (22.7) is satisfied.

22.4 Application of the Contraction Mapping Principle

First, we consider the special case $\beta = 0$.

Theorem 6. *For any $M > 0$ the Hammerstein integral equation (22.3)–(22.5), with $\beta = 0$, has a unique solution $u^* \in B(M)$ provided that*

$$\mu < e^{-M} (1 + M)^{-1}. \tag{22.8}$$

Proof. For $u \in B(M)$,

$$\|Tu\| = \|\mu K N u\| \leq |\mu| \|K\| \|Nu\|$$

$$= \mu \sup_{x \in [0,1]} |f(x, u(x))| \quad \text{since} \quad \|K\| = 1 \quad \text{and} \quad \mu > 0$$

$$= \mu \sup_{x \in [0,1]} |u(x) \exp u(x)| \leq \mu M e^M,$$

so that T is invariant on $B(M)$ provided $\mu \leq e^{-M}$.

Now, $B(M)$ is convex and N is Fréchet differentiable with Fréchet derivative at $u \in C[0,1]$ given by

$$(DN)(u) : C[0,1] \longrightarrow C[0,1],$$

$$[(DN)(u)]v(x) = -e^{u(x)}(1 + u(x))v(x).$$

For all $u \in B(M)$,

$$\|(DN)(u)\| = \sup_{x \in [0,1]} |e^{u(x)}(1 + u(x))| \leq e^M(1 + M).$$

Thus, N is Lipschitz in $B(M)$ with Lipschitz constant $L = e^M(1 + M)$. By Lemma 3, T is Lipschitz in $B(M)$ with a Lipschitz constant

$$\gamma = |\mu|L = \mu e^M(1 + M).$$

For T to be a contraction, we need $\gamma < 1$ so that we require condition (22.8) to hold in order to apply Theorem 2.

Let $\mu_0(M) := e^{-M}(1 + M)^{-1}$. For $\beta = 0$ and any $M > 0$, $\mu < \mu_0(M)$ (i.e. μ satisfies (22.8)) guarantees the existence of a unique solution in $B(M)$ of (22.3)–(22.5). Note that $\mu_0(M)$ is monotonically decreasing in M with $\sup_{M>0} \mu_0(M) = 1$. Since, with $\beta = 0$, it is obvious that the trivial solution $u = 0 \in B(M)$ is always a solution of (22.3)–(22.5), we can deduce from Fig. 1 that no other solutions can exist in the region bounded by the axes and the curve $\mu = \mu_0(\|u\|)$. For any $M > 0$, there is a unique solution u^* with $\|u^*\| \leq M$ for all $\mu \leq \mu_0(M)$. Since the trivial solution $u = 0$ exists, there can be no other solution in the rectangle shown. As M varies from 0 to ∞, these rectangles fill the area between the axes and the curve $\mu = \mu_0(\|u\|)$. We summarize this discussion in the following theorem.

Theorem 7. *Consider equations (22.3)–(22.5) with $\beta = 0$ and $\mu > 0$ a bifurcation parameter. The trivial solution $u = 0 \in C[0,1]$ is a solution for all $\mu > 0$ and there are no nontrivial solutions that bifurcate from this trivial solution while $\mu < 1$.*

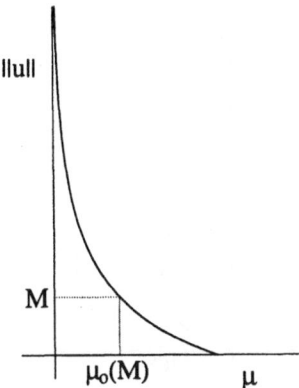

Fig. 1. $\mu = \mu_0(\|u\|)$.

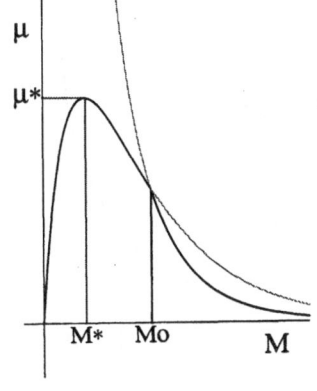

Fig. 2. $\mu = \mu_\beta(M)$.

Theorem 8. *($\beta \neq 0$) For any $M > 0$, the Hammerstein integral equation (22.3)–(22.5) has a unique solution $u^* \in B(M)$ provided*

$$\mu < \min\{Me^{-M}(\beta + M)^{-1}, e^{-M}(|\beta - 1| + M)^{-1}\} =: \mu_\beta(M) . \qquad (22.9)$$

Proof. Arguing as in Theorem 6, we have for any $u \in B(M)$,

$$\|Tu\| = \sup_{x \in [0,1]} |Tu(x)| \leq M \text{ provided } \mu \leq Me^{-M}(\beta + M)^{-1}.$$

$B(M)$ is convex and $N : B(M) \longrightarrow B(M)$ defined by $Nu = (\beta - u)e^u$, is Fréchet differentiable with the Fréchet derivative at any point $u \in C[0,1]$ given by

$$(DN)(u) : C[0,1] \longrightarrow C[0,1],$$

$$[(DN)(u)]v(x) = e^{u(x)}(\beta - 1 - u(x))v(x).$$

For all $u \in B(M)$, $\|(DN)(u)\| \leq e^M(|\beta - 1| + M)$, so that N is Lipschitz in $B(M)$ with Lipschitz constant $L = e^M(|\beta - 1| + M)$. Hence, by Lemma 2, T is Lipschitz in $B(M)$ with a Lipschitz constant $\gamma = |\mu|L = \mu(|\beta - 1| + M)e^M$. For T to be a contraction, we require $\gamma < 1$. Thus, combining the two restrictions on μ we obtain the stated condition (22.9). Note here that, unlike the case $\beta = 0$, the trivial function $u = 0$ is not a solution.

In Fig. 2 the curve $\mu = \mu_\beta(M)$ is shown as the appropriate combination of the curves $\mu = Me^{-M}(\beta + M)^{-1}$ and $\mu = e^{-M}(|\beta - 1| + M)^{-1}$. These curves intersect at $M = M_0 = \frac{1}{2}(1 - |\beta - 1| + \sqrt{(|\beta - 1| - 1)^2 + 4\beta})$. It is easily shown that M_0 is greater than $M^* = \frac{1}{2}(-\beta + \sqrt{\beta^2 + 4\beta})$, the location of the maximum turning point of $\mu = Me^{-M}(\beta + M)^{-1}$ so that for all $\beta > 0$ the curves are indeed related as shown. Thus,

$$\sup_{M > 0} \mu_\beta(M) = M^* e^{-M^*}(\beta + M^*)^{-1} =: \mu^*.$$

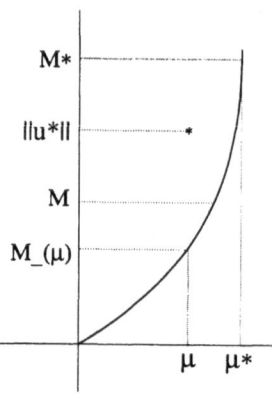

Fig. 3. $\|u\| = M_-(\mu)$.

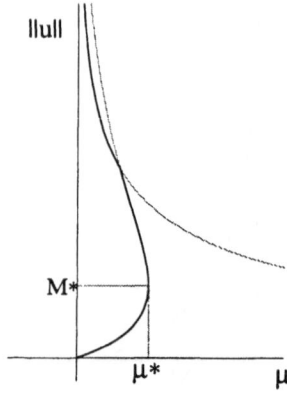

Fig. 4. $\mu = \mu_\beta(\|u\|)$.

For $\mu < \mu^*$ there exists a unique solution u^* of (22.3)–(22.5) in $B(M^*)$. For each $\mu < \mu^*$, $\|u^*\| \leq M_-(\mu)$, where $M_-(\mu)$ is the smaller function defined implicitly by $\mu = \mu_\beta(\|u\|)$. We prove this by contradiction. Suppose for some $\mu < \mu^*$ the unique solution u^* satisfies $\|u^*\| > M_-(\mu)$, as shown in Fig. 3. Choose M such that $M_-(\mu) < M < \|u^*\|$. Then by Theorem 8, (22.3)–(22.5) has a unique solution $v^* \in B(M)$. But then u^* and v^* are two distinct solutions in $B(M^*)$, which is a contradiction.

Also for $\mu < \mu^*$ any other solution of (22.3)–(22.5) must have norm greater than $M_+(\mu)$, where $M_+(\mu)$ is the larger function defined implicitly by $\mu = \mu_\beta(\|u\|)$. This is easily proved in the same way as for the case $\beta = 0$.

In conclusion, we see that no solution can exist in the area between the curve $\mu = \mu_\beta(\|u\|)$ and the $\|u\|$-axis of Fig. 4 and the unique solution in $B(M)$ for $\mu < \mu^*$ guaranteed by Theorem 8 must be below the curve.

References

1. A.B. Poore, A tubular chemical reactor model, in *A Collection of Nonlinear Model Problems Contributed to the Proceeding of the AMS-SIAM*, 1989, 28–31.

2. R. Heinemann and A. Poore, The effect of activation energy on tubular reactor multiplicity, *Chem. Engrg. Sci.* **37** (1982), 128–131.

3. R. Heinemann and A. Poore, Multiplicity, stability, and oscillatory dynamics of the tubular reactor, *Chem. Engrg. Sci.* **36** (1989), 1411–1419.

4. M. Lovo and V. Balakotaiah, Multiplicity features of adiabatic autothermal reactors, *AIChE J.* **38** (1992), 101–115.

5. D.H. Griffel, *Applied Functional Analysis*, Wiley & Sons, New York, 1981.

23 About Localized Boundary-Domain Integro-Differential Formulations for a Quasilinear Problem with Variable Coefficients

Sergey E. Mikhailov

23.1 Introduction

Application of the boundary integral equation (BIE) method (boundary element method) to the solution of linear boundary value problems (BVPs) for partial differential equations (PDEs) has been intensively developed over recent decades. Using a fundamental solution of an auxiliary linear PDE, a nonlinear BVP can be reduced to a nonlinear boundary-domain integral or integro-differential equation (BDIE or BDIDE) (see, e.g., [1,2,3]). However, a fundamental solution necessary for the reduction is usually highly nonlocal, which generally leads after discretization to a system of nonlinear algebraic equations with a fully populated matrix. Moreover, the fundamental solution is generally not available if the coefficients of the auxiliary PDE are not constant.

To prevent these difficulties, some *localized* parametrixes are constructed and used in this paper developing the approach of [4], to reduce nonlinear (quasilinear) BVPs with variable coefficients to quasilinear *localized* BDIDEs. This results, after discretization, in sparsely populated systems of quasilinear algebraic equations. Some techniques of the parametrix localization are discussed and the corresponding nonlinear LBDIDEs are introduced. A mesh-based algorithm for the localized equations discretization is described.

23.2 Stationary Nonlinear Heat Transfer Problem in Inhomogeneous Body and Parametrix

For illustration of the general approach we consider a BVP of stationary nonlinear heat transfer in an isotropic inhomogeneous 2D or 3D body Ω, with a prescribed temperature $\bar{u}(x)$ on a closed part $\partial_D\Omega$ of the boundary $\partial\Omega$ and prescribed heat flux $\bar{t}(x)$ on the remaining open part $\partial_N\Omega$ of $\partial\Omega$,

$$[L(u)u](x) = f(x), \quad x \in \Omega, \tag{23.1}$$

$$u(x) = \bar{u}(x), \quad x \in \partial_D\Omega, \tag{23.2}$$

$$[T(u)u](x) = \bar{t}(x), \quad x \in \partial_N\Omega, \tag{23.3}$$

where Ω is an open domain, $u(x)$ is an unknown temperature, $[L(\lambda)u](x) := \frac{\partial}{\partial x_i}\left[a(\lambda(x), x)\frac{\partial u(x)}{\partial x_i}\right]$ is a linear differential operator, $[T(\lambda)u](x) := a(\lambda(x), x)\partial u(x)/\partial n(x)$ is a linear surface flux operator and $a(\lambda, x) > C > 0$ is a variable thermo-conductivity coefficient depending on a function $\lambda(x)$, $f(x)$ is a known distributed heat source, $n(x)$ is an outward normal vector to the boundary $\partial\Omega$, $\bar{u}(x)$ and $\bar{t}(x)$ are known functions. Summation in repeated indices is supposed from 1 to 2 in the 2D and from 1 to 3 in the 3D case unless stated otherwise. Note that the well-known Kirchhoff transform cannot reduce this problem to a linear one since $a(u, x)$ depends not only on u but also on x.

Then the Green formula for the differential operator $L(u)$ has the form

$$\int_\Omega \{uL(u)v - vL(u)u\}\, d\Omega = \int_{\partial\Omega} \{uT(u)v - vT(u)u\}\, d\Gamma. \tag{23.4}$$

For partial differential operators with variable coefficients, like $L(\lambda)$ in (23.1), a fundamental solution is usually not available in an explicit form or the form is too expensive for numerical applications. However, a *parametrix* $P(\lambda, x, y)$ is often available instead, which is a function of x and y, which depends also on λ, satisfying equation $L_x(\lambda)\, P(\lambda, x, y) = \delta(x-y) + R(\lambda, \nabla\lambda, x, y)$ where the remainder term $R(\lambda, \nabla\lambda, x, y)$ as function of $x \in \Omega$ has not more than a weak (integrable) singularity with respect to x. One can check that a parametrix for (23.1) is given by the fundamental solution to the same equation but with the "frozen" coefficient $a = a(\lambda(y), y)$, that is, $P(\lambda, x, y) = F_\Delta(x, y)/a(\lambda(y), y)$, where $F_\Delta(x, y)$ is the fundamental solution for the Laplace operator. Denoting $|x - y| = \sqrt{(x_i - y_i)(x_i - y_i)}$, we have,

$$\text{2D case}: \quad P(\lambda, x, y) = \frac{\ln|x - y|}{2\pi a(\lambda(y), y)}, \tag{23.5}$$

$$R(\lambda, \nabla\lambda, x, y) = \frac{x_i - y_i}{2\pi a(\lambda(y), y)|x - y|^2}\left[\frac{\partial a(\lambda, x)}{\partial \lambda}\frac{\partial \lambda(x)}{\partial x_i} + \frac{\partial a(\lambda, x)}{\partial x_i}\right]_{\lambda=\lambda(x)};$$

$$\text{3D case}: \quad P(\lambda, x, y) = \frac{-1}{4\pi a(\lambda(y), y)|x - y|}, \tag{23.6}$$

$$R(\lambda, \nabla\lambda, x, y) = \frac{x_i - y_i}{4\pi a(\lambda(y), y)|x - y|^3}\left[\frac{\partial a(\lambda, x)}{\partial \lambda}\frac{\partial \lambda(x)}{\partial x_i} + \frac{\partial a(\lambda, x)}{\partial x_i}\right]_{\lambda=\lambda(x)}.$$

Using the parametrix $P(u, x, y)$ as $v(x)$ in the Green formula (23.4), one can reduce BVP (23.1)–(23.3) to a BDIDE, which can in turn be reduced after some discretization to a system of nonlinear algebraic equations and solved numerically. The system will include unknowns not only at the boundary but also at internal points. Moreover, since the commonly used parametrixes, see e.g. (23.5), (23.6), are highly nonlocal, the matrix of the system will be fully populated and this prevents the use of special methods for sparsely populated systems. To avoid this difficulty, we present below some ideas of constructing *localized* parametrixes and consequently *localized* BDIDEs (LBDIDEs).

23.3 Localized Parametrix and LBDIDPs

A parametrix is not unique and all parametrixes for a differential operator $L(\lambda)$ have the same singularity at $x = y$ but can differ at other points. Thus we can perturb an available (not localized) parametrix $P^0(\lambda, x, y)$ to localize it. Particularly, we can consider $P_\omega(\lambda, x, y) = \chi(x, y)P^0(\lambda, x, y)$, where $\chi(x, y)$ is a cut-off function, such that $\chi(y, y) = 1$ and $\chi(x, y) = 0$ at points x not belonging to the closure of an open localization domain $\omega(y)$ (a neighborhood of y). Then $P_\omega(\lambda, x, y)$ has the same singularity as $P^0(\lambda, x, y)$ at $x = y$ but is localized (nonzero) only on $\omega(y)$. Further we have, $L_x(\lambda)P_\omega = L_x(\lambda)[\chi P^0(\lambda)] = L_x(\lambda)P^0(\lambda) + L_x(\lambda)[(1-\chi)P^0(\lambda)] = \delta(x-y) + R_\omega(\lambda, \nabla\lambda, x, y)$, $R_\omega(\lambda, \nabla\lambda, x, y) = R^0(\lambda, \nabla\lambda, x, y) + L_x(\lambda)[(1 - \chi)P^0(\lambda)]$. Consequently R_ω will have the necessary properties of the remainder, that is, $P_\omega(\lambda, x, y)$ is also a parametrix at least if χ is smooth enough. Let the domain $\omega(y) \ni y$ be an open neighbourhood of a point y and $\chi(x, y)$ be piecewise constant, $\chi(x, y) = \{1$ if $x \in \omega(y)$, 0 if $x \notin \omega(y)\}$. Then $P_\omega(\lambda, x, y) = \chi(x, y)P^0(\lambda, x, y) = \{P^0(\lambda, x, y)$ if $x \in \omega(y)$, 0 if $x \notin \omega(y)\}$ is a discontinuous localized parametrix. Substituting $P_\omega(u, x, y)$ for $v(x)$ and taking $u(x)$ as a solution to (23.1) in the Green formula for intersection of Ω with $\omega(y)$, we arrive at the integral equality localized on $\bar\omega(y) \cap \bar\Omega$,

$$c(y)u(y) - \int_{\bar\omega(y)\cap\partial\Omega} u(x)T_x(u)P_\omega(u, x, y)d\Gamma(x)$$

$$+ \int_{\bar\omega(y)\cap\partial\Omega} P_\omega(u, x, y)T(u)u(x)d\Gamma(x)$$

$$- \int_{\Omega\cap\partial\omega(y)} u(x)T_x(u)P_\omega(u, x, y)d\Gamma(x)$$

$$+ \int_{\Omega\cap\partial\omega(y)} P_\omega(u, x, y)T(u)u(x)d\Gamma(x)$$

$$+ \int_{\omega(y)\cap\Omega} R_\omega(u, \nabla u, x, y)u(x)d\Omega(x)$$

$$= \int_{\omega(y)\cap\Omega} P_\omega(u, x, y)f(x)d\Omega(x). \tag{23.7}$$

Here $c(y) = 1$ if $y \in \Omega$, $c(y) = 0$ if $y \notin \bar\Omega$, $c(y) = \alpha(y)/(2\pi)$ if $y \in \partial\Omega$ in 2D, $c(y) = \alpha(y)/(4\pi)$ if $y \in \partial\Omega$ in 3D and $\alpha(y)$ is an interior space angle at a corner point y of the boundary $\partial\Omega$.

Substitution of boundary conditions (23.2) and (23.3) in (23.7), introduction of a new variable $t(x) = Tu(x)$ for the unknown flux at $x \in \partial\Omega_D$ and use of (23.7) at $y \in \Omega \cup \partial\Omega$ reduce BVP (23.1)–(23.3) to the LBDIDE for $u(x)$, $x \in \Omega\cup\partial_N\Omega$ and $t(x) = Tu(x)$, $x \in \partial_D\Omega$. The equation is integro-differential, since it includes an unknown flux $T(u)u(x)$ on $\Omega \cap \partial\omega(y)$ as well as dependence on the gradient ∇u in R_ω. This implies the LBDIDE is to be complemented by the Dirichlet boundary condition (23.2) at $y \in \partial_D \Omega$ thus reducing BVP (23.1)–(23.3) to a localized boundary-domain integro-differential problem (t), which will be called LBDIDP(t).

Otherwise, we can substitute (23.2) and (23.3) into integral equality (23.7) but leave T as the differential operator, acting on u at $\partial_D \Omega$, and use (23.7) only at $y \in \Omega \cup \partial \Omega_N$. Complementing the LBDIDE with the Dirichlet boundary condition (23.2) at $y \in \partial_D \Omega$ reduces BVP (23.1)–(23.3) to a localized boundary-domain integro-differential problem (T) for $u(x)$, $x \in \Omega \cup \partial_N \Omega$, called LBDIDP(T).

To get rid of the fourth integral including $T(u)u$ on $\Omega \cap \partial\omega(y)$ in (23.7), one can construct a localized parametrix $P_\omega(u, x, y)$ vanishing on $\Omega \cap \partial\omega(y)$.

The Green function for a corresponding BVP with "frozen" constant coefficients and without junior derivative terms in the differential operator L on $\omega(y)$ was employed in [5,6] as a parametrix $P_\omega^0(x, y)$ vanishing on $\partial\omega(y)$. However, the Green function is available in an analytical form only for sufficiently simple shapes of the localization domain $\omega(y)$, e.g. for a ball. It seems to be simpler and more universal to construct a proper localized parametrix as $P_\omega(\lambda, x, y) = \chi(x, y) P^0(\lambda, x, y)$, where $\chi(x, y)$ is a continuous in $x \in \Omega$ cut-off function, which is smooth in $\bar\omega(y)$ and equal to zero both on the boundary and outside of $\omega(y)$, whereas P^0 is an available parametrix (e.g., a fundamental solution for a corresponding differential operator with "frozen" coefficients).

To simplify the integral representation even further by getting rid of the remaining integral along $\partial\omega(y)$, one can employ a smooth in $x \in \bar\Omega$ cut-off function $\chi(x, y)$, which vanishes on $\partial\omega(y)$ together with its normal derivative in x. Then the same holds true also for the parametrix $P_\omega(\lambda, x, y) = \chi(x, y) P^0(\lambda, x, y)$. For such a parametrix, third and forth integrals disappear on the left-hand side of (23.7). Some examples of internally or globally smooth cut-off functions are presented in [4].

To reduce LBDIDPs (t) or (T) to a sparsely populated system of quasi-linear algebraic equations e.g. by the collocation method, one has to employ a local interpolation or approximation formula for the unknown function $u(x)$.

23.4 Mesh-Based Discretization

Suppose the domain Ω is covered by a mesh of closures of volume elements e_k with nodes set up at the corners, edges, faces, and/or inside the elements. Let J be the total number of nodes x^i ($i = 1, 2, ..., J$), from which there are J_D nodes on $\partial_D \Omega$. One can use each node x^i as a collocation point for an LBDIDE with a localization domain $\omega(x^i)$. Let the union of closures of the volume elements that intersect with $\omega(x^i)$ be called the *total* localization domain $\tilde\omega(x^i)$, Fig. 1. Then the closure $\bar\omega(x^i) \cap \bar\Omega$ belongs to $\tilde\omega(x^i)$. If $\omega(x^i)$ is sufficiently small, then $\tilde\omega(x^i)$ consists only of the elements adjacent to the collocation point x^i. If $\omega(x^i)$ is ab initio chosen as consisting only of the elements adjacent to the collocation point x^i, then $\tilde\omega(x^i) = \bar\omega(x^i)$.

Let $\tilde{u}_{\tilde{\omega}(x^i)}$ be the array of the function values $u(x^j)$ at the node points $x^j \in \tilde{\omega}(x^i)$ and $J_{\tilde{\omega}(x^i)}$ be the number of the node points.

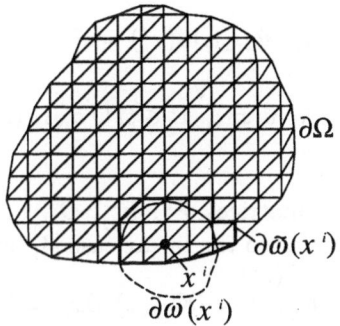

Fig. 1. A localization domain $\omega(x^i)$ and a total localization domain $\tilde{\omega}(x^i)$ associated with a collocation point x^i of a body Ω for a mesh-based discretization.

Let $u(x) = \sum_j u(x^j)\phi_{kj}(x)$ be a continuous piece-wise smooth interpolation of $u(x)$ at any point $x \in \Omega$ along the values $u(x^j)$ at the node points x^j belonging to the same element $\bar{e}_k \subset \Omega$ as x, and the shape functions $\phi_{kj}(x)$ be localized on e_k. Collecting the interpolation formulas for all $x \in \tilde{\omega}(x^i)$, we have

$$u(x) = \sum_{x^j \in \tilde{\omega}(x^i)} u(x^j)\Phi_{ij}(x), \quad \Phi_{ij}(x) = \begin{cases} \phi_{kj}(x) & \text{if } x, x^j \in \bar{e}_k \subset \tilde{\omega}(x^i), \\ 0 & \text{otherwise}, \end{cases}$$

$$\frac{\partial u(x)}{\partial x_q} = \sum_{x^j \in \tilde{\omega}(x^i)} u(x^j)\frac{\partial \Phi_{ij}(x)}{\partial x_q}, \tag{23.8}$$

$$\frac{\partial \Phi_{ij}(x)}{\partial x_q} = \begin{cases} \frac{\partial \phi_{kj}(x)}{\partial x_q} & \text{if } x, x^j \in \bar{e}_k \subset \tilde{\omega}(x^i), \\ 0 & \text{otherwise}. \end{cases}$$

Consequently, $\Phi_{ij}(x) = \frac{\partial \Phi_{ij}(x)}{\partial x_q} = 0$ if $x^j \notin \tilde{\omega}(x^i)$. We can also use a local interpolation of $t(x) = [T(u)u](x)$ along only boundary nodes belonging to $\tilde{\omega}(x^i) \cap \partial_D\Omega$, $t(x) = \sum_{x^j \in \tilde{\omega}(x^i) \cap \partial_D\Omega} t(x^j)\Phi'_{ij}(x)$, $x \in \bar{\omega}(x^i) \cap \partial_D\Omega$. Here $\Phi'_{ij}(x)$ are the shape functions on the boundary obtained similar to $\Phi_{ij}(x)$ in (23.8).

After substitution of the interpolations, in LBDIDP(t) and taking into account (23.2), we arrive at the following system of J quasilinear algebraic equations for J unknowns: $u(x^j)$, $x^j \in \Omega \cup \partial_N\Omega$ and $t(x^j) = (Tu)(x^j)$, $x^j \in \partial_D\Omega$,

$$c^0(x^i)u(x^i) + \sum_{x^j \in \Omega \cup \partial_N\Omega} K^0_{ij}(\tilde{u}_{\tilde{\omega}(x^i)})u(x^j) + \sum_{x^j \in \partial\Omega} Q_{ij}(\tilde{u}_{\tilde{\omega}(x^i)})t(x^j)$$

$$= \mathcal{F}^0(\tilde{u}_{\tilde{\omega}(x^i)}, x^i) - \sum_{x^j \in \partial_D\Omega} K^0_{ij}(\tilde{u}_{\tilde{\omega}(x^i)})\bar{u}(x^j), \quad x^i \in \bar{\Omega} \text{ (no sum on } i). \tag{23.9}$$

Instead, one can arrive at the system of only $J - J_D$ quasilinear algebraic equations for $J - J_D$ unknowns $u(x^j)$, $x^j \in \Omega \cup \partial_N \Omega$, if one substitutes interpolation formulae (23.8) in LBDIDP(T),

$$c(x^i)u(x^i) + \sum_{x^j \in \Omega \cup \partial_N \Omega} K_{ij}(\tilde{u}_{\tilde{\omega}(x^i)})u(x^j) = \mathcal{F}(\tilde{u}_{\tilde{\omega}(x^i)}, x^i)$$

$$- \sum_{x^j \in \partial_D \Omega} K_{ij}(\tilde{u}_{\tilde{\omega}(x^i)})\bar{u}(x^j), \quad x^i \in \Omega \cup \partial_N \Omega \text{ (no sum on } i). \quad (23.10)$$

The right-hand sides and matrices $\mathcal{F}^0$, K_{ij}^0, Q_{ij} in (23.9) and $\mathcal{F}$, K_{ij} in (23.10) are calculated similar to the linear case [4] but now depend also on the unknowns $u(x^j)$, $x^j \in \tilde{\omega}(x^i)$.

From the definitions, $\Phi_{ij}(x) = \partial\Phi_{ij}(x)/\partial x_q = T\Phi_{ij}(x) = \Phi'_{ij}(x) = 0$ and consequently $K_{ij}^0 = Q_{ij} = K_{ij} = 0$ if $x^j \notin \tilde{\omega}(x^i)$. Thus, each of the equations in systems (23.9) and (23.10), has not more than $J_{\tilde{\omega}(x^i)}\mathcal{L}J$ nonzero entries, i.e. the systems are sparsely populated. For example, the typical number of entries in each equation is $J_{\tilde{\omega}(x^i)} = 7$ for the 2D case and triangle mesh presented in Fig. 1.

In addition to the mesh-based discretisation described here, a meshless discretization of the quasilinear LBDIDPs (t) and (T) can be also done similar to the linear case described in [4].

References

1. C.A. Brebbia, J.C.F. Telles, and L.C. Wrobel, *Boundary Element Techniques*, Springer-Verlag, Berlin, 1984.

2. P.K. Banerjee, *Boundary Element Methods in Engineering*, McGraw-Hill, London, 1994.

3. X.-W. Gao and T.G. Davies, *Boundary Element Programming in Mechanics*, Cambridge University Press, Cambridge, 2002.

4. S.E. Mikhailov, Localized boundary-domain integral formulations for problems with variable coefficients, *Engrg. Anal. with Boundary Elements* **26** (2002), 681–690.

5. T. Zhu, J.-D. Zhang, and S.N. Atluri, A local boundary integral equation (LBIE) method in computational mechanics, and a meshless discretization approach, *Comput. Mech.* **21** (1998), 223–235.

6. T. Zhu, J.-D. Zhang, and S.N. Atluri, A meshless numerical method based on the local boundary integral equation (LBIE) to solve linear and non-linear boundary value problems, *Engrg. Anal. with Boundary Elements* **23** (1999), 375–389.

24 Uniqueness for Inverse Inhomogeneous Transmission Problems in Lipschitz Domains

Dorina Mitrea

24.1 Introduction

Let Ω be a bounded Lipschitz domain in $\mathbb{R}^3$ with connected complement and outward unit normal n. Denote by $H^{r,p}$, $r \in \mathbb{R}$, the usual Sobolev type spaces on Ω, by $L^p(\partial\Omega)$ the space of pth power integrable functions on $\partial\Omega$, and by $L_1^p(\partial\Omega)$ the collection of functions $f \in L^p(\partial\Omega)$ whose tangential gradient, $\nabla_{\tan}f$, belongs to $L^p(\partial\Omega)$. The scale of Besov spaces $B_s^{p,p}(\partial\Omega)$, $0 < s < 1$, is then obtained by the method of real interpolation between $L^p(\partial\Omega)$ and $L_1^p(\partial\Omega)$. We denote by $u|_{\partial\Omega_{\pm}}$ the restriction of u to the boundary in the pointwise non-tangential limit sense from inside or outside of Ω. The non-tangential maximal operator acting on a function $u : \Omega \to \mathbb{R}$, is given at a boundary point x by

$$\mathcal{N}(u)(x) := \sup\{|u(y)|; \ y \in \Omega, \ |x - y| \le 2\operatorname{dist}(y, \partial\Omega)\}.$$

The inhomogeneous transmission boundary value problem for the Helmholtz equation consists of finding a pair of functions u, v satisfying

$$(\Delta + k^2\rho(x))u(x) = 0 \quad \text{in} \quad \mathbb{R}^3\backslash\overline{\Omega},$$
$$(\Delta + k^2q(x))v(x) = 0 \quad \text{in} \quad \Omega,$$
$$u \in H^{2,p}(B_R\backslash\overline{\Omega}) + \{w; \mathcal{N}(\nabla w) \in L^p(\partial\Omega), \exists\, w|_{\partial\Omega_-}, \exists\, \nabla w|_{\partial\Omega_-}\},$$
$$v \in H^{2,p}(\Omega) + \{w; \mathcal{N}(\nabla w) \in L^p(\partial\Omega), \exists\, w|_{\partial\Omega_+}, \exists\, \nabla w|_{\partial\Omega_+}\}, \qquad (24.1)$$
$$u|_{\partial\Omega_-} - v|_{\partial\Omega_+} = f \in L_1^p(\partial\Omega),$$
$$\langle(\nabla u)|_{\partial\Omega_-}, n\rangle - \mu\langle(\nabla v)|_{\partial\Omega_+}, n\rangle - \lambda u|_{\partial\Omega_-} = g \in L^p(\partial\Omega),$$
$$\lim_{r\to\infty} r\left(\tfrac{\partial u}{\partial r} - iku\right) = 0, \quad r = |x|,$$

where B_R is the ball of radius R centered at the origin, $k \in \mathbb{R}_+$, $\mu > 0$, and $\lambda \in L^\infty(\partial\Omega)$, $\rho \in L^\infty(\mathbb{R}^3)$, $q \in L^\infty(\Omega)$, are complex-valued functions. Throughout the paper we shall assume that $\operatorname{Im}\lambda \le 0$, $\operatorname{Im}q \ge 0$, $\operatorname{Im}\rho \ge 0$, $\rho \equiv 1$ outside a ball of radius R_0.

This work was partially supported by a UMC Summer Research Fellowship.

The above problem models the scattering of inhomogeneous acoustic time-harmonic waves by a penetrable bounded obstacle Ω. The incident plane wave $u^i(x) = e^{ik\langle x,d\rangle}$, $x \in \mathbb{R}^3$, with $d \in S^2$ the propagation direction, will produce a (radiating) scattered wave u^s in the exterior of Ω and a transmitted wave v in Ω. The waves u^s and v are annihilated by the inhomogeneous Helmholtz operators $\Delta + k^2\rho(x)$ and $\Delta + k^2q(x)$, respectively, and verify the so-called conductive boundary conditions

$$(u^i + u^s) = v \quad \text{and} \quad \langle \nabla(u^i + u^s), n\rangle = \mu\langle\nabla v, n\rangle + \lambda(u^i + u^s) \quad \text{on} \quad \partial\Omega.$$

In particular, the problem (24.1) reduces precisely to the above for $u := u^s$ and $f := -u^i|_{\partial\Omega_-}$, $g := -\frac{\partial u^i}{\partial n} + \lambda u^i|_{\partial\Omega_-}$ (as usual, $\frac{\partial u^i}{\partial n}$ denotes $\langle\nabla u^i, n\rangle$).

The main result of this paper is concerned with the inverse obstacle problem associated with (24.1) for scatterers with Lipschitz boundaries. Recall that the scattered wave u^s has the asymptotic behavior

$$u^s(x) = \frac{e^{ik|x|}}{|x|}\left[u_\infty\left(\frac{x}{|x|}\right) + \mathcal{O}\left(\frac{1}{|x|}\right)\right] \quad \text{as} \quad |x| \to \infty,$$

where u_∞ is the *far-field pattern* of u^s; see, e.g., [1].

Theorem 1. *Suppose two conductive scatterers occupy the interiors of two bounded Lipschitz domains Ω_1, Ω_2 in $\mathbb{R}^3$, with connected complements and inhomogeneities given by q_i, $i = 1, 2$. Also, we assume that the inhomogeneity of $\mathbb{R}^3\backslash\Omega_j, j = 1, 2$ is ρ.*

Suppose that, for a fixed $k > 0$ and any incident direction $d \in S^2$, the two far-field patterns for Ω_1 and Ω_2 corresponding to the incident plane wave $e^{ik\langle x,d\rangle}$ coincide. Then, $\Omega_1 = \Omega_2$.

It is well known that well-definiteness of the inverse problem is based on the well-posedness of the direct problem (24.1). The latter is considered in Section 2 where we show that (24.1) is uniquely solvable in Lipschitz domains for p in a neighborhood of 2. The proof of Theorem 1 is dealt with in Section 3.

In the case of a smooth domain, similar results have been proved in [2]. In order to treat the case of Lipschitz domains the ideas in [2] have to be refined considerably. In the process, we make essential use of mapping properties for layer potentials in non-smooth domains and of Rellich type identities (for controlling the local behavior of solutions near the boundary). This approach was used successfully in the proof of the uniqueness for the inverse transmission problem in the case of the homogeneous Helmholtz equation on Lipschitz domains; see [3].

24.2 The Direct Problem

Theorem 2. *Let Ω be a bounded Lipschitz domain in $\mathbb{R}^3$. Then there exists $\delta = \delta(\partial\Omega, k, \mu) > 0$ so that (24.1) is uniquely solvable for each $2 - \delta < p < 2 + \delta$. Moreover the solution depends continuously on the boundary data.*

Proof. Uniqueness in the case $p = 2$ (and thus for $p \geq 2$) can be proved based on the arguments in Theorem 3.12 in [1].

Recall $\Phi_k(x, y) := -\frac{1}{4\pi}\frac{e^{ik|x-y|}}{|x-y|}$, $x \neq y$, the fundamental solution for $\Delta + k^2$ in $\mathbb{R}^3$. Associated with Φ_k we have the classical single and double layer potential operators:

$$\mathcal{S}_k f(x) := \int_{\partial\Omega} \Phi_k(x, y) f(y)\, d\sigma(y), \quad \mathcal{D}_k f(x) := \int_{\partial\Omega} \frac{\partial \Phi_k(y, x)}{\partial n(y)} f(y)\, d\sigma(y),$$

for $f \in L^p(\partial\Omega)$, $1 < p < \infty$, and $x \in \mathbb{R}^3 \setminus \partial\Omega$. If $x \in \partial\Omega$ these integrals are considered in the principal value sense, and we denote them by $S_k f$, and $K_k f$, respectively. Recall the jump relations

$$\mathcal{D}_k f|_{\partial\Omega_\pm} = (\pm\tfrac{1}{2}I + K_k)f, \quad \frac{\partial S_k f}{\partial n}\Big|_{\partial\Omega_\pm} = (\mp\tfrac{1}{2}I + K_k^*)f,$$

where K_k^* is the formal adjoint of K_k. For more on mapping properties for these layer potentials see [4]. For a fixed $R \geq R_0$, $1 < p < \infty$ define

$$T_{k,R}u := k^2 \Pi_{k, B_R\setminus\bar{\Omega}}[(1 - \rho)u], \quad T_{k,\Omega}v := k^2 \Pi_{k,\Omega}[(1 - q)v], \quad (24.2)$$

where $\Pi_{k,D}u(x) := \iint_D \Phi_k(x, y)u(y)\, dy$, is the Newtonian type layer potential for D a domain in $\mathbb{R}^3$, $u \in L^p(D)$ and $x \in \mathbb{R}^3$. The composition of the operators in (24.2) with the trace operator $Tr : H^{s,p}(\Omega) \to B^{p,p}_{s-\frac{1}{p}}(\partial\Omega)$, will be denoted by $T_{k,R}$ and $T_{k,\Omega}$. At this point we observe that if for each p close enough to 2, we can determine $u \in L^p(B_R\setminus\bar{\Omega})$, $v \in L^p(\Omega)$, $\psi \in L^p_1(\partial\Omega)$, $\phi \in L^p(\partial\Omega)$, such that

$$u = T_{k,R}(u) + \mathcal{D}_k\psi + \mathcal{S}_k\phi \quad \text{in} \quad B_R\setminus\bar{\Omega},$$

$$v = T_{k,\Omega}(v) + \tfrac{1}{\mu}\mathcal{D}_k\psi + \tfrac{1}{\mu}\mathcal{S}_k\phi \quad \text{in} \quad \Omega,$$

$$T_{k,R}(u) - T_{k,\Omega}(v) + \left[-\tfrac{1}{2}(1 + \tfrac{1}{\mu})I + (1 - \tfrac{1}{\mu})K_k\right]\psi$$
$$+ (1 - \tfrac{1}{\mu})S_k\phi = f \quad \text{on} \quad \partial\Omega,$$

$$\frac{\partial T_{k,R}}{\partial n}(u) - \lambda T_{k,R}u - \mu\frac{\partial T_{k,\Omega}}{\partial n}(v) + \phi$$
$$- \lambda(-\tfrac{1}{2}I + K_k)\psi - \lambda S_k\phi = g \quad \text{on} \quad \partial\Omega,$$

then there exists $\tilde{u}$ a suitable extension of u to $\mathbb{R}^3\backslash\bar{\Omega}$, such that $(\tilde{u}, v)$ is a solution of (24.1). In this way, matters are reduced to proving the invertibility of the operator U mapping $L^p(B_R\backslash\bar{\Omega}) \oplus L^p(\Omega) \oplus L_1^p(\partial\Omega) \oplus L^p(\partial\Omega)$ onto itself for p sufficiently close to 2, where

$$U(u, v, \psi, \phi) = (0, 0, f, g), \quad U = A + B,$$

$$A := \begin{pmatrix} -I & 0 & 0 & 0 \\ 0 & -I & 0 & 0 \\ 0 & 0 & -\frac{\mu+1}{2\mu}I + \frac{\mu-1}{\mu}K_{k_0} & 0 \\ 0 & 0 & -\lambda(-\frac{1}{2}I + K_k) & I \end{pmatrix},$$

$$B := \begin{pmatrix} T_{k,R} & 0 & D_k & S_k \\ 0 & T_{k,\Omega} & \frac{1}{\mu}D_k & \frac{1}{\mu}S_k \\ T_{k,R} & -T_{k,\Omega} & \frac{\mu-1}{\mu}(K_k - K_{k_0}) & \frac{\mu-1}{\mu}S_k \\ \frac{\partial T_{k,R}}{\partial n} - \lambda T_{k,R} & -\mu\frac{\partial T_{k,\Omega}}{\partial n} & 0 & -\lambda S_k \end{pmatrix};$$

here k_0 is some fixed complex number with $\operatorname{Im} k_0 > |\operatorname{Re} k_0|$. The operators A and B above are considered from the space $L^p(B_R\backslash\bar{\Omega})\oplus L^p(\Omega)\oplus L_1^p(\partial\Omega)\oplus L^p(\partial\Omega)$ onto itself.

Invoking the spectral properties of the double layer potential for the Helmholtz operator with a regular wave number, cf. Theorem 7.4 in [5] (observe that $\eta := \left|\frac{\mu+1}{2(\mu-1)}\right| > \frac{1}{2}$ for $\mu \neq 1$), we see that the operator $-\frac{\mu+1}{2\mu}I + \frac{\mu-1}{\mu}K_{k_0}$ is invertible from $L_1^p(\partial\Omega)$ onto itself, provided p is close enough to 2. This in concert with the fact that B is compact for all $1 < p < \infty$, further implies that U is a Fredholm operator with index zero for p in a suitable neighborhood of 2. In the case $p = 2$, the injectivity of U follows from the uniqueness of (24.1) and the jump relations for the layer potentials. Hence, U is an isomorphism for $p = 2$. Now based on interpolation results, we conclude that the operator U is an isomorphism for $|p - 2| < \delta$, some $\delta > 0$ small enough. This clearly entails the existence of a solution for (24.1) for p near 2.

We are left with proving uniqueness for (24.1) for $p - \delta < p < 2$. The reasoning is similar to the one used for the homogeneous case (see [3]) and we omit it. The fact that the unique solution to (24.1) depends continuously on the boundary data follows from the proof of the existence part.

Remark. A close look at the above proof reveals that the operator U is also an isomorphism from $L^p(B_R\backslash\bar{\Omega}) \oplus L^p(\Omega) \oplus L^p(\partial\Omega) \oplus L^p(\partial\Omega)$ onto itself, for p close to 2. In particular, by interpolation, we have that U is an isomorphism from $L^p(B_R\backslash\bar{\Omega}) \oplus L^p(\Omega) \oplus B_s^{p,p}(\partial\Omega) \oplus L^p(\partial\Omega)$, for any $0 < s < 1$, with corresponding norm estimates for the solution to (24.1). These norm estimates can be further used in conjunction with the explicit form of u and v and the mapping properties of layer potentials, in order to bound the $H^{1,p}$ norms of u and v by appropriate norms of the boundary data. In particular, it follows that

$$\|v\|_{H^{1,2}(\Omega)} \leq C(\|f\|_{B^{2,2}_{1/2}(\partial\Omega)} + \|g\|_{L^2(\partial\Omega)}). \qquad (24.3)$$

24.3 The Inverse Problem

For the proof of Theorem 1 we employ ideas similar to the ones used in [3] for the proof of the uniqueness for the transmission problem for the homogeneous Helmholtz equation. As such, we will sketch the main ideas and for details refer the interested reader to [3]. Let Ω_1, Ω_2, q_1, q_2, be as in the hypotheses of Theorem 1. First, by standard arguments (see [1]) one can conclude that if the hypothesis of Theorem 1 holds true, then the scattered waves u_1^s, u_2^s for Ω_1 and Ω_2, corresponding to the incoming waves $\Phi_k(\cdot, x_0)$ coincide in $G := \mathbb{R}^3 \setminus \overline{\Omega_1 \cup \Omega_2}$, for each fixed $x_0 \in G$.

Seeking a contradiction, suppose that $\Omega_1 \neq \Omega_2$. Then, there exists a point $z \in \partial\Omega_1$ with $z \notin \overline{\Omega}_2$ and $z \in \partial G$. Considering a sequence of points $x_j \in G$, which converges non-tangentially to z such that the distance from x_j to $\partial\Omega_1$ is comparable to $\frac{1}{j}$, for each incident wave $\Phi_k(\cdot, x_j)$ we obtain a unique solution (u_j^l, v_j^l) for Ω_l, $l = 1, 2$, such that $u_j^1 = u_j^2$ in G. Fix $r > 0$ small enough so that $B := B(z, r)$, the ball of radius r centered at z, does not intersect $\overline{\Omega}_2$. Then the sequence $\{u_j^1\}_j$ is bounded in $C^1(\overline{B} \setminus \Omega_1)$. Setting $w_j := v_j^1 - \phi_{k^*}(\cdot, x_j)$ in Ω_1, for some $k^* > 0$, $k^* \neq k$, and using the boundary conditions verified by (u_j^1, v_j^1), it turns out that

$$\{w_j\}_j, \quad \{\nabla_{\tan} w_j\}_j \quad \text{are bounded in } L^\infty(\partial\Omega_1 \cap \overline{B}), \qquad (24.4)$$

$$\frac{\partial w_j}{\partial n}(x) \text{ is of order of } |x - x_j|^{-2} \text{ for } x \in \partial\Omega_1 \cap \overline{B} \text{ as } j \to \infty. \qquad (24.5)$$

At this point, arguing as in [3], it can be proved (based on the Rellich identity in [6]) that there exists $\Gamma_0 \subset \Gamma_1 \subseteq B \cap \partial\Omega_1$ such that for each $\epsilon > 0$, one can find constants $C, C_\epsilon > 0$, both independent of j, for which

$$\int_{\Gamma_0} |\frac{\partial w_j}{\partial n}|^2 d\sigma \leq C_\epsilon \int_{\Gamma_1} |\nabla_{\tan} w_j|^2 d\sigma + \epsilon \int_{\Gamma_1} |\frac{\partial w_j}{\partial n}|^2 d\sigma$$

$$+ C \int\int_{\Omega_1} (|v_j^1|^2 + |\nabla v_j^1|^2 + |\phi_{k^*}(\cdot, x_j)|^2 + |\nabla\phi_{k^*}(\cdot, x_j)|^2). \qquad (24.6)$$

Now (24.5) implies that the left side of (24.6) and the second integral in the right side of (24.6) are of the order of j^2 as $j \to \infty$. The essential thing is that ϵ can be chosen sufficiently small so that the second integral in the right side of (24.6) can be absorbed in the left side, yielding there a term that is still of the order of j^2 as $j \to \infty$. Also, from (24.4) we have that the first integral in the right side of (24.6) is $\mathcal{O}(1)$ as $j \to \infty$. In particular, if the solid integral in (24.6) is $\mathcal{O}(j)$ as $j \to \infty$, we arrive at a contradiction (as far as the growth rate estimates of the terms in (24.6) are concerned). Using the definition of $\phi_{k^*}(\cdot, x_j)$, in order to show that the solid integral in (24.6) is indeed $\mathcal{O}(j)$ as $j \to \infty$, it suffices to prove that $\|v_j^1\|_{H^{1,2}(\Omega_1)}^2 = \mathcal{O}(j)$ as $j \to \infty$. With this in mind, define

$$\tilde{w}_j := v_j^1 - \tfrac{1}{\mu}\phi_{k^*}(\cdot, x_j), \qquad h_j := \left[\tfrac{(k^*)^2}{\mu} - \tfrac{k^2 q_1}{\mu}\right]\phi_{k^*}(\cdot, x_j)$$

in Ω_1. Then, $(u_j^1, \tilde{w}_j - \Pi_{k,\Omega_1}(h_j))$ is a solution of the version of (24.1) corresponding to the domain Ω_1, the inhomogeneity q_1, and the boundary data $f_j + \Pi_{k,\Omega_1}(h_j)|_{\partial\Omega_1}$, $g_j + \mu\frac{\partial\Pi_{k,\Omega_1}}{\partial n}(h_j)$, where

$$f_j := -\phi_k(\cdot, x_j) + \frac{1}{\mu}\phi_{k^*}(\cdot, x_j),$$

$$g_j := -\frac{\partial\phi_k(\cdot, x_j)}{\partial n} + \frac{\partial\phi_{k^*}(\cdot, x_j)}{\partial n} + \lambda\phi_k(\cdot, x_j).$$

At this point, the estimate (24.3) applied to $\tilde{w}_j$ yields the desired growth rate for $\|\tilde{w}_j\|_{H^{1,2}(\Omega_1)}$. With this the proof of Theorem 1 is complete.

References

1. D. Colton and R. Kress, *Inverse Acoustic and Electromagnetic Scattering Theory*, Springer-Verlag, Berlin, 1992.

2. A. Kirsch and L. Päivärinta, On recovering obstacles inside inhomogeneities, *Math. Methods Appl. Sci.* **21** (1998), 619–651.

3. D. Mitrea and M. Mitrea, Uniqueness for inverse conductivity and transmission problems in the class of Lipschitz domains, *Comm. Partial Differential Equations* **23** (1998), 1419–1448.

4. D. Mitrea, M.Mitrea, and J. Pipher, Vector potential theory on nonsmooth domains in $\mathbb{R}^3$ and applications to electromagnetic scattering, *J. Fourier Anal. Appl.* **3** (1997), 131–192.

5. B. Jawerth and M. Mitrea, Higher dimensional scattering theory on C^1 and Lipschitz domains, *Amer. J. Math.* **117** (1995), 929–963.

6. J. Nečas, *Les Méthodes Directes en Théorie des Équations Élliptiques*, Academia, Prague, 1967.

25 Analytic Solution for an Enhanced Theory of Bending of Plates

Radu Mitric and Christian Constanda

25.1 Notation and Prerequisites

Let S be a domain in $\mathbb{R}^2$ bounded by a simple closed C^2-curve ∂S, and let $h_0 = \text{const}$ be such that $0 < h_0 \ll \text{diam}\, S$. By a thin plate we understand an elastic body that occupies the region $\bar{S} \times [-h_0/2, h_0/2]$; here h_0 is called the plate thickness. We denote by $x = (x_1, x_2)$ a generic point in $\mathbb{R}^2$, and write $z = x_1 + ix_2 \in \mathbb{C}$, $\partial_\alpha = \partial/\partial x_\alpha$, $\alpha = 1, 2$, and $\partial_z = \partial/\partial z$. Also, we denote by S^+ and S^- the domains interior and exterior to ∂S, respectively.

In what follows we consider the problem of bending of an elastic plate with transverse shear deformation and transverse normal strain, accounted for through displacements of the form

$$U_\alpha = x_3 u_\alpha(x_1, x_2), \quad U_3 = u_{31}(x_1, x_2) + x_3^2 u_{32}(x_1, x_2), \quad \alpha = 1, 2.$$

The general form of a rigid displacement is $u = Fk$, where $F(x)$ is the (4×3)-matrix of columns $(1, 0, -x_1, 0)^{\mathrm{T}}$, $(0, 1, -x_2, 0)^{\mathrm{T}}$ and $(0, 0, 1, 0)^{\mathrm{T}}$, k is an arbitrary constant (3×1)-vector, and the superscript T indicates matrix transposition.

In [1] we considered the system of equilibrium equations in terms of displacements, with the Dirichlet and Neumann boundary conditions. We reduced these boundary value problems to singular integral equations on the contour of the domain and solved them in spaces of smooth functions. At the end of this initial stage we were not able to elucidate the physical meaning of certain analytic constraints imposed on the asymptotic behavior of the solutions. This question is answered here, where we construct the complete integral of the system through complex variable treatment.

25.2 Complex Representation of the Stresses

We consider the homogeneous system corresponding to the equations of equilibrium, which, in terms of moments and forces, is

$$N_{\alpha\beta,\beta} - N_{3\alpha} = 0,$$
$$N_{3\beta,\beta} = 0, \tag{25.1}$$
$$M_{3\beta,\beta} - 2M_{33} = 0,$$

and the compatibility conditions

$$N_{21,1} + h^2(N_{31,12} - N_{32,11}) = (1 - \sigma)N_{11,2} - \sigma(N_{22,2} + M_{33,2}),$$

$$N_{21,2} + h^2(N_{32,12} - N_{31,22}) = (1 - \sigma)N_{22,1} - \sigma(N_{11,1} + M_{33,1}),$$

$$(M_{31} - h^2 N_{31})_{,2} = (M_{32} - h^2 N_{32})_{,1},$$

$$[(1 - \sigma)M_{33} - \sigma(N_{11} + N_{22})]_{,\alpha} = 5h^{-2}(M_{3\alpha} - h^2 N_{3\alpha}),$$

(25.2)

where $\sigma = \lambda/(3\lambda + 2\mu)$, t_{ij} are the stresses, and

$$N_{\alpha\beta} = \frac{1}{h_0} \int_{-h_0/2}^{h_0/2} x_3 t_{\alpha 3} \, dx_3, \quad N_{3\alpha} = \frac{1}{h_0} \int_{-h_0/2}^{h_0/2} t_{\alpha 3} \, dx_3,$$

$$M_{3\alpha} = \frac{1}{h_0} \int_{-h_0/2}^{h_0/2} x_3^2 t_{\alpha 3} \, dx_3, \quad M_{33} = \frac{1}{h_0} \int_{-h_0/2}^{h_0/2} x_3 t_{33} \, dx_3.$$

We investigate the homogeneous system corresponding to (25.1), (25.2) and construct its analytic solution in S. By computation, we find that

$$N_{11} = -\text{Im}[4h^2\eta_{,zz} + \tilde{k}\bar{z}(\Omega_1'' + \Omega_2'') - \tilde{k}\Omega_2' + (2 - \tilde{k})\Omega_1'] - \tilde{\sigma}\rho_{,22},$$

$$N_{22} = +\text{Im}[4h^2\eta_{,zz} + \tilde{k}\bar{z}(\Omega_1'' + \Omega_2'') + \tilde{k}\Omega_1' - (2 - \tilde{k})\Omega_2'] - \tilde{\sigma}\rho_{,11},$$

$$N_{21} = -\text{Re}[4h^2\eta_{,zz} + \tilde{k}\bar{z}(\Omega_1'' + \Omega_2'') + \Omega_1' - \Omega_2'] - 2\tilde{\sigma}\text{Im}\,\rho_{,zz},$$

$$N_{31} = -\text{Im}[2\eta_{,z} + (1 + \tilde{k})(\Omega_1'' + \Omega_2'')],$$

$$N_{32} = -\text{Re}[2\eta_{,z} + (1 + \tilde{k})(\Omega_1'' + \Omega_2'')],$$

(25.3)

$$M_{31} = -\text{Im}[2h^2\eta_{,z} + \tfrac{4}{5}h^2(1 + 2\tilde{k})(\Omega_1'' + \Omega_2'')] - 2\text{Re}\,\rho_{,z},$$

$$M_{32} = -\text{Re}[2h^2\eta_{,z} + \tfrac{4}{5}h^2(1 + 2\tilde{k})(\Omega_1'' + \Omega_2'')] - 2\text{Im}\,\rho_{,z},$$

$$M_{33} = -\frac{5}{4h^2}\left(1 + \frac{1}{\tilde{k}}\right)\rho,$$

where Ω_α are arbitrary analytic functions in S, $\eta(z, \bar{z})$ and $\rho(z, \bar{z})$ are arbitrary real solutions in S of the equations

$$\Delta\eta - \frac{1}{h^2}\eta = 0 \quad \text{and} \quad \Delta\rho - \frac{1}{k^2}\rho = 0,$$

(25.4)

respectively, and

$$k^2 = h^2(\lambda + 2\mu)/[20(\lambda + \mu)], \quad \tilde{k} = (1 - 2\sigma)/(3 - 2\sigma),$$

$$\sigma = \lambda/(3\lambda + 2\mu), \quad \tilde{\sigma} = \sigma/[2(1 - \sigma)].$$

25.3 The Displacement Boundary Value Problem

Consider the Dirichlet boundary condition

$$u_\alpha = \tilde{u}_\alpha, \quad u_{3\alpha} = \tilde{u}_{3\alpha} \quad \text{on } \partial S.$$

We introduce the complex functions

$$\Gamma = u_1 + iu_2, \quad \Theta = u_{31}, \quad \tilde{\Theta} = u_{32},$$
$$\Phi = N_{11} - N_{22} + 2iN_{21}, \quad \Psi = N_{11} + N_{22},$$
$$\Lambda = N_{31} + iN_{32}, \quad \Sigma = M_{31} + iM_{32}, \quad \Pi = M_{33}.$$

Replacing N_{ij}, M_{3i} in terms of displacements in (25.3), we arrive at

$$\Phi = 4h^2 \mu \Gamma_{,\bar{z}}, \quad \Psi = 4h^2(\lambda + \mu)\mathrm{Re}\Gamma_{,z} + 4h^2 \lambda \tilde{\Theta},$$
$$\Lambda = \mu(\Gamma + 2\Theta_{,\bar{z}}) + 2h^2 \mu \tilde{\Theta}_{,\bar{z}}, \quad \Sigma = h^2 \mu(\Gamma + 2\Theta_{,\bar{z}}) + \tfrac{18}{5} h^4 \mu \tilde{\Theta}_{,\bar{z}},$$
$$\Pi = 2h^2 \lambda \mathrm{Re}\Gamma_{,z} + 2h^2(\lambda + 2\mu)\tilde{\Theta}.$$

Setting

$$\Omega_\alpha = \omega'_\alpha, \quad \psi = -\frac{2}{\mu}\eta, \quad \Omega = \frac{i\tilde{k}}{2h^2 \mu}(\omega'_1 + \omega'_2), \quad \omega = \frac{i}{2h^2 \mu}(\omega_1 - \omega_2),$$

we now obtain

$$\Gamma = i\psi_{,\bar{z}} + z\bar{\Omega}' + \Omega + \bar{\omega}' + l + \frac{\lambda}{4h^2 \mu(\lambda + \mu)}\rho_{,\bar{z}},$$

$$\Theta = \mathrm{Re}\left[2h^2 \frac{9\lambda + 8\mu}{\lambda + 2\mu}\Omega' - \bar{z}\Omega - \omega - l\bar{z} + m + \frac{9\lambda + 10\mu}{8h^2 \mu(\lambda + \mu)}\rho\right],$$

$$\tilde{\Theta} = \mathrm{Re}\left(-\frac{2\lambda}{\lambda + 2\mu}\Omega' - \frac{5}{4h^4 \mu}\rho\right),$$

with $l \in \mathbb{C}$ and $m \in \mathbb{R}$. The matrix defined by the terms containing l and m represents a rigid displacement so we can ignore these terms, considering them incorporated in Ω and ω, respectively. Thus, the displacement boundary value problem reduces to finding a solution ψ of (25.4)$_1$, a solution ρ of (25.4)$_2$, and analytic functions Ω and ω in S such that Γ, Θ, and $\tilde{\Theta}$ take prescribed values on ∂S. With these functions we then construct

$$\Phi = 4h^2 \mu[i\psi_{,\bar{z}\bar{z}} + z\bar{\Omega}'' + \bar{\omega}'' + \frac{\lambda}{4h^2 \mu(\lambda + \mu)}\rho_{,zz}],$$

$$\Psi = 4h^2 \mu \frac{3\lambda + 2\mu}{\lambda + 2\mu}(\Omega' + \bar{\Omega}') - \frac{5\lambda}{h^2(\lambda + 2\mu)}\rho,$$

$$\Lambda = i\mu\psi_{,\bar{z}} + 16h^2\mu\frac{\lambda+\mu}{\lambda+2\mu}\bar{\Omega}'',$$

$$\Sigma = ih^2\mu\psi_{,\bar{z}} + \frac{8}{5}h^4\mu\frac{9\lambda+10\mu}{\lambda+2\mu}\Omega'' - 2\rho_{,\bar{z}},$$

$$\Pi = -\frac{10(\lambda+\mu)}{h^2(\lambda+2\mu)}\rho.$$

25.4 Bounded Multiply Connected Domain

From the last system we deduce that $\Phi, \Psi, \Lambda, \Sigma, \Pi$ are single valued if and only if $\rho, \Omega'', \psi, \Omega'', \mathrm{Re}\,(\bar{z}\Omega+\omega)$ and $z\bar{\Omega}'+\Omega+\bar{\omega}'$ are single valued. Since S is multiply connected, by integrating the single-valued functions Ω'', ω'' over S, we obtain a multiple-valued result. Suppose that ∂S consists of $n+1$ disjoint simple curves of which one is $S_0 = \partial S_0$, and that $S_0 \supset \bigcup_{k=1}^{n} \partial S_k$. We choose arbitrary points z_k inside ∂S_k. Then, according to [2],

$$\Omega = \frac{1}{2\pi i}\sum_{k=1}^{n}(c_k z + d_k)\log(z - z_k) + \tilde{\Omega},$$

$$\omega = \frac{1}{2\pi i}\sum_{k=1}^{n}(p_k z + q_k)\log(z - z_k) + \tilde{\omega},$$

where $c_k, d_k, p_k, q_k \in \mathbb{C}, k = 1, \ldots, n$, and $\tilde{\Omega}$ and $\tilde{\omega}$ are analytic (so single valued) functions in S. Then, the single valuedness of the potentials imply that

$$\mathrm{Re}\,c_k = 0, \quad d_k + \bar{p}_k = 0, \quad \mathrm{Re}\,q_k = 0.$$

Traversing ∂S once anticlockwise, we obtain the resultant force and moment on ∂S in the form

$$\mathcal{N}_k = -\int_{\partial S_k} N_3 ds = -\int_{\partial S_k} (N_{31}\nu_1 + N_{32}\nu_2)ds = -16h^2\mu\frac{\lambda+\mu}{\lambda+2\mu}\,\mathrm{Im}\,c_k,$$

$$\mathcal{M}_k = -\int_{\partial S_k}[-N_2 + x_2 N_3 + i(N_1 - x_1 N_3)]ds = -2h^2\mu\left(\frac{1}{\bar{z}}d_k - \bar{p}_k\right),$$

and we conclude that

$$\Omega = -c\sum_{k=1}^{n}(z\mathcal{N}_k - i\mathcal{M}_k)\log(z - z_k) + \tilde{\Omega},$$

$$\omega = -c\sum_{k=1}^{n}(iz\bar{\mathcal{M}}_k + s_k)\log(z - z_k) + \tilde{\omega},$$

where $c = (\lambda + 2\mu)[32\pi h^2\mu(\lambda + \mu)]^{-1}$.

25.5 Unbounded Multiply Connected Domain

Suppose that ∂S has expanded to infinity. We introduce the notation

$$N = \sum_{k=1}^{n} \mathcal{N}_k, \quad M = \sum_{k=1}^{n} \mathcal{M}_k, \quad s = \sum_{k=1}^{n} s_k.$$

Then, by a similar argument to that used in [2], we find that

$$\Omega = -c(Nz - iM)\log z + \sum_{n=-\infty}^{+\infty} a_n z^n,$$

$$\omega = -c(i\bar{M}z + s)\log z + \sum_{n=-\infty}^{+\infty} b_n z^n,$$

where $a_n, b_n \in \mathbb{C}$.

Using the asymptotics of ψ and ρ, we deduce that these functions and their derivatives vanish when $|x - y| \to \infty$, so the far-field pattern of the solution depends only on the structure of Ω and ω (it is independent on ψ and ρ).

Then, by computation, we find that $\Phi, \Psi, \Lambda, \Sigma,$ and Π are bounded at infinity if and only if

$$N = 0, \quad a_n = 0 \ (n \geq 3), \quad b_n \geq 0 \ (n \geq 3),$$

and that for a finite energy solution, that is,

$$\Gamma = O(1), \quad \Theta = O(\ln z), \quad \bar{\Theta} = O(1),$$

we must have

$$M = 0, \quad a_1 = i\alpha \ (\alpha \in \mathbb{R}), \quad b_2 = 0, \quad b_1 = \bar{a}_0. \tag{25.5}$$

Under these assumptions, it follows that

$$\Omega = \sum_{n=-\infty}^{-1} a_n z^n, \quad \omega = a \ln z + b + \sum_{n=-\infty}^{-1} b_n z^n. \tag{25.6}$$

For $a = b$ in (25.6), the expansion of u coincides with the class $\mathcal{A}$, in which we can solve the exterior Dirichlet, Neumann, and Robin problems. Also, we obtain the asymptotic behavior

$$\Phi = O(|z|^{-2}), \quad \Psi = O(|z|^{-2}), \quad \Lambda = O(|z|^{-3}), \quad \Sigma = O(|z|^{-3}),$$
$$\Pi = O(e^{-|z|}), \quad \Gamma = O(|z|^{-1}), \quad \Theta = O(\ln |z|), \quad \tilde{\Theta} = O(|z|^{-2}).$$

Therefore, the Betti formula holds in the exterior domain, so the condition that $u \in \mathcal{A}$, which was shown to be sufficient for the solvability of the exterior Neumann problem, is also necessary if we want a unique solution. Removing the restriction $a = b$ in (25.6) means that the regular solution is unique up to an arbitrary vertical translation.

25.6 Physical Significance of the Restrictions

If $|z| \to \infty$, then

$$\Phi_\infty = 8h^2 \mu \bar{b}_2, \quad \Psi_\infty = 8h^2 \mu \frac{3\lambda + 2\mu}{\lambda + 2\mu} \operatorname{Re} a_1,$$
$$\Lambda_\infty = \Sigma_\infty = \Pi_\infty = 0,$$

which means that the bending and twisting moments are uniformly distributed, while the transverse shear force, Σ, and Π vanish at infinity. Condition (25.5) is equivalent to

$$\Phi_\infty = 0, \quad \Psi_\infty = 0, \quad \varepsilon_\infty = 0,$$

where $\varepsilon = \varepsilon_1 + i\varepsilon_2$ and $\varepsilon_\alpha = \frac{1}{2} u_\alpha - u_{31,\alpha} - x_3^2 u_{32,\alpha}$ is the rotation in the vertical x_α-coordinate plane.

In conclusion, an analytic solution of our problem is of finite energy if and only if the corresponding bending and twisting moments, transverse shear force, Σ, Π, and the rotations in the vertical coordinate planes vanish at infinity. Then, according to a previous remark, $\mathcal{A}$ is the class of all finite energy solutions of our problem that contain no vertical translation.

References

1. R. Mitric and C. Constanda, An enhanced theory of bending of plates, in *Integral Methods in Science and Engineering*, Birkhäuser, Boston, 2002, 191–196.

2. N.I. Muskhelishvili, *Some basic problems in the mathematical theory of elasticity*, 3rd ed., P. Noordhoff, Groningen, 1949.

26 On Stabilization of Solutions of Elliptic Equations Containing Bessel Operators

Andrey B. Muravnik

26.1 Introduction

Singular differential equations containing Bessel operators frequently occur in modern models of mathematical physics and their theory is a rapidly developing research area (see [1] and references therein). This paper is devoted to stabilization of solutions of elliptic equations containing singularities of the specified kind.

The phenomenon called *stabilization* is well known for *parabolic* equations; it means the existence of a finite limit of the solution as $t \to \infty$. However, there are well-posed non-isotropic *elliptic* boundary value problems in unbounded domains in which we can talk about stabilization in the following sense: the solution has a finite limit as a selected *spatial* variable tends to infinity.

In this paper we investigate nonclassical boundary value problems of the above kind for singular elliptic equations, and find a necessary and sufficient condition of stabilization of solution in terms of weighted means of the boundary value function. We investigate the linear case and quasilinear equations with so-called nonlinearities of the Burgers–Kardar–Parisi–Zhang type (see [2] and [3]). Equations with such nonlinearities arise, for example, in modeling of directed polymers and interface growth.

26.2 Preliminaries

Let $k \overset{\text{def}}{=} 2\nu + 1$ be a positive parameter, and let

$$B_y \overset{\text{def}}{=} \frac{1}{y^{2\nu+1}} \frac{\partial}{\partial y}\left(y^{2\nu+1}\frac{\partial}{\partial y}\right) = \frac{\partial^2}{\partial y^2} + \frac{2\nu+1}{y}\frac{\partial}{\partial y},$$

$$T_y^h f(y) \overset{\text{def}}{=} \frac{\Gamma(\nu+1)}{\sqrt{\pi}\,\Gamma(\nu+\frac{1}{2})} \int_0^\pi f\left(\sqrt{y^2+h^2-2yh\cos\theta}\,\right) \sin^{2\nu}\theta\, d\theta$$

The author was partially supported by INTAS, grant 00-136, and RFBR, grant 02-01-312. He is indebted to S.I. Pohozhaev and A.L. Skubacheskii for their attention to this work.

be the Bessel operator acting with respect to y and its corresponding generalized translation operator. Also, let $j_\nu(y) \stackrel{\text{def}}{=} (2/y)^\nu \Gamma(\nu+1) J_\nu(y)$ be the normalized (in the uniform sense) Bessel function of the first kind.

We investigate the case of the positive parameter at the singularity of the Bessel operator, so it is assumed that $\nu > -\frac{1}{2}$.

In what follows, $\mathbb{R}^{n+1}_+$ denotes the half-space $\{\xi \in \mathbb{R}^n; \eta > 0\}$, $B_+(A)$ denotes the semi-ball $\{|\xi|^2 + \eta^2 < a^2; \eta > 0\}$, and $S_+(A)$ denotes the hemisphere $\{|\xi|^2 + \eta^2 = A^2; \eta > 0\}$.

26.3 Linear Case

We consider the problem

$$\sum_{j=1}^n \frac{\partial^2 u}{\partial x_j^2} + B_y u + \frac{\partial^2 u}{\partial z^2} = 0, \quad x \in \mathbb{R}^n, \ y \geq 0, \ z > 0, \tag{26.1}$$

$$u|_{z=0} = \varphi(x,y), \quad x \in \mathbb{R}^n, \ y \geq 0, \tag{26.2}$$

$$\frac{\partial u}{\partial y}\bigg|_{y=0} = 0, \quad x \in \mathbb{R}^n, \ z > 0. \tag{26.3}$$

Here φ is continuous and bounded.

Let us denote $\dfrac{2\Gamma(\frac{n+k}{2}+1)}{\pi^{\frac{n+1}{2}}\Gamma(\frac{k+1}{2})}$ as C and define on $\mathbb{R}^n \times (0,+\infty) \times (0,+\infty)$ the following function:

$$u(x,y,z) \stackrel{\text{def}}{=} Cz \int\limits_0^\infty \int\limits_{\mathbb{R}^n} \frac{\eta^k}{(|\xi|^2 + \eta^2 + z^2)^{\frac{n+k}{2}+1}} T_\eta^y \varphi(x-\xi, \eta) \, d\xi \, d\eta. \tag{26.4}$$

The function u is well defined because integral (26.4) converges absolutely for any (x,y,z) from $\mathbb{R}^n \times (0,+\infty) \times (0,+\infty)$ (since $|T_\eta^y \varphi(\cdot)| \leq \sup |\varphi|$).

The following assertion holds.

Theorem 1. *Function (26.4) is the unique bounded solution of (26.1)–(26.3).*

A brief scheme of the principal steps of the proof is as follows.

First we substitute (26.4) into (26.1) in order to prove that (26.4) satisfies (26.1) formally. Then we have to prove that the formal substitution is legitimate. To do that, we consider

$$\int\limits_0^\infty \int\limits_{\mathbb{R}^n} \eta^k \frac{az^2 + b|\xi|^2 + c\eta^2}{(|\xi|^2 + \eta^2 + z^2)^{\frac{n+k}{2}+3}} T_\eta^y \varphi(x-\xi, \eta) \, d\xi \, d\eta$$

(formal differentiation above leads to terms like this); its absolute value does not exceed $\mathrm{const} \int\limits_0^\infty \int\limits_{\mathbb{R}^n} (|\xi|^2 + \eta^2 + z^2)^{-(\frac{n}{2}+2)} d\xi \, d\eta$. The latter integral converges because $z > 0$. Thus, $u(x,y,z)$ is a classical solution of (26.1).

Further, the function $T_\eta^y f(\eta)$ is even with respect to variable y (see, e.g., [1], p. 35), therefore, (26.4) also satisfies condition (26.3).

Now let us prove that condition (26.2) is fulfilled too. Let $x_0 \in \mathbb{R}^n$, $y_0 > 0$. Then for any positive z,

$$u(x_0, y_0, z) - \varphi(x_0, y_0)$$

$$= C \int\limits_0^\infty \int\limits_{\mathbb{R}^n} \frac{z\eta^k}{\left(|\xi|^2 + \eta^2 + z^2\right)^{\frac{n+k}{2}+1}} \left[T_{y_0}^\eta \varphi(x_0 - \xi, y_0) - \varphi(x_0, y_0)\right] d\xi d\eta$$

$$= C\left(\int\limits_{B_+(A)} + \int\limits_{\mathbb{R}_+^{n+1} \setminus B_+(A)} \right) \overset{\text{def}}{=} I_1 + I_2,$$

where A is a positive parameter.

Let $\varepsilon > 0$. First we choose A so large that $|I_2| < \varepsilon/2$ for any positive z. It is possible because of convergence of the whole integral and boundedness of φ. Now we fix A and consider I_1. $T_{y_0}^{z\eta}\varphi(x_0 - z\xi, y_0) \overset{z \to +0}{\longrightarrow} \varphi(x_0, y_0)$ uniformly with respect to $(\xi, \eta) \in B_+(A)$, so one can take z_0 so small that $|I_1| < \varepsilon/2$ for any $z \in (0, z_0)$, $(\xi, \eta) \in B_+(A)$.

Finally, $u(x, y, z)$ is bounded, therefore, by the maximum principle (see, e.g., [4]), there are no more bounded solutions of (26.1)–(26.3).

The following assertion is the main result about the behavior of $u(x, y, z)$ as $z \to \infty$ is

Theorem 2. *Let* $l \in (-\infty, +\infty), x \in \mathbb{R}^n, y \geq 0$. *Then*

$$u(x, y, z) \overset{z \to \infty}{\longrightarrow} l \iff \lim_{r \to \infty} \frac{n+k+1}{r^{n+k+1}} \int\limits_{B_+(r)} \eta^k \varphi(\xi, \eta) d\xi d\eta = \frac{\pi^{\frac{n}{2}} \Gamma\left(\frac{k+1}{2}\right)}{\Gamma\left(\frac{n+k+1}{2}\right)} l.$$

The proof consists of the following main steps.

We start our consideration from the initial value of the solution and use a corollary of the Wiener–Tauber theorem (see, e.g., [5], 1003–1004]) to prove that the limiting relation

$$\lim_{r \to \infty} \frac{1}{r^{n+k+1}} \int\limits_{B_+(r)} \eta^k \varphi(\xi, \eta) \, d\xi \, d\eta = 0 \qquad (26.5)$$

is a necessary condition for the limiting relation

$$u(0, 0, z) \overset{z \to \infty}{\longrightarrow} 0. \qquad (26.6)$$

To prove that (26.5) implies (26.6), we represent $u(0, 0, z)$ as

$$\tilde{C} \int\limits_0^\infty \frac{\rho^{n+k+2} g(z\rho) d\rho}{\left(\rho^2 + 1\right)^{\frac{n+k}{2}+2}} = \tilde{C}\left(\int\limits_0^A + \int\limits_A^\infty \right) \overset{\text{def}}{=} I_1 + I_2,$$

where A is a positive parameter and

$$g(r) = \frac{C}{r^{n+k+1}} \int\limits_{B_+(r)} \eta^k \varphi(\xi, \eta) d\xi d\eta.$$

Then I_1 and I_2 are estimated similarly to the proof of Theorem 1. This proves Theorem 2 for $x = y = l = 0$.

To extend it for arbitrary $x = a \in \mathbb{R}^n$, $y = b > 0$, we represent $u(a, b, z)$ as $v(0, 0, z)$, where $v(x, y, z)$ is the bounded solution of (26.1)–(26.3) with the boundary-value function $\psi(x, y) \stackrel{\text{def}}{=} T_y^b \varphi(a - x, y)$. Then we use the result of [6], that for any (a, b) from $\mathbb{R}_+^{n+1}$,

$$\frac{1}{r^{n+k+1}} \left| \int\limits_{B_+(r)} \eta^k \psi(\xi, \eta) \, d\xi \, d\eta - \int\limits_{B_+(r)} \eta^k \varphi(\xi, \eta) d\xi \, d\eta \right| \xrightarrow{r \to \infty} 0.$$

The final step of the proof is its extension for the case $l \neq 0$: we consider problem (26.1)–(26.3) with the boundary-value function $\psi(x, y) - l$ and use the linearity of the equation and the uniqueness proved in Theorem 1.

26.4 Nonlinear Case

Hereafter we will denote $\Delta_x + B_y$ by Δ_B. Consider the equation

$$\Delta_B u + g(u)|\nabla u|^2 + \frac{\partial^2 u}{\partial z^2} = 0, \quad x \in \mathbb{R}^n, \ y \geq 0, \ z > 0 \qquad (26.7)$$

with a continuous coefficient g.

The following assertion is valid.

Theorem 3. *There exists a unique bounded solution of (26.7), (26.2), and (26.3).*

We use the method of proof explained in [7]: on $\mathbb{R}^1$ we define the function

$$f(s) \stackrel{\text{def}}{=} \int\limits_0^s e^{\int_0^x g(\tau)d\tau} \, dx, \qquad (26.8)$$

which is strictly monotone and satisfies $g(s) = f''(s)/f'(s)$.

Let us denote $f[\varphi(x, y)]$ by $\psi(x, y)$. Then $\psi(x, y)$ is also continuous and bounded, hence (see Theorem 1), there exists a unique bounded solution of the problem

$$\Delta_B v + \frac{\partial^2 v}{\partial z^2} = 0, \quad x \in \mathbb{R}^n, \ y \geq 0, \ z > 0,$$

$$v\big|_{z=0} = \psi(x, y), \quad x \in \mathbb{R}^n, \ y \geq 0,$$

$$\frac{\partial v}{\partial y}\bigg|_{y=0} = 0, \quad x \in \mathbb{R}^n, \ z > 0.$$

Since f is strictly monotone, on $[f(m), f(M)]$ we can define the function f^{-1} with values in $[m, M]$. Denote $f^{-1}[v(x, y, z)]$ by $u(x, y, z)$. Then direct substitution shows that u satisfies (26.7), (26.2), and (26.3). The proof of uniqueness uses the strict positivity of f'.

The investigation of the behavior of the solution of problem (26.7), (26.2), and (26.3) as $z \to \infty$ is also based on the ansatz (26.8): function $v(x, y, z) \overset{\text{def}}{=} f[u(x, y, z)]$ is introduced and Theorem 2 is applied. It yields the following assertion.

Theorem 4. *Let $A \in (-\infty, +\infty)$, $u(x, y, z)$ be the bounded solution of problem (26.7), (26.2), and (26.3). Then for any (x, y) in $\overline{\mathbb{R}^{n+1}_+}$,*

$$u(x, y, z) \overset{z \to \infty}{\longrightarrow} A \iff \frac{(n + k + 1)\Gamma\left(\frac{n+k+1}{2}\right)}{\pi^{\frac{n}{2}}\Gamma\left(\frac{k+1}{2}\right)r^{n+k+1}} \int_{B_+(r)} \eta^k f[\varphi(\xi, \eta)]d\xi d\eta \overset{r \to \infty}{\longrightarrow} f(A),$$

where f is defined by (26.8).

26.5 Singular Coefficient at the Nonlinearity

Let $\alpha > -1$; we will look for positive solutions of the equation

$$\Delta_B u + \frac{\partial^2 u}{\partial z^2} + \frac{\alpha}{u}|\nabla u|^2 = 0, \quad x \in \mathbb{R}^n, \ y \geq 0, \ z > 0. \tag{26.9}$$

On $(0, \infty)$ we define the function $f(s)$ as $s^{\alpha+1}$. Then $f'(s) > 0$ on $(0, \infty)$ and $f''(s)/f'(s) = \alpha/s$.

Acting as in the previous case (see also [7]), we arrive at the following assertion.

Theorem 5. *Let $\varphi(x, y)$ be a nonzero, continuous, bounded, and non-negative function. Then there exists a unique bounded positive solution of problem (26.9), (26.2), and (26.3).*

Theorem 6. *Let $A \geq 0$, let $\varphi(x, y)$ be a continuous, bounded, nonnegative, and nonzero function, and let $u(x, y, z)$ be the bounded positive solution of problem (26.9), (26.2), and (26.3). Then for any (x, y) from $\overline{\mathbb{R}^{n+1}_+}$,*

$$u(x, y, z) \overset{z \to \infty}{\longrightarrow} A \text{ if and only if }$$

$$\frac{(n + k + 1)\Gamma\left(\frac{n+k+1}{2}\right)}{\pi^{\frac{n}{2}}\Gamma\left(\frac{k+1}{2}\right)r^{n+k+1}} \int_{B_+(r)} \eta^k \varphi^{\alpha+1}(\xi, \eta)d\xi d\eta \overset{r \to \infty}{\longrightarrow} A^{\alpha+1}.$$

In the same way, theorems on the unique solvability and the stabilization of the solution are proved for the case when $g(s) = \frac{C}{s^\beta}$, $\beta \in (0, 1)$, in equation (26.7). In this case $f(s)$ is defined as in Section 26.4, i.e.,

$$f(s) = \int\limits_0^s e^{\frac{C}{1-\beta}x^{1-\beta}} dx. \tag{26.10}$$

Thus, for any real C and any $\beta \in (0,1)$, Theorem 5 is valid for problem (26.7), (26.2), and (26.3) and Theorem 4 (with f defined by (26.10)) is valid for the bounded positive solution of that problem.

The proofs are the same as above apart with the exception of the following point. We have to prove that the constructed solution u is positive (which in the case of equation (26.9) follows directly from the explicit form of f). If $m \overset{\text{def}}{=} \inf \varphi > 0$, then the required positiveness is evident. Let $m = 0$. Then $f(m) = 0$ too (see (26.10)), hence, $v(x,y,z) > f(m)$, by virtue of the positiveness of v. Since f is strictly monotone (note that (26.10) is a particular case of (26.8)), this means that $u(x,y,z) > m = 0$.

References

1. I.A. Kipriyanov, *Singular Elliptic Boundary Problems*, Nauka, Moscow, 1997 (Russian).

2. M. Kardar, G. Parisi, and Y.-C. Zhang, Dynamic scaling of growing interfaces, *Phys. Rev. Lett.* **56** (1986), 889–892.

3. E. Medina, T. Hwa, M. Kardar, and Y.-C. Zhang, Burgers equation with correlated noise: renormalization group analysis and applications to directed polymers and interface growth, *Phys. Rev.* **A39** (1989), 3053–3075.

4. A. Huber, On the uniqueness of generalized axially symmetric potentials, *Ann. of Math.* **60** (1954), 351–358.

5. N. Dunford and J.T. Schwartz, *Linear Operators, Part 2*, Interscience Publishers, New York, 1963.

6. A.B. Muravnik, On stabilization of solution of a singular problem, in *Boundary Value Problems for Nonclassical Equations of Mathematical Physics,* Institute of Mathematics of the Siberian Division USSR Academy of Sciences, 1987, 99–104 (Russian).

7. V.N. Denisov and A.B. Muravnik, On stabilization of the solution of the Cauchy problem for quasilinear parabolic equations, *Differential Equations* **38** (2002), 369–374.

27 New Zonal, Spectral Solutions for Compressible Navier–Stokes Partial Differential Equations

Adriana Nastase

27.1 Introduction

The paper proposes new zonal, spectral solutions for the partial differential equations (PDE) of the three-dimensional stationary, compressible Navier–Stokes layer (NSL), without any simplifications, as given in [1] and [2]. These solutions are useful for the computation of the flow over flattened, flying configurations (FC). Let us introduce a new spectral coordinate, namely

$$\eta = (x_3 - Z(x_1, x_2))/\delta(x_1, x_2). \tag{27.1}$$

Here $Z(x_1, x_2)$ is the equation of the FC, $\delta(x_1, x_2)$ is the thickness of the NSL, and the coordinate x_3 is measured perpendicular to the planform of the FC. The spectral forms of the axial, lateral, and vertical velocity components u_δ, v_δ, and w_δ, of the density function $R = \ln \rho$, and of the absolute temperature T and their nine boundary conditions, at the NSL-edge ($\eta = 1$), are (see [3]–[6])

$$u_\delta = u_e \sum_{i=1}^{N} u_i \eta^i, \quad v_\delta = v_e \sum_{i=1}^{N} v_i \eta^i, \quad w_\delta = w_e \sum_{i=1}^{N} w_i \eta^i, \tag{27.2a,b,c}$$

$$R = R_w + (R_e - R_w) \sum_{i=1}^{N} r_i \eta^i, \quad T = T_w + (T_e - T_w) \sum_{i=1}^{N} t_i \eta^i. \tag{27.2d,e}$$

The spectral forms (27.2a–e) automatically satisfy the nonslip conditions at the FC wall ($\eta = 0$). The boundary conditions at the NSL-edge ($\eta = 1$) are here written in an original explicit form (see [3]–[6]), namely

$$u_{N-2} = \alpha_{0,N-2} + \sum_{i=1}^{N-3} \alpha_{i,N-2}\, u_i, \quad v_{N-2} = \alpha_{0,N-2} + \sum_{i=1}^{N-3} \alpha_{i,N-2}\, v_i, \tag{27.3a,b}$$

$$u_{N-1} = \alpha_{0,N-1} + \sum_{i=1}^{N-3} \alpha_{i,N-1}\, u_i, \quad v_{N-1} = \alpha_{0,N-1} + \sum_{i=1}^{N-3} \alpha_{i,N-1}\, v_i,$$

$$(27.3\text{c,d})$$

$$u_N = \alpha_{0,N} + \sum_{i=1}^{N-3} \alpha_{i,N}\, u_i, \quad v_N = \alpha_{0,N} + \sum_{i=1}^{N-3} \alpha_{i,N}\, v_i,$$

$$(27.3\text{e,f})$$

$$w_N = \gamma_{0,N} + \sum_{i=1}^{N-1} \gamma_{i,N}\, w_i, \quad \sum_{i=1}^{N} r_i = 1, \quad \sum_{i=1}^{N} t_i = 1.$$

$$(27.3\text{g,h,i})$$

Here R_w and T_w are the prescribed values of R and T at the wall and u_e, v_e, w_e, R_e, and T_e are the values of u, v, w, R, and T at the NSL-edge, derived from an inviscid reinforced potential solver used here as outer flow (instead of the parallel undisturbed flow used by Prandtl in his boundary layer theory) over the modified FC, obtained after the solidification of the NSL on the FC, and u_i, v_i, w_i, r_i, and t_i are the free spectral coefficients, determined by satisfying the NSL's PDEs. The physical equation of ideal gas for the pressure p and an exponential law for the viscosity μ versus T are used further:

$$p \equiv R_g \rho T = R_g e^R \left[T_w + (T_e - T_w) \sum_{i=1}^{N} t_i \eta^i \right], \qquad (27.4)$$

$$\mu = \mu_\infty \left[\frac{T}{T_\infty} \right]^{n_1}. \qquad (27.5)$$

Here R_g is the universal gas constant, T_∞ is the absolute temperature of the undisturbed flow, and n_1 is the viscosity exponent.

27.2 Derivatives of the Velocity Components

The first and second derivatives of the velocity components u_δ, v_δ, and w_δ, which occur in all the NSL's PDEs, namely the impulse, temperature, and continuity, can easily and accurately be computed and are linear and homogeneous functions versus the spectral coefficients u_i, v_i, and w_i. The first and the second derivatives of u_δ and their proportionality coefficients a_{1i} and $a_{1i,j}$ are given here as an example:

$$\frac{\partial u_\delta}{\partial x_1} = \sum_{i=1}^{N} a_{1i}\, u_i, \quad \frac{\partial u_\delta}{\partial x_2} = \sum_{i=1}^{N} a_{2i}\, u_i, \quad \frac{\partial u_\delta}{\partial x_3} = \sum_{i=1}^{N} a_{3i}\, u_i,$$

$$(27.6\text{a,b,c})$$

$$a_{1i} = \left\{ \frac{\partial u_e}{\partial x_1} \eta - \frac{iu_e}{\delta} \left[\frac{\partial Z}{\partial x_1} + \eta \frac{\partial \delta}{\partial x_1} \right] \right\} \eta^{i-1}, \qquad (27.7a)$$

$$a_{2i} = \left\{ \frac{\partial u_e}{\partial x_2} \eta - \frac{iu_e}{\delta} \left[\frac{\partial Z}{\partial x_2} + \eta \frac{\partial \delta}{\partial x_2} \right] \right\} \eta^{i-1}, \qquad (27.7b)$$

$$a_{3i} = \frac{iu_e}{\delta} \eta^{i-1}, \qquad (27.7c)$$

$$\frac{\partial^2 u_\delta}{\partial x_1^2} = \sum_{i=1}^{N} a_{11,i} u_i, \quad \frac{\partial^2 u_\delta}{\partial x_2^2} = \sum_{i=1}^{N} a_{22,i} u_i, \quad \frac{\partial^2 u_\delta}{\partial x_3^2} = \sum_{i=1}^{N} a_{33,i} u_i,$$

$$(27.8a,b,c)$$

$$a_{11,i} = \left\{ \left[(i-1) \left(\frac{\partial \eta}{\partial x_1} \right)^2 + \eta \frac{\partial^2 \eta}{\partial x_1^2} \right] i u_e + 2i\eta \frac{\partial \eta}{\partial x_1} \frac{\partial u_e}{\partial x_1} + \eta^2 \frac{\partial^2 u_e}{\partial x_1^2} \right\} \eta^{i-2},$$

$$(27.9a)$$

$$a_{22,i} = \left\{ \left[(i-1) \left(\frac{\partial \eta}{\partial x_2} \right)^2 + \eta \frac{\partial^2 \eta}{\partial x_2^2} \right] i u_e + 2i\eta \frac{\partial \eta}{\partial x_2} \frac{\partial u_e}{\partial x_2} + \eta^2 \frac{\partial^2 u_e}{\partial x_2^2} \right\} \eta^{i-2},$$

$$(27.9b)$$

$$a_{33,i} = (i-1) \eta^{i-2} \left(\frac{\partial \eta}{\partial x_3} \right)^2 u_e, \qquad (27.9c)$$

$$\frac{\partial^2 u_\delta}{\partial x_1 \partial x_2} = \sum_{i=1}^{N} a_{12,i} u_i, \qquad (27.10a)$$

$$\frac{\partial^2 u_\delta}{\partial x_1 \partial x_3} = \sum_{i=1}^{N} a_{13,i} u_i, \quad \frac{\partial^2 u_\delta}{\partial x_2 \partial x_3} = \sum_{i=1}^{N} a_{23,i} u_i, \qquad (27.10b,c)$$

$$a_{12,i} = \left\{ \left[(i-1) \frac{\partial \eta}{\partial x_1} \frac{\partial \eta}{\partial x_2} + \eta \frac{\partial^2 \eta}{\partial x_1 \partial x_2} \right] i u_e + i\eta \left[\frac{\partial \eta}{\partial x_2} \frac{\partial u_e}{\partial x_1} + \frac{\partial \eta}{\partial x_1} \frac{\partial u_e}{\partial x_2} \right] \right.$$
$$\left. + \eta^2 \frac{\partial^2 u_e}{\partial x_1 \partial x_2} \right\} \eta^{i-2}, \qquad (27.11a)$$

$$a_{13,i} = \left\{ \left[(i-1) \frac{\partial \eta}{\partial x_1} \frac{\partial \eta}{\partial x_3} + \eta \frac{\partial^2 \eta}{\partial x_1 \partial x_3} \right] u_e + \eta \frac{\partial \eta}{\partial x_3} \frac{\partial u_e}{\partial x_1} \right\} i \eta^{i-2},$$

$$(27.11b)$$

$$a_{23,i} = \left\{ \left[(i-1) \frac{\partial \eta}{\partial x_2} \frac{\partial \eta}{\partial x_3} + \eta \frac{\partial^2 \eta}{\partial x_2 \partial x_3} \right] u_e + \eta \frac{\partial \eta}{\partial x_3} \frac{\partial u_e}{\partial x_2} \right\} i \eta^{i-2}.$$

$$(27.11c)$$

The first and the second derivatives of v_δ and w_δ are obtained in the same manner. All these derivatives are linear and homogeneous functions versus the velocity's spectral coefficients. If now the spectral velocities given in the formulas (27.2a-c) and the first and second derivatives of these velocities computed above in the formulas (27.5a-c), (27.8a-c) and (27.10a-c) are introduced in the PDEs of impulse, the following quadratical algebraic system (QAS), with slightly variable coefficients, is obtained:

$$
\sum_{i=1}^{N-3} u_i \left[\sum_{j=1}^{N-3} \left(\bar{A}_{ijk}^{(1)} u_j + \bar{B}_{ijk}^{(1)} v_j \right) + \sum_{j=1}^{N-1} \bar{C}_{ijk}^{(1)} w_j \right]
$$

$$
= \bar{D}_k^{(1)} + \sum_{i=1}^{N-3} \left(\bar{A}_{ik}^{(1)} u_i + \bar{B}_{ik}^{(1)} v_i \right) + \sum_{i=1}^{N-1} \bar{C}_{ik}^{(1)} w_i, \qquad (27.12a)
$$

$$
\sum_{i=1}^{N-3} v_i \left[\sum_{j=1}^{N-3} \left(\bar{A}_{ijk}^{(2)} u_j + \bar{B}_{ijk}^{(2)} v_j \right) + \sum_{j=1}^{N-1} \bar{C}_{ijk}^{(2)} w_j \right]
$$

$$
= \bar{D}_k^{(2)} + \sum_{i=1}^{N-3} \left(\bar{A}_{ik}^{(2)} u_i + \bar{B}_{ik}^{(2)} v_i \right) + \sum_{i=1}^{N-1} \bar{C}_{ik}^{(2)} w_i, \qquad (27.12b)
$$

$$
\sum_{i=1}^{N-3} w_i \left[\sum_{j=1}^{N-3} \left(\bar{A}_{ijk_1}^{(3)} u_j + \bar{B}_{ijk_1}^{(3)} v_j \right) + \sum_{j=1}^{N-1} \bar{C}_{ijk_1}^{(3)} w_j \right]
$$

$$
= \bar{D}_{k_1}^{(3)} + \sum_{i=1}^{N-3} \left(\bar{A}_{ik_1}^{(3)} u_i + \bar{B}_{ik_1}^{(3)} v_i \right) + \sum_{i=1}^{N-1} \bar{C}_{ik_1}^{(3)} w_i. \qquad (27.12c)
$$

The coefficients of this QAS depend on the density, temperature, viscosity and pressure, which are computed by using the continuity's and the temperature's PDE, an exponential law and the physical equation of gas, as given below.

27.3 Determination of the Spectral Coefficients of the Density Function and of the Temperature

The continuity equation is written in an original form by using the density function $R = ln\rho$. This equation, which is non-linear in ρ, is linear in R. If the relations (27.2a–d) are introduced in the continuity equation and the relation (27.3h) and the collocation method are used, the spectral coefficients r_i of R are obtained only as functions of the velocity's spectral coefficients u_i, v_i, w_i , by solving a linear algebraic system, namely

$$
\sum_{i=1}^{N} g_{ip} r_i = \gamma_p \quad (p = 1, 2, \ldots, N), \qquad (27.13)
$$

because the coefficients g_{ip} and γ_p depend only on the velocity's spectral coefficients. Similarly, if the relations (27.2a–e) are used, the viscosity μ, computed with the exponential law (27.5), and the pressure p, obtained from the physical equation of gas (27.4), are introduced in the temperature's PDE, the relation (27.3i) and the collocation method are also used, the spectral coefficients t_i of the absolute temperature T are expressed only as functions of the spectral coefficients of the velocity by solving of a transcendental algebraic system, namely:

$$\sum_{i=1}^{N} h_{ip} t_i + h_{0p} \left(T^{n_1}\right)_p = \theta_p \quad (p = 1, 2, ..., N), \qquad (27.14)$$

because the coefficients h_{ip}, h_{0p}, and θ_p depend also only on the velocity's spectral coefficients.

27.4 An Iterative Method to Solve a Quadratic Algebraic System

Here we propose our own method, which reduces the solving of QAS of impulse (27.12a–c) to the iterative solving of a cascade of linear algebraic systems (LAS). For this purpose, we make the notation

$$x_i = \begin{cases} u_i, & 1 \le i \le N - 3, \\ v_i, & N - 2 \le i \le 2N - 6, \\ w_i, & 2N - 5 \le i \le 3N - 7, \end{cases}$$

and, for the coefficients of (27.12a,b,c),

$$a_{ijk}^{(1)} = \begin{cases} \bar{A}_{ijk}^{(1)} \\ \bar{B}_{ijk}^{(1)} \\ \bar{C}_{ijk}^{(1)} \end{cases}, \quad a_{ijk}^{(2)} = \begin{cases} \bar{A}_{ijk}^{(2)} \\ \bar{B}_{ijk}^{(2)} \\ \bar{C}_{ijk}^{(2)} \end{cases}, \quad a_{ijk}^{(3)} = \begin{cases} \bar{A}_{ijk_1}^{(3)} \\ \bar{B}_{ijk_1}^{(3)} \\ \bar{C}_{ijk_1}^{(3)} \end{cases},$$

where the top line is for $1 \le k \le N - 3$, the second one for $N - 2 \le k \le 2N - 6$, and the third one for $2N - 5 \le k \le 3N - 7$, with $k_1 = 3N - 7 - 2k$.

The impulse equations can be written in the classical form of a QAS, namely

$$\sum_{i=1}^{M} \left[\sum_{j=1}^{M} a_{ij}^{(k)} x_i x_j + 2a_{i,M+1}^{(k)} x_i \right] + a_{M+1,M+1}^{(k)} = 0. \qquad (27.15)$$

Here the coefficients are $a_{ij}^{(k)} = \left(a_{ijk}^{(1)}, a_{ijk}^{(2)}, a_{ijk}^{(3)}\right)^T$ (i.e. the transposed matrix). Furthermore, with the notation

$$X_1 = x_1^2, \quad X_2 = x_2^2, \quad ..., \quad X_M = x_M^2 \qquad (27.16)$$

introduced in the QAS (27.15), we formally obtain the linear algebraic system

$$\sum_{i=1}^{M} A_{ii}^{(k)} X_i = R_i. \tag{27.17}$$

Here the free term R_k on the right-hand side of each k-th equation ($k = 1, 2, \ldots, M$) of the LAS (27.17) contains the free term, the linear terms, and the mixed products of the form $x_i x_j$ ($i \neq j$) of the k-th equation of the QAS (27.12a–c). This LAS has a unique solution, if the characteristic determinant is not zero. In the first iteration, the values of R_i are guessed and the new values of X_i are obtained by solving the LAS (27.17). The new values of x_i, obtained by using the relations (27.16), are used for the actualization of the values of R_i. The values of X_i, in the second iteration, are obtained by solving the LAS (27.17) with actualized coefficients R_i. In conclusion, the proposed hybrid numerical-analytic zonal, spectral solutions for the NSL's PDEs have the following advantages: they are more accurate and need less computer time than the full-numerical methods, because they do not need grids, the derivatives of the velocities are exactly computed, and the equations are exactly satisfied, in an arbitrarily chosen number of points. These NSL solutions are reinforced, due to the outer potential flow, because they satisfy the asymptotic behaviors along their singular lines like leading edges, junction lines wing/fuselage, wing/leading edge's flaps, and at infinity. For the supersonic flow, the conditions on the characteristic surface are also satisfied.

References

1. A.D. Young, *Boundary Layers*, Blackwell, London, 1989.

2. H. Schlichting, *Boundary Layer Theory*, McGraw-Hill, 1979.

3. A. Nastase, Spectral solutions for the Navier-Stokes equations and the shape's optimal design, in *Proc. of ECCOMAS 2000*, Barcelona, Spain.

4. A. Nastase, Zonal, spectral solutions for Navier-Stokes partial differential equations, Sonderheft GAMM 2001, ETH Zürich, Switzerland.

5. A. Nastase, Zonal, spectral solutions for the three-dimensional, compressible, Navier-Stokes partial differential equations, in *Proc. Internat. Symposium 100th Birthday of Elie Carafoli*, Bucharest, 2001.

6. A. Nastase, Comparison of zonal, spectral solutions for compressible boundary layer and Navier–Stokes equations, in *Integral Methods in Science and Engineering*, P. Schiavone, C. Constanda, and A. Mioduchowski (eds.), Birkhäuser, Boston, 2002.

28 Crack Problems and Boundary Variational Inequalities

David Natroshvili

28.1 Introduction

Signorini contact problems as well as crack problems have been extensively studied for homogeneous elastic bodies (see [1],[2],[3] and the references therein). The main tool to investigate these problems is the theory of spatial variational inequalities.

The purpose of the present paper is to formulate three-dimensional *crack problems* in terms of Signorini conditions and to reduce these to variational inequalities which live on the two–dimensional boundary of a domain occupied by a homogeneous elastic *anisotropic* body. Clearly this approach reduces the dimension of the problem by 1.

We show uniqueness and existence for the solution of the *boundary variational inequality* (BVI) and further formulate a Galerkin boundary element procedure along with an a priori error estimate.

A similar approach for the Signorini problem in the *isotropic* case was considered in [4], where a unilateral boundary value problem is reduced to a system consisting of a BVI and a singular boundary integral equation.

28.2 Formulation of the Problem and Uniqueness Theorem

Let an elastic homogeneous anisotropic body in the natural configuration occupy a bounded connected region $\overline{\Omega^+} = \Omega^+ \cup \partial\Omega^+$ of the three-dimensional space $\mathbb{R}^3$, surrounded by a closed surface $S := \partial\Omega^+$; By $\Omega^- = \mathbb{R}^3 \setminus \overline{\Omega^+}$ we denote the corresponding exterior domain which is supposed to be simply connected. Further we assume that the body in question contains an interior crack along an open simply connected surface $\Sigma_c \subset \Omega^+$ which is a part of some two–dimensional, closed, non-self-intersecting surface $\Sigma \subset \Omega^+$.

Denote $\Omega_c^+ := \Omega^+ \setminus \overline{\Sigma_c}$. Let S be divided into two disjoint parts S_D and S_N: $S = \overline{S_D} \cup \overline{S_N}$.

In what follows for simplicity we assume that S, Σ, S_D, S_N, Σ_c, ∂S_D, ∂S_N, and $\partial \Sigma_c$, are C^∞–smooth. Moreover, S_D, S_N, and Σ_c have positive measures.

Denote by $u = (u_1, u_2, u_3)^\top$ the displacement vector, by $\varepsilon_{ij}(u) = 2^{-1}(\partial_j u_i + \partial_i u_j)$ the strain tensor, by $\sigma_{ij}(u) = c_{kjpq}\varepsilon_{pq}(u)$ the stress tensor, by $T(\partial, n) := [c_{kjpq} n_j \partial_q]_{3 \times 3}$ the stress operator, and by $T(\partial, n)u$ a stress vector. The superscript $\top$ denotes transposition and $\partial := (\partial_1, \partial_2, \partial_3)$ with $\partial_j = \partial/\partial x_j$, $j = 1, 2, 3$. Here and in what follows we employ the Einstein summation convention. For the elasticity constants we assume the usual symmetry conditions $c_{kjpq} = c_{pqkj} = c_{jkpq}$.

In what follows $H^r(\mathcal{M})$ stands for the Sobolev–Slobodetski space on $\mathcal{M}$, while $\tilde{H}^r(S_N)$ (resp. $\tilde{H}^r(\Sigma_c)$) denotes the subspace of functions from $H^r(S)$ (resp. $H^r(\Sigma)$) with a support in $\overline{S_N}$ (resp. $\overline{\Sigma_c}$).

Problem (P). Find a vector-function $u \in [H^1(\Omega_c^+)]^3$ satisfying
(i) the equilibrium equations in Ω_c^+ (in the distributional sense)

$$A(\partial)u(x) = 0, \quad x \in \Omega_c^+, \tag{28.1}$$

(ii) the Dirichlet condition on S_D,

$$[u(x)]^+ = 0, \quad x \in S_D, \tag{28.2}$$

(iii) the Neumann condition on S_N,

$$[T(\partial_x, n(x))\, u(x)]^+ = g(x), \quad x \in S_N, \tag{28.3}$$

(iv) the Signorini type conditions on the crack surface Σ_c,

$$[u(x) \cdot n(x)]^+ - [u(x) \cdot n(x)]^- \le 0, \tag{28.4}$$

$$[T(\partial_x, n(x))u(x) \cdot n(x)]^+ = [T(\partial_x, n(x)]u(x) \cdot n(x)]^- \le 0, \tag{28.5}$$

$$[T(\partial_x, n(x))u(x) - n(x)\,(T(\partial_x, n(x))u(x) \cdot n(x))]^\pm = 0, \tag{28.6}$$

$$\left\langle [T(\partial_x, n(x))u(x) \cdot n(x))]^+, \ [u(x) \cdot n(x)]^+ - [u(x) \cdot n(x)]^- \right\rangle_{\Sigma_c} = 0, \tag{28.7}$$

where $A(\partial)$ is the matrix differential operator of linearized elastostatics: $A(\partial) = [A_{kp}(\partial)]_{3 \times 3}$, $A_{kp}(\partial) = c_{kjpq}\partial_j\,\partial_q$, the symbols $[\,\cdot\,]^\pm$ denote the traces on S and Σ_c, $n(x)$ is the outward normal vector to S and $\Sigma (\supset \Sigma_c)$, $\langle\,\cdot\,,\,\cdot\,\rangle_\mathcal{M}$ are duality brackets between the spaces $H^{-r}(\mathcal{M})$ and $\tilde{H}^r(\mathcal{M})$, and $g \in H^{-1/2}(S_N)$ is a given vector-function.

Equation (28.1) corresponds to the equilibrium state of the elastic body. Condition (28.2) shows that the body is fixed along the subsurface S_D, while (28.3) shows that the body is subjected to some assigned surface stresses on S_N.

The Signorini type conditions (called "ambiguous boundary conditions" in the original terminology of Signorini [1]) on the crack surface describe a generalized contact interaction along Σ_c without friction since the tangential stresses vanish on Σ_c due to (28.5). Condition (28.4) shows that there is no penetration along Σ_c, while (28.7) along with (28.4) and (28.6) implies that at each point of Σ_c either there is a gap where normal stresses vanish, or there is a contact (the contact zones on Σ_c are called *regions of coincidence* (cf. [1]). We have the following uniqueness

Theorem 1. *Problem (P) has at most one solution.*

28.3 Reduction to a Boundary Variational Inequality

Let us introduce the cones

$$K(\Omega_c) := \{u \in [H^1(\Omega_c)]^3 \; : \; [u]^+_{S_D} = 0 \text{ and } [u \cdot n]^+_{\Sigma_c} - [u \cdot n]^-_{\Sigma_c} \leq 0\},$$

$$K(S, \Sigma_c) := \{(\varphi', \varphi'') \in [\tilde{H}^{1/2}(S_N)]^3 \times [\tilde{H}^{1/2}(\Sigma_c)]^3 \; : \; \varphi'' \cdot n \leq 0 \text{ on } \Sigma_c\}.$$

Denote by $\Gamma(\cdot)$ the fundamental matrix of the operator $A(\partial)$ (see, e.g., [5] and [6]). Let $\mathcal{M}$ be some two-dimensional smooth manifold (with or without boundary) and introduce the single-layer and double-layer potentials

$$V_{\mathcal{M}}(h)(x) := \int_{\mathcal{M}} \Gamma(x - y)\, h(y)\, dS_y, \quad x \in \mathbb{R}^3 \setminus \overline{\mathcal{M}},$$

$$W_{\mathcal{M}}(h)(x) := \int_{\mathcal{M}} [T(\partial_y, n(y))\, \Gamma(y - x)]^\top h(y)\, dS_y, \quad x \in \mathbb{R}^3 \setminus \overline{\mathcal{M}}.$$

The properties of these potentials and the corresponding boundary integral and pseudodifferential operators

$$\mathcal{H}_{\mathcal{M}}\, h(x) := \int_{\mathcal{M}} \Gamma(x - y)\, h(y)\, dS_y,$$

$$\mathcal{K}^*_{\mathcal{M}}\, h(x) := \int_{\mathcal{M}} [T(\partial_y, n(y))\Gamma(y - x)]^\top h(y)\, dS_y,$$

$$\mathcal{K}_{\mathcal{M}}\, h(x) := \int_{\mathcal{M}} T(\partial_x, n(x))\Gamma(x - y)\, h(y)\, dS_y,$$

$$\mathcal{L}_{\mathcal{M}}\, h(x) := \lim_{\Omega^\pm \ni z \to x \in \mathcal{M}} T(\partial_z, n(x)) \int_{\mathcal{M}} [T(\partial_y, n(y))\Gamma(y - z)]^\top h(y)\, dS_y$$

in the Hölder ($C^{k+\alpha}$), Bessel potential (H^s_p) and Besov ($B^s_{p,q}$) spaces are studied in, e.g., [5], [6], [7], [8], [9], [10], [11].
Further let

$$\mathcal{N}_1\, h := [\mathcal{H}_S]^{-1}[-2^{-1}I + \mathcal{K}^*_S]\, h, \quad h \in [H^{1/2}(S)]^3,$$

$$\mathcal{N}_2\, h := [\mathcal{H}_S]^{-1}[W_{\Sigma_c}(h)]_S, \quad h \in [\tilde{H}^{1/2}(\Sigma_c)]^3,$$

$$\mathcal{N}_3\, h := [TW_S(h) - TV_S(\mathcal{N}_1\, h)]_{\Sigma_c}, \quad h \in [H^{1/2}(S)]^3,$$

$$\mathcal{N}_4\, h := \mathcal{L}_{\Sigma_c} h - [TV_S(\mathcal{N}_2\, h)]_{\Sigma_c}, \quad h \in [\tilde{H}^{1/2}(\Sigma_c)]^3,$$

and

$$\mathcal{N} := \begin{bmatrix} \mathcal{N}_1 & \mathcal{N}_2 \\ \mathcal{N}_3 & \mathcal{N}_4 \end{bmatrix}_{6 \times 6},$$

where I is the unit 3×3 matrix, $[\cdot]_\mathcal{M}$ is the restriction operator to $\mathcal{M}$, and $[\mathcal{H}_S]^{-1}$ is the operator inverse to the invertible operator $\mathcal{H}_S : H^{-1/2}(S) \to H^{1/2}(S)$.

Theorem 2. *Let $\tilde{H}^s(S_N, \Sigma_c) := [\tilde{H}^s(S_N)]^3 \times [\tilde{H}^s(\Sigma_c)]^3$. The operator $\mathcal{N}$ has the mapping property*

$$\mathcal{N} : \tilde{H}^{1/2}(S_N, \Sigma_c) \to H^{-1/2}(S_N, \Sigma_c).$$

Moreover, let $\Phi = (\varphi', \varphi'')$, $\Psi = (\psi', \psi'') \in \tilde{H}^{1/2}(S_N, \Sigma_c)$. The bilinear form

$$P(\Phi, \Psi) := \langle \mathcal{N}_1\varphi' + \mathcal{N}_2\varphi'' , \psi' \rangle_{S_N} + \langle \mathcal{N}_3\varphi' + \mathcal{N}_4\varphi'' , \psi'' \rangle_{\Sigma_c}$$

is bounded and coercive in $\tilde{H}^{1/2}(S_N, \Sigma_c)$.

Let us consider the following *boundary variational inequality*: Find $\Phi = (\varphi', \varphi'') \in K(S, \Sigma_c)$ such that

$$P(\Phi, \Psi - \Phi) \geq \mathcal{F}(\Psi - \Phi) \quad \text{for all} \quad \Psi = (\psi', \psi'') \in K(S, \Sigma_c), \qquad (28.8)$$

where

$$\mathcal{F}(\cdot) = \langle g, \cdot \rangle_{S_N} \quad \text{with} \quad g \in H^{-1/2}(S_N) \qquad (28.9)$$

is a bounded linear functional on $\tilde{H}^{1/2}(S_N)$.

It can be shown that Problem (P) is equivalent to the BVI (28.8) in the following sense:

(i) if $\Phi = (\varphi', \varphi'') \in K(S, \Sigma_c)$ is a solution to (28.8), then the vector-function

$$u(x) = W_S(\varphi')(x) - V_S(\mathcal{N}_1\varphi' + \mathcal{N}_2\varphi'')(x) + W_{\Sigma_c}(\varphi'')(x), \quad x \in \Omega_c^+,$$

solves Problem (P);

(ii) if u solves Problem (P), then $\varphi' = [u]_S^+$ and $\varphi'' = [u]_{\Sigma_c}^+ - [u]_{\Sigma_c}^-$ is a solution of the BVI (28.8).

Theorem 3. *The BVI (28.8) is uniquely solvable.*

In view of the equivalence mentioned above, the following assertion holds.

Corollary 4. *Problem (P) is uniquely solvable.*

It can be shown that the BVI (28.8) is equivalent to the following *spatial variational inequality*: Find $u \in K(\Omega_c)$ such that

$$B(u, v - u) \geq \mathcal{F}([v]_{S_N}^+) \quad \text{for all} \quad v \in K(\Omega_c), \qquad (28.10)$$

where $\mathcal{F}(\cdot)$ is given by (28.9) and

$$B(u, v) := \int_{\Omega_c} c_{kjpq}\, \varepsilon_{kj}(u)\, \varepsilon_{pq}(v)\, dx.$$

In turn (28.10) is equivalent to the minimization problem on $K(\Omega_c)$ for the functional

$$J(u) := 2^{-1} B(u, u) - \mathcal{F}([u]_{S_N}^+).$$

Note that

$$B(u, u) \geq \delta_0 \int_{\Omega_c} \varepsilon_{kj}(u)\, \varepsilon_{kj}(u)\, dx$$

with some $\delta_0 = \text{const} > 0$.

28.4 Error Estimates for Galerkin Approximations

Assume that $\tilde{H}_h^{1/2}(S_N, \Sigma_c)$ is a finite dimensional subspace of $\tilde{H}^{1/2}(S_N, \Sigma_c)$ given by a boundary discretization, and let

$$K_h(S, \Sigma_c) := \{\Psi_h \in \tilde{H}_h^{1/2}(S_N, \Sigma_c) : \Psi_h \in K(S, \Sigma_c)\}$$

be a convex closed nonempty subset of $\tilde{H}_h^{1/2}(S_N, \Sigma_c)$.

An element $\Phi_h \in K_h(S, \Sigma_c)$ defines an *approximate solution* of the BVI (28.8) if

$$P(\Phi_h, \Psi_h - \Phi_h) \geq \mathcal{F}(\Psi_h - \Phi_h) \quad \text{for all} \quad \Psi_h \in K_h(S, \Sigma_c). \tag{28.11}$$

The existence and uniqueness theorems for the approximate solutions to the BVI (28.11) are due to convex analysis [12].

The convergence and asymptotic error estimates for a corresponding boundary element Galerkin approximation is given by the following

Theorem 5. *Let $\Phi \in K(S, \Sigma_c)$ be a solution to the BVI (28.8) and $\Phi_h \in K_h(S, \Sigma_c)$ be a solution to the BVI (28.11).*

Then we have the a priori error estimate

$$\|\Phi - \Phi_h\|_{\tilde{H}^{1/2}(S_N, \Sigma_c)}^2 \leq c_0 \inf_{\Psi_h \in K_h(S, \Sigma_c)} \{\|\Phi - \Psi_h\|_{\tilde{H}^{1/2}(S_N, \Sigma_c)}^2$$
$$+ P(\Phi, \Psi_h - \Phi) - \mathcal{F}(\Psi_h - \Phi)\},$$

where c_0 is a positive constant independent of Φ, Φ_h, and g.

References

1. G. Fichera, *Existence Theorems in Elasticity*, Handbuch der Physik, Bd. 6/2, Springer–Verlag, Heidelberg, 1973.

2. A.M. Khludnev and V.A. Kovtunenko, *Analysis of Cracks in Solids*, WIT Press, Southampton, Boston, 2000.

3. J.-L. Lions and G. Stampacchia, Variational inequalities, *Comm. Pure Appl. Math.* **20** (1976), 493–519.

4. H. Han, A boundary element procedure for the Signorini problem in three-dimensional elasticity, *Numerical Mathematics* **3** (1994), 104–117.

5. D. Natroshvili, *Investigation of boundary value and initial-boundary value problems of the mathematical theory of anisotropic elasticity and thermoelasticity by means of potential methods*, Doctoral Thesis, A. Razmadze Math. Inst., Tbilisi, 1984, 1–325 (Russian).

6. W. McLean, *Strongly Elliptic Systems and Boundary Integral Equations*, Cambridge University Press, Cambridge, 2000.

7. M. Costabel and W.L. Wendland, Strong ellipticity of boundary integral operators, *J. Reine Angew. Math.* **372** (1986), 34–63.

8. G. Eskin, *Boundary Value Problems for Elliptic Pseudodifferential Equations*, Amer. Math. Soc., Providence, R.I., 1981.

9. R. Duduchava, D. Natroshvili and E. Shargorodsky, Basic boundary value problems of thermoelasticity for anisotropic bodies with cuts, *Georgian Math. J.* **2** (1995), I, 123–140; II, 259–276.

10. D. Natroshvili, O. Chkadua and E. Shargorodsky, Mixed boundary value problems of the anisotropic elasticity, *Proc. I. Vekua Inst. Appl. Math. Tbilisi State Univ.* **39** (1990), 133–181.

11. V.D. Kupradze, T.G. Gegelia, M.O. Basheleishvili, and T.V. Burchuladze, *Three-Dimensional Problems of the Mathematical Theory of Elasticity and Thermoelasticity*, North–Holland, Amsterdam-New York-Oxford, 1979.

12. R. Glowinski, J.-L. Lions, and R. Tremolieres, *Numerical Analysis of Variational Inequalities*, North–Holland, Amsterdam, 1981.

29 Spline Approximations for Weakly Singular Volterra Integro-Differential Equations

Inga Parts and Arvet Pedas

29.1 Introduction

We consider a linear Volterra integro-differential equation

$$y'(t) = p(t)y(t) + q(t) + \int_0^t K(t,s)y(s)ds, \quad 0 \le t \le T, \qquad (29.1)$$

with $0 < T < \infty$, and with given initial condition

$$y(0) = y_0, \quad y_0 \in \mathbb{R} \equiv (-\infty, \infty). \qquad (29.2)$$

We assume that $p, q \in C^{m,\nu}(0,T]$, $K \in W^{m,\nu}(\Delta_T)$, where $m \in \mathbb{N} \equiv \{1, 2, \ldots\}$, and $\nu \in \mathbb{R}$, $\nu < 1$. Here $C^{m,\nu}(0,T]$, $m \in \mathbb{N}$, $\nu < 1$, is defined as the collection of all continuous functions $x : [0,T] \to \mathbb{R}$ which are m times continuously differentiable on $(0,T]$ and such that the estimate

$$\left| x^{(k)}(t) \right| \le c \begin{cases} 1 & \text{if } k < 1 - \nu, \\ 1 + |\log t| & \text{if } k = 1 - \nu, \\ t^{1-\nu-k} & \text{if } k > 1 - \nu \end{cases} \qquad (29.3)$$

holds with a constant $c = c(x)$ for all $t \in (0,T]$ and $k = 1, \ldots, m$. The set $W^{m,\nu}(\Delta_T)$, $m \in \mathbb{N}$, $\nu < 1$, $\Delta_T = \{(t,s) \in \mathbb{R}^2 : 0 \le t \le T, 0 \le s < t\}$, consists of m times continuously differentiable functions $K : \Delta_T \to \mathbb{R}$ satisfying

$$\left| \left(\frac{\partial}{\partial t} \right)^i \left(\frac{\partial}{\partial t} + \frac{\partial}{\partial s} \right)^j K(t,s) \right| \le c \begin{cases} 1 & \text{if } \nu + i < 0, \\ 1 + |\log(t-s)| & \text{if } \nu + i = 0, \\ (t-s)^{-\nu-i} & \text{if } \nu + i > 0, \end{cases} \qquad (29.4)$$

with a constant $c = c(K)$ for all $(t,s) \in \Delta_T$ and all non-negative integers i and j such that $i + j \le m$. It follows from (29.4) (with $i = j = 0$, $0 \le \nu < 1$) that the kernel $K(t,s)$ of equation (1) may possess a weak

---

This work was supported by the Estonian Science Foundation (Research Grant No. 4410).

singularity as $s \to t$. In the case $\nu < 0$, the kernel $K(t,s)$ is bounded on Δ_T, but its derivatives may be singular as $s \to t$. Often the kernel K of equation (1) has the form $K \equiv K_\alpha(t,s) = \kappa(t,s)(t-s)^{-\alpha}$, $0 < \alpha < 1$, or $K \equiv K_0(t,s) = \kappa(t,s)\log(t-s)$, where κ is an m times continuously differentiable function on $\overline{\Delta}_T = \{(t,s) : 0 \le s \le t \le T\}$. Clearly, $K_\alpha \in W^{m,\alpha}(\Delta_T)$ and $K_0 \in W^{m,0}(\Delta_T)$.

If $p, q \in C^{m,\nu}(0,T]$, $K \in W^{m,\nu}(\Delta_T)$, $m \in \mathbb{N}$, $\nu \in \mathbb{R}$, $\nu < 1$, $y_0 \in \mathbb{R}$, then it follows from [1,2] that the initial-value problem (1), (2) has a unique solution $y \in C^{m+1,\nu-1}(0,T]$. Moreover, the derivatives $y''(t), \ldots, y^{(m)}(t)$ of $y(t)$ are typically nonsmooth at $t = 0$ (see (29.3), $m \ge 2$) even if $p(t)$ and $q(t)$ are m-smooth on the whole interval $[0,T]$. Therefore it is quite complicated to construct high order approximation methods for the numerical solution of such equations (see, for example, [1–5]).

In the sequel we construct for the initial-value problem (1), (2) a collocation method using polynomial splines on various grids. We discuss the convergence rate of proposed schemes and present some numerical illustrations. The main results of the paper extend the corresponding results of [1,2] and are formulated in Theorems 1–3.

29.2 Collocation Method

For given $N \in \mathbb{N}$, let $\Pi_N = \{t_0, t_1, \ldots, t_N : 0 = t_0 < t_1 < \ldots < t_N = T\}$ be a partition (a grid) of the interval $[0,T]$ (for ease of notation we suppress the index N in $t_j = t_j^{(N)}$ indicating the dependence of the grid points on N). A sequence of partitions for $[0,T]$ is called quasi-uniform if there exists a constant θ independent of N so that

$$\max_{j=1,\ldots,N}(t_j - t_{j-1}) / \min_{j=1,\ldots,N}(t_j - t_{j-1}) \le \theta, \quad n \in \mathbb{N}. \tag{29.5}$$

If the grid points are given by

$$t_j = T(j/N)^r, \quad j = 0, 1, \ldots, N, \tag{29.6}$$

then Π_N is called a graded grid; in the present context the so-called grading exponent $r \in \mathbb{R}$ will always satisfy $r \ge 1$. It follows from (29.5) and (29.6) that for both the quasi-uniform grid and the graded grid

$$h_N \equiv \max_{j=1,\ldots,N}(t_j - t_{j-1}) \to 0 \quad \text{as} \quad N \to \infty. \tag{29.7}$$

For given integers $m \ge 0$ and $-1 \le d \le m - 1$, let $S_m^{(d)}(\Pi_N)$ be the spline space of piecewise polynomial functions on the grid Π_N:

$$S_m^{(d)}(\Pi_N) = \{u : u|_{[t_{j-1},t_j]} \equiv u_j \in \pi_m, \quad j = 1, \ldots, N;$$
$$u_j^{(k)}(t_j) = u_{j+1}^{(k)}(t_j), \quad 0 \le k \le d, \ j = 1, \ldots, N-1\}.$$

Here π_m denotes the set of polynomials of degree not exceeding m and $u|_{[t_{j-1},t_j]}$ is the restriction of u to the subinterval $[t_{j-1}, t_j]$. Note that the elements of $S_m^{(-1)}(\Pi_N) = \{u : u|_{[t_{j-1},t_j]} \in \pi_m, j = 1, \ldots, N\}$ may have jump discontinuities at the interior grid points $t_1, \ldots, t_{N-1}$.

Using the notation $y' = z$ and (2), we may rewrite equation (1) as a linear Volterra integral equation of the second kind with respect to z:

$$z(t) = f(t) + p(t) \int_0^t z(s)ds + \int_0^t K(t,s)\left(\int_0^s z(\tau)d\tau\right)ds, \qquad (29.8)$$

where

$$f(t) = q(t) + y_0 p(t) + y_0 \int_0^t K(t,s)ds, \quad t \in [0,T].$$

For given $m, N \in \mathbb{N}$ we look for an approximation v to the solution z of equation (29.8) in the space $S_{m-1}^{(-1)}(\Pi_N)$ determining $v \equiv v^{(N,m)} \in S_{m-1}^{(-1)}(\Pi_N)$ from the following conditions:

$$v_j(t_{jk}) = f(t_{jk}) + p(t_{jk}) \int_0^{t_{jk}} v(s)ds + \int_0^{t_{jk}} K(t_{jk}, s)\left(\int_0^s v(\tau)d\tau\right)ds,$$
$$k = 1, \ldots, m; \; j = 1, \ldots, N.$$
$$(29.9)$$

Here $v_j \equiv v|_{[t_{j-1},t_j]}$ is the restriction of v to $[t_{j-1}, t_j]$, $j = 1, \ldots, N$, and the collocation points

$$t_{jk} = t_{j-1} + \eta_k(t_j - t_{j-1}) \quad (k = 1, \ldots, m; \; j = 1, \ldots, N) \qquad (29.10)$$

are completely characterized by the given grid Π_N and by the collocation parameters $\eta_1, \ldots, \eta_m$ which do not depend on j and N and satisfy

$$0 \leq \eta_1 < \ldots < \eta_m \leq 1. \qquad (29.11)$$

Having determined the approximation v for $z = y'$, we can also determine the approximation u for y, the solution of the initial-value problem (1), (2), setting

$$u(t) = y_0 + \int_0^t v(s)ds, \qquad t \in [0,T]. \qquad (29.12)$$

Remark 1. The choice of collocation points (29.10) with $\eta_1 = 0$, $\eta_m = 1$ in (29.11) actually implies that the resulting collocation approximation v belongs to the smoother polynomial spline space $S_{m-1}^{(0)}(\Pi_N)$. Note also that $v \in S_{m-1}^{(-1)}(\Pi_N)$ implies that $u \in S_m^{(0)}(\Pi_N)$, and $v \in S_{m-1}^{(0)}(\Pi_N)$ implies that $u \in S_m^{(1)}(\Pi_N)$.

Remark 2. Conditions (29.9) form a system of equations whose exact form is determined by the choice of a basis in $S_{m-1}^{(-1)}(\Pi_N)$ (or in $S_{m-1}^{(0)}(\Pi_N)$

if $\eta_1 = 0$, $\eta_m = 1$). For instance, in each subinterval $[t_{j-1}, t_j]$ $(j = 1, \ldots, N)$ we may use the representation $v_j(t_{j-1} + \tau(t_j - t_{j-1})) = \sum_{k=1}^{m} c_{jk} L_k^{(m-1)}(\tau)$, $\tau \in [0, 1]$, where $L_k^{(m-1)}(\tau)$ denotes the kth Lagrange fundamental polynomial of degree $m - 1$ associated with the parameters $0 \le \eta_1 < \cdots < \eta_m \le 1$, that is $L_k^{(m-1)}(\tau) = \prod_{i \ne k}^{m} (\tau - \eta_i)/(\eta_k - \eta_i)$, $\tau \in [0, 1]$. The collocation conditions (29.9) then lead to a linear system of equations for the coefficients $c_{jk} \equiv c_{jk}^{(N)} = v_j(t_{jk})$, $k = 1, \ldots, m$; $j = 1, \ldots, N$.

Remark 3. Method (29.9), (29.12) where we have discretized the integral equation (29.8) is equivalent to the collocation method applied directly to the initial-value problem (29.1), (29.2). In the latter form the collocation method in a more particular case has been examined in [3]–[5].

Theorem 1. *Let $p, q \in C^{m,\nu}(0, T]$, $K \in W^{m,\nu}(\Delta_T)$, $m \in \mathbb{N}$, $\nu \in \mathbb{R}$, $\nu < 1$, $y_0 \in \mathbb{R}$, and assume that the underlying grid sequence (Π_N) satisfies (29.7). Then, for all sufficiently large $N \in \mathbb{N}$ and for every choice of parameters $0 \le \eta_1 < \cdots < \eta_m \le 1$ with $\eta_1 > 0$ or $\eta_m < 1$, the equations (29.12) and (29.9) determine unique approximations $u \in S_m^{(0)}(\Pi_N)$ and $v \in S_{m-1}^{(-1)}(\Pi_N)$ (with $v|_{[t_{j-1}, t_j]} = (u|_{[t_{j-1}, t_j]})'$, $j = 1, \ldots, N$) to the solution y of the initial-value problem (29.1), (29.2) and its derivative y', respectively. If $\eta_1 = 0$, $\eta_m = 1$, then $u \in S_m^{(1)}(\Pi_N)$ and $v = u' \in S_{m-1}^{(0)}(\Pi_N)$. The collocation error $e^{(k)} \equiv u^{(k)} - y^{(k)}$ for $k = 0$ and $k = 1$ satisfies*

$$\|e^{(k)}\|_\infty \le c \begin{cases} h_N^m & \text{for} \quad m < 1 - \nu, \\ h_N^m(1 + |\log h_N|) & \text{for} \quad m = 1 - \nu, \\ h_N^{1-\nu} & \text{for} \quad m > 1 - \nu. \end{cases}$$

Here h_N is given by (29.7), c is a constant not depending on N and

$$\|e^{(k)}\|_\infty = \max_{j=1,\ldots,N} \Big(\max_{t \in [t_{j-1}, t_j]} |u_j^{(k)}(t) - y^{(k)}(t)| \Big), \quad u_j = u|_{[t_{j-1}, t_j]}. \quad (29.13)$$

Theorem 2. *Let p, q, K and y_0 be subject to the conditions stated in Theorem 1. Moreover, assume that the underlying grid sequence (Π_N) is quasi-uniform (i.e. satisfies (29.5)). Then, for all sufficiently large $N \in \mathbb{N}$, in the notation of Theorem 1, the collocation error $e^{(k)} \equiv u^{(k)} - y^{(k)}$ for $k = 0$ and $k = 1$ satisfies the following estimates:*
  1) if $m < 2 - \nu - k$, then $\|e^{(k)}\|_\infty \le c N^{-m}$;
  2) if $m = 2 - \nu - k$, then $\|e^{(k)}\|_\infty \le c N^{-m}(1 + \log N)$;
  3) if $m > 2 - \nu - k$, then $\|e^{(k)}\|_\infty \le c N^{-(2-\nu-k)}$.
Here c is a constant not depending on N and $\|e^{(k)}\|_\infty$ is given by (29.13).

Theorem 3. *Let p, q, K and y_0 be subject to the conditions stated in Theorem 1. Moreover, assume that the underlying grid sequence (Π_N) is graded (i.e. satisfies (29.6)). Then, for all sufficiently large $N \in \mathbb{N}$, in the*

*notation of Theorem 1, the collocation error $e^{(k)} \equiv u^{(k)} - y^{(k)}$ for $k = 0$
and $k = 1$ satisfies the following estimates:*
  1) if $m < 2 - \nu - k$, then $\|e^{(k)}\|_\infty \le c\, N^{-m}$ for $r \ge 1$;
  2) if $m = 2 - \nu - k$, then

$$\|e^{(k)}\|_\infty \le c \begin{cases} N^{-m}(1 + \log N) & for \quad r = 1, \\ N^{-m} & for \quad r > 1; \end{cases}$$

3) if $m > 2 - \nu - k$, then

$$\|e^{(k)}\|_\infty \le c \begin{cases} N^{-r(2-\nu-k)} & for \quad 1 \le r < m/(2 - \nu - k), \\ N^{-m}(1 + \log N)^{1-k} & for \quad r = m/(2 - \nu - k), \\ N^{-m} & for \quad r > m/(2 - \nu - k). \end{cases}$$

Here c is a constant not depending on N and $\|e^{(k)}\|_\infty$ is given by (29.13).

29.3 Numerical Experiments

In this section we test the convergence behavior numerically. We consider
the initial-value problem (1), (2) where

$$p(t) = -1, \quad q(t) = \frac{7}{5}t^{\frac{2}{5}} + t^{\frac{7}{5}} + t^{\frac{9}{5}} \int_0^1 (1 - \tau)^{-\frac{3}{5}} \tau^{\frac{7}{5}} d\tau,$$

$$K(t, s) = -(t - s)^{-\frac{3}{5}}, \quad T = 1, \ y_0 = 0.$$

In this case the exact solution of problem (1), (2) is $y(t) = t^{7/5}$. We see that
$p, q \in C^{m,\nu}(0, T]$, $K \in W^{m,\nu}(\Delta_T)$ for $\nu = 3/5$ and for any $m \in \mathbb{N}$. This
problem is solved numerically by method (29.9) by using the corresponding
Lagrange basis for $S_{m-1}^{(-1)}(\Pi_N)$ (see Remark 2). All integrals occurring in
method (29.9) are evaluated analytically. Some of the results obtained are
presented in Table 1 for $m = 2$ and $N = 4, 8, 16, 32, 64$. In fact, in Table 1
for different grids Π_N^1, $\Pi_{N,5}^{(0)}$, $\Pi_{N,5}^{(1)}$, Π_N^r ($r > 1$), the error (compare (29.13))

$$\varepsilon_N^{(l)} = \{\max |u^{(l)}(\tau_{jk}) - y^{(l)}(\tau_{jk})| : k = 1, \ldots, 9; j = 1, \ldots, N\},$$

and the ratio $\varrho_N^{(l)} = \varepsilon_{N/2}^{(l)}/\varepsilon_N^{(l)}$ for $l = 0$ and $l = 1$ are given. In order to
calculate the error (29.13) we have taken $t = \tau_{jk}$ where $\tau_{jk} = t_{j-1} + k(t_j - t_{j-1})/10$, $k = 1, \ldots, 9$; $j = 1, \ldots, N$. Further, Π_N^r ($r \ge 1$) is the graded
grid Π_N with grid points (29.6) and $\Pi_{N,5}^{(0)}$, $\Pi_{N,5}^{(1)}$ are the quasi-uniform grids
Π_N, with $\theta = 5$ in (29.5), defined as follows: $\Pi_{N,\theta}^{(l)} = \{t_0, \ldots, t_N : t_j = 2jT_l/N, j = 0, \ldots, N/2; t_{j+(N/2)} = T_l + 2j(T - T_l)/N, j = 1, \ldots, N/2\}$,
$T_l = \theta^{1-l}T/(\theta + 1)$, $l = 0, l = 1$. The numerical results are obtained with
the collocation parameters $\{\eta_1 = 1/4, \eta_2 = 3/4\}$ (see (29.11) with $m = 2$).

N	Π_N^1 $\varepsilon_N^{(0)}$	$\varrho_N^{(0)}$	$\Pi_{N,5}^{(0)}$ $\varepsilon_N^{(0)}$	$\varrho_N^{(0)}$	$\Pi_{N,5}^{(1)}$ $\varepsilon_N^{(0)}$	$\varrho_N^{(0)}$	$\Pi_N^r, r=1.1$ $\varepsilon_N^{(0)}$	$\varrho_N^{(0)}$	$\Pi_N^r, r=1.8$ $\varepsilon_N^{(0)}$	$\varrho_N^{(0)}$
4	3.8E-3		7.2E-3		1.9E-3		3.2E-3		1.2E-3	
8	1.5E-3	2.5	3.0E-3	2.4	5.1E-4	3.7	1.1E-3	2.8	2.9E-4	4.2
16	5.9E-4	2.6	1.2E-3	2.5	1.4E-4	3.6	4.0E-4	2.9	7.0E-5	4.1
32	2.3E-4	2.6	4.6E-4	2.6	4.9E-5	2.9	1.4E-4	2.9	1.7E-5	4.0
64	8.6E-5	2.6	1.8E-4	2.6	1.9E-5	2.6	4.8E-4	2.9	4.4E-6	4.0

N	Π_N^1 $\varepsilon_N^{(1)}$	$\varrho_N^{(1)}$	$\Pi_{N,5}^{(0)}$ $\varepsilon_N^{(1)}$	$\varrho_N^{(1)}$	$\Pi_{N,5}^{(1)}$ $\varepsilon_N^{(1)}$	$\varrho_N^{(1)}$	$\Pi_N^r, r=1.1$ $\varepsilon_N^{(1)}$	$\varrho_N^{(1)}$	$\Pi_N^r, r=5.1$ $\varepsilon_N^{(1)}$	$\varrho_N^{(1)}$
4	5.8E-2		6.4E-2		4.1E-2		5.6E-2		2.7E-2	
8	4.7E-2	1.2	5.5E-2	1.2	3.1E-2	1.3	4.4E-2	1.3	7.4E-3	3.6
16	3.7E-2	1.3	4.4E-2	1.3	2.4E-2	1.3	3.3E-2	1.3	1.8E-3	4.1
32	2.8E-2	1.3	3.4E-2	1.3	1.8E-2	1.3	2.5E-2	1.3	4.5E-4	4.1
64	2.1E-2	1.3	2.6E-2	1.3	1.4E-2	1.3	1.8E-2	1.4	1.1E-4	4.1

Table 1.

From Theorems 2 and 3 for $m = 2$ and $\nu = 3/5$ we can derive the following convergence results. In case of quasi-uniform grids the ratio $\varrho_N^{(0)}$ ought to be approximately $2.6(\approx 2^{2-\nu})$, and the ratio $\varrho_N^{(1)}$ ought to be approximately $1.3(\approx 2^{1-\nu})$. In case of graded grids the ratio $\varrho_N^{(0)}$ ought to be approximately $2.9(\approx 2^{r(2-\nu)})$ for $r = 1.1$ $(1 \le r < \frac{m}{2-\nu})$ and $4(= 2^m)$ for $r = 1.8$ $(r > \frac{m}{2-\nu})$, and the ratio $\varrho_N^{(1)}$ ought to be approximately $1.4(\approx 2^{r(1-\nu)})$ for $r = 1.1$ $(1 \le r < \frac{m}{1-\nu})$ and $4(= 2^m)$ for $r = 5.1$ $(r > \frac{m}{1-\nu})$. From Table 1 we can see that the numerical results are in good agreement with the theoretical estimates of Theorems 2 and 3.

References

1. H. Brunner, A. Pedas, and G. Vainikko, Piecewise polynomial collocation methods for linear Volterra integro-differential equations with weakly singular kernels, *SIAM J. Numer. Anal.* **39** (2001), 957–982.

2. H. Brunner, A. Pedas, and G. Vainikko, A spline collocation method for linear Volterra integro-differential equations with weakly singular kernels, *BIT* **41** (2001), 891–900.

3. H. Brunner and P.J. van der Houwen, *The Numerical Solution of Volterra Equations*, North-Holland, Amsterdam, 1986.

4. T. Tang, A note on collocation methods for Volterra integro-differential equations with weakly singular kernels, *IMA J. Numer. Anal.* **13** (1993), 93–99.

5. Q.Y. Hu, Geometric meshes and their application to Volterra integro-differential equations with singularities, *IMA J. Numer. Anal.* **18** (1998), 151–164.

30 Hybrid Laplace and Poisson Solvers. II: Robin BCs

Fred R. Payne

30.1 Introduction

Direct formal integration (DFI; Payne [1]) is an analytic-numerical method that has been uniformly successful in solving a large number of nonlinear (NL) and linear physical problems. Applications to Prandtl boundary layer, Navier–Stokes, turbulence, cavity flow, aerodynamics, chaos, Tricomi transsonics and Euler problems have been described (cf. Payne [2] through [8]). Other successes include solid state physics, predator-prey systems, and flight and orbital mechanics. The handling of parabolic and elliptic PDEs is straightforward. One hyperbolic PDE (Tricomi) has been treated. DFI has three stages:

1. Formally integrate DEs along one or more trajectories thereby converting the DEs to Volterra-type IEs or IDEs of second kind.
2. Study all forms (IEs, IDEs, DEs) for new insights into the problem.
3. Solve the equations a) analytically by hand near initial points to discover the solution behavior there and b) numerically upon computer for details.

This work solved 14 Laplace and Poisson problems. In Part-I of this series (cf. Payne [6]) Dirichlet BCs were treated in six cases. Eight Robin BCs are solved here.

30.2 DFI Elliptics

The 2D Poisson equation, on the unit square, is twice successively y-integrated on $[0, y(\to 1)]$ to convert the PDE system to 2nd kind Volterra IDE types. Hence, with $Uxx + Uyy = f(x, y)$, it follows that

$$U_y(x, y) = U_y(x, 0) - \left(\int_0^y U(x, s)\, ds \right)_{xx} + \int_0^y f(x, s)\, ds, \quad (30.1)$$

$$U(x, y) = U(x, 0) + yU_y(x, 0)$$

$$- \left(\int_0^y [y - s]U(x, s)\, ds \right)_{xx} + \int_0^y [y - s]f(x, s)\, ds. \quad (30.2)$$

The Lovitt [9] "lag factor" $[y-s]$, due to repeated integrals, yields Volterras faltung or convolution (cf. p.23 of [10]). Equations (30.1) or (30.2) can be used for Poisson or Laplace. Equation (30.2) has the advantage of negating the usual Volterra–Picard iterations for any two-point quadrature. These forms do require "sweeping" but relaxation is automatic. DFI/ NAD method (x-integrate (30.2) twice) eliminates all derivatives since the $(n-1)$ derivative is not here. A few NAD Euler tests required one sweep (cf. Payne [5]). Robin BCs result in the undesirable form of (30.1) for the derivative field. However, expanding $U(x,y)$ in a formal, integral equation avoids this by: $U(x,y) = U(x,0) + \int_0^y U_y(x,s)\,ds$, to recast (30.1) into equation (30.3) below with advantage:

$$U_y(x,y) = U_y(x,0) - yU_{xx}(x,0)$$

$$- \left(\int_0^y [y-s]U_y(x,s)\,ds \right)_{xx} + \int_0^y f(x,s)\,ds. \qquad (30.3)$$

All algols ((30.1),(30.2), and (30.3)) have the same, usable structure; however, the forms in (30.2) and (30.3) preclude Volterra iterations via Lovitt decoupling. Neumann and Robin are best suited by (30.1) or (30.3) and Dirichlet by (30.2). All DFI stencils are of a "hammerhead" type; i.e. the interpolation grid has a flat-end type of geometry. Thus, let o denote known ("old") values and f denote the value to be computed; the DFI Lovitt stencil is

$$\begin{matrix} & f & \\ 0 & 0 & 0\,, \\ 0 & 0 & 0 \end{matrix}$$

where there will be three columns for 2nd order central differences, five columns for 4th order central differences, etc.

Lovitt decoupling of f from itself is a huge advantage; f is determined from the BC alone for the first step. The calculation proceeds by a guess for the shooter (secant) to match f at the far boundary. The "shooter" is the normal derivative at the $y = 0$ boundary for Dirichlet (30.2) or the 2nd x-derivative (30.3) for Robin. DFI induces optimum computer numerics here as in many cases of NL Physics, 1981–2002, with no failures. College sophomores can solve a Riccati NL ODE since only basic calculus and the concept of iteration are required (personal experience, 1996–2000). A goal of this paper is to induce the reader to try DFI (a diskette of sample codes and outputs is available upon request). The three basic DFI modes contained in the formalism are:
1) "SIMPLEX", the only mode here, uses a single trajectory which results in the following conversion: ODE → pure IE and PDE → hybrid IDE
2) "DUPLEX" which uses two or more trajectories $(x, y, \ldots)$.
3) "NAD" (Natural Anti-Derivative) integrates over all coordinates and thus eliminates all derivatives, converting to pure integral equations.

30.3 Test Cases

All test cases were run on a square grid $= [0, 1] \times [0, 1]$:

Elliptic function	Dirichlet BC	Robin BC
$x^2 - y^2$	X	X
$x^3 y - x y^3$	X	X
$x^2 y - x y^3 / 3$	X	X
$x^2 + y^2$	X	X
$x^2 + y + y$	X	X
$x^3 + y^3$		X
$x^2 y$		X
$x^3 y + x y^3$		X

Numeric validations used as "debuggers" included:
  1. Global sum, RMS of $\nabla^2 U = 0$? (or $f(x, y)$?. Is the DE satisfied? Used for all BC.)
  2a. Global sum, RMS of: $[U_{exact} - U_{calculated}]^2 = 0$? (for Dirichlet BC)
  2b. Global sum, RMS of: $[\partial_y U_{exact} - \partial_y U_{calculated}]^2 = 0$? (for Robin BC)

30.4 DFI Tutorials

Valuable experience with DFI can be gained by following some simple ODE examples (cf. Payne [6] for details). A version of Riccati's equation has been solved by sophomores (personal experience, 1996–2000). A more complex equation modeling boundary layer flow: $u''(y) = u^2 - 1$, where u is the fluid velocity field subject to $u(0) = 0$ ["no slip"] and $u(y \rightarrow \infty) \rightarrow 1$ [matching the free stream velocity] has been solved by seniors (1986–2000). Two integrations yield the DFI form

$$u(y) = y u(0) - y^2/2 + \int_0^y (y - s) u^2(s)\, ds \quad \text{[no iteration possible]}.$$

This shows DFI power when coupled with Lovitt, an iteration free scheme.
N.B. Lovitt form: $\int_0^y dz \int_0^z u^2(s)\, ds = \int_0^y (y - s) u^2(s)\, ds$ (n-integrations yield $[y - s]^{n-1}/[n - 1]!$ factor).

30.5 Robin BC Cases

Again, one formal y-integration of Laplace/Poisson yields

$$U_y(x, y) = U_y(x, 0) - \left(\int_0^y U(x, s)\, ds \right)_{xx} + \int_0^y f(x, s)\, ds. \quad (30.1R)$$

All Robin BC set U_y on the top/bottom walls and U on the side-walls. A second, numerical integration in (1R) [of U_y to get U] leads to complex codes. A simpler way is to expand U as below:

$$U(x,y) = U(x,0) + \int_0^y U_y(x,s)\,ds. \tag{30.4}$$

Insert (30.4) into (30.1R); the result is like (30.2), which is the Lovitt–Dirichlet form

$$U_y(x,y) = Uy(x,0) + yU_{xx}(x,0)$$
$$- \left(\int_0^y (y-s)U_y(x,s)\,ds \right)_{xx} + \int_0^y f(x,s)\,ds. \tag{30.3R}$$

The "shooter" is the 2nd x-derivative, $U_{xx}(x,0)$. Three types of initial fields were used: 1) exact solution, 2) bi-linear interpolation and 3) zeroing the field. The last simulates unsteady problems with fixed BCs, with diffusion in "time" (each sweep serves as a time step). Errors followed closely a $1/e$ decay (improvement) per sweep, confirming the simulation. Error measures used for Dirichlet and Robin, as "debuggers" and validators, were: Max = Maximum point-wise PDE error; RMS = Global mean square of PDE RMS; RMSU_y = Mean square of DFI U_y-derivative versus exact value; U = average point-wise DFI solution difference from exact U [32 sweeps unless noted].

30.6 Results of Robin Cases

N.B.: Errors are in powers of 10; zero = no error.

Case	Max	RMS	RMS U_y	U	Comments
$x^2 - y^2$	zero	-16	-15	-9	
$x^3 y - xy^3$	-20	-6	-6	-5	
$x^2 y - y^3/3$	-10	-6	-10	-10	[16 sweeps]
$x^2 + y$	zero	-6	-17	-11	[99 sweeps]
$x^2 + y^2$	zero	-6	-26	-30	[16 sweeps]
$x^3 + y^3$	zero	-7	-14	-9	[64 sweeps]
$x^2 y$	zero	-8	-26	-13	[110 sweeps]
$x^3 y + xy^3$	-9	-7	-11	-10	

30.7 Discussion

Aspect Ratio $(AR = \Delta x/\Delta y)$ is similar to "relaxation factor" in SOR. Many values were exercised, from $AR = 1$ to 512 and y-grid sizes from 64 to 4096 on the unit square. Optimum here was $AR = 128$ for 1024/4096 y-intervals, giving good accuracy. $AR = 64$ often failed to converge, usually due to shooter failure. Conjectured is a behavior like that of column instability; the DFI stencil is either three columns wide (2nd order CD) or five columns (4th order CD); see sketch above of DFI stencil. As the DFI columns grow in height, for some AR, instability sets in. For Robin, U was constructed via $U(x, y+h) = U(x,y) + hU_y(x, y+h)$. A more complex averaging of two U_y seems "overkill" in view of $h \leq 1/1024(= .000977)$

but could be useful in some problems. Many decimal (0.01), rather than binary, step sizes ran into trouble due to accumulated round-off error. In Robin cases it sufficed to interpolate linearly U_y on the side walls and U on the bottom wall to start ("bootstrap") Robin numerics. DFI and the smoothing (diffusive and averaging) qualities of the Laplacian accelerate the calculations. "Micro-" Picard Iteration is DFIs second "life blood" [Lovitt is the first] since it improves Volterra iteration (if needed) so DFI speeds are competitive. This scheme iterates until converged on the first interval and then forms a new IVP on the next interval which it solves in the same way to convergence. One then repeats this "step-by-step" process until global convergence. Numerical tests reduce run times by about 90–95% over standard Picard. This and DFI generate a hierarchy which includes all DE problems:

  1. IVP = a sequence of "mini" IVPs
  2. BVP = a limit of a sequence of [a sequence of "mini" IVPs]
  3. ABVP = a limit of a limit of a sequence of [a sequence of "mini" IVPs]

For ABVP one goes "far enough" so the computer numbers do not change (or the change is acceptable). Here, Lovitt precludes iteration.

30.8 Conclusions

DFI is a very simple and powerful technique, with wide applicability. This might prompt the question: "Why was DFI found only so recently?" While Joseph Louiville [11] did something like DFI, conjectures include: 1) "Micro-" Picard is much faster than standard Picard; 2) Academic mathematics usually emphasizes DEs but neglects IEs.

30.9 Closure

The DFI solver is quite simple for Dirichlet boundary conditions; one merely "shoots" for the normal derivative at one boundary. Robin is a bit more involved but yet another formal integration reduces it to close similarity to Dirichlet. In both cases, a major decision is a good first estimate for the "shooter" so it can match the BC on the other side of the computational domain. DFI has no failures, 1980–2002. From a fundamental view, integral methods are inherently and organically more compatible to digital machines. For example, if more accuracy is desired, Romberg extrapolation is quite easy to incorporate; simply add one more "DO" loop. Up to seven levels [accuracy = $O(h^1 4)$] requires only 10–20 new lines of code.

DFI was designed for nonlinear problems in viscous fluid flows. Complete success with elliptics and parabolics is demonstrated. One hyperbolic (Tricomi transonics) case indicates these operators will also yield to DFI.

A serendipity is that DFI forms provide new insights into any problem due to multiple but equivalent mathematical formulations. Consider Laplace/Poisson with Robin BC. From the above three forms, $(1, 2, 3)$, leading solution terms are:

(1R) $U_y(x, y) = U + y(x, 0) + \cdots$,
(2R) $U(x, y) = U(x, 0) + yU_y(x, 0) + \cdots$,
(3R) $U_y(x, y) = U_y(x, 0) - yU_{xx}(x, 0) + \cdots$,

which follow from y-integrations; x-integrations yield three similar forms.

Numericists find this ideal: eight (6 IDE, PDE, IE) mathematical descriptions of and numeric schemes for the same problem. To switch integration trajectories, merely switch two FORTRAN "Do Loop" indices, changing six lines of code of a total of 250-350 lines.

References

1. F.R. Payne, Lecture Notes: 1981 AIAA Symposium, Arlington, TX (unpublished).

2. F.R. Payne, A simple conversion of two-point BVP, in *Trends in the Theory and Practice of Nonlinear Analysis*, North-Holland, Amsterdam, 1985, 377–385.

3. F.R. Payne, Direct formal integration (DFI): a global alternative to FDM/FEM, in *Integral Methods in Science and Engineering,* Hemisphere, Washington, 1986, 62–73.

4. F.R. Payne, A triad of solutions for 2-D Navier–Stokes: global, semilocal and local, in *Integral Methods in Science and Engineering*, Hemisphere, New York, 1991, 352–359.

5. F.R. Payne, Euler and inviscid Burger high-accuracy solutions, in *Nonlinear Problems in Aerospace and Aviation*, vol. 2, European Conference Publications, Cambridge, 2001, 601–606.

6. F.R. Payne, Hybrid Laplace and Poisson solvers. Part I: Dirichlet BCs, in *Integral Methods in Science and Engineering* , Birkhäuser, Boston, 2002, 203–209.

7. F.R. Payne and K.R. Payne, New facets of DFI, a DE solver for all seasons, in *Integral Methods in Science and Engineering*, Vol. 2 Pitman Res. Notes Math. Ser. **375**, Longman, Harlow, 1997, 176–180.

8. F.R. Payne and K.R. Payne, Linear and sublinear Tricomi via DFI, in *Integral Methods in Science and Engineering*, Res. Notes Math. Ser. **418**, Chapman & Hall/CRC, Boca Raton, FL, 2000, 268–273.

9. W.V. Lovitt, *Linear Integral Equations*, McGraw-Hill, Dover, 1950.

10. F.G. Tricomi, *Integral Equations*, Dover Publications, Inc., New York, 1985.

11. J. Liouville, Sur le dévelopment des fonctions ou parties de fonctions en séries. Sur la théorie des équations differentielles linéairs et les dévelopments des fonctions en séries, *J. Math.* **2** (1837), 16–22; **3** (1838), 561–614.

31 Vibrating Systems with Many Concentrated Masses: on the Low Frequencies and the Local Problem

Eugenia Pérez

31.1 Introduction

We consider the vibrations of a body occupying a domain Ω of $\mathbb{R}^n$, $n = 2, 3$, that contains many small regions of high density near the boundary, so-called *concentrated masses*. We refer to [1]–[5] for a previous study of these vibrating systems, and to [6]–[8] for vibrating systems with one single concentrated mass. We study the asymptotic behavior, as $\varepsilon \to 0$, of the eigenelements $(\lambda^\varepsilon, u^\varepsilon)$ of the corresponding spectral problem (31.2).

Let Ω be any bounded domain in $\mathbb{R}^n$, $n = 2, 3$, with a Lipschitz boundary $\partial\Omega$. Let Σ and Γ_Ω be non-empty parts of the boundary, such that $\partial\Omega = \bar{\Sigma} \cup \bar{\Gamma}_\Omega$, and Σ is assumed to be in contact with $\{x_n = 0\}$. Let ε and η be two small parameters such that $\varepsilon \mathcal{L} \eta$ and $\eta = \eta(\varepsilon) \to 0$ as $\varepsilon \to 0$.

For $n = 2$, let B be the semi-circle $B = \{(y_1, y_2) \ / \ y_1^2 + y_2^2 < 1, y_2 < 0\}$ in the auxiliary space $\mathbb{R}^2$ with coordinates y_1, y_2. For $n = 3$, let B be the half-ball $B = \{(y_1, y_2, y_3) \ / \ y_1^2 + y_2^2 + y_3^2 < 1, y_3 < 0\}$ in the auxiliary space $\mathbb{R}^3$ with coordinates y_1, y_2, y_3. Let ∂B be the boundary of B, $\partial B = \bar{T} \cup \bar{\Gamma}$, where T is the part lying on $\{y_n = 0\}$. Let B^ε (and similarly T^ε, Γ^ε) denote its homothetic εB (εT, $\varepsilon \Gamma$). Let B_k^ε (and similarly T_k^ε, Γ_k^ε) denote the domain obtained by translation of the previous B^ε (T^ε, Γ^ε) centered at the point $\tilde{x}_k$ of Σ at distance η between them. k is a parameter ranging from 1 to $N(\varepsilon)$, $k \in \mathbf{N}$. $N(\varepsilon)$ denotes the number of B_k^ε contained in Ω; $N(\varepsilon)$ is of order $O(\frac{1}{\eta})$ when $n = 2$ and $O(\frac{1}{\eta^2})$ when $n = 3$. The parameter α denotes the value

$$\alpha = \lim_{\varepsilon \to 0} \frac{-1}{\eta \ln \varepsilon} \text{ when } n = 2 \text{ and } \alpha = \lim_{\varepsilon \to 0} \frac{\varepsilon}{\eta^2} \text{ when } n = 3. \quad (31.1)$$

We consider the eigenvalue problem

$$-\Delta u^\varepsilon = \rho^\varepsilon \lambda^\varepsilon u^\varepsilon \text{ in } \Omega,$$

$$u^\varepsilon = 0 \text{ on } \Gamma_\Omega \cup \bigcup T^\varepsilon, \quad \frac{\partial u^\varepsilon}{\partial n} = 0 \text{ on } \Sigma - \overline{\bigcup T^\varepsilon}, \quad (31.2)$$

This work was partly supported by DGES: BFM2001-1266.

where $\rho^\varepsilon = \rho^\varepsilon(x)$ is the density function defined as

$$\rho^\varepsilon(x) = \frac{1}{\varepsilon^m} \text{ if } x \in \bigcup B^\varepsilon, \quad \rho^\varepsilon(x)1 \text{ if } x \in \Omega - \overline{\bigcup B^\varepsilon},$$

the symbol $\bigcup$ is extended, for fixed ε, to all the regions B_k^ε contained in Ω, and the parameter m is a real number, $m > 2$ (see [1]–[3] for different values of the parameter m, boundary conditions, and shapes of the domains).

As is well known, problem (31.2) has a discrete spectrum. For fixed ε, let $\{\lambda_i^\varepsilon\}_{i=1}^\infty$ be the sequence of eigenvalues of (31.2), converging to ∞, with the classical convention of repeated eigenvalues. It has been proved (see [1]–[3]) that they satisfy the estimates $C\varepsilon^{m-2} \le \lambda_i^\varepsilon \le C_i\varepsilon^{m-2}$, where C is a constant independent of ε and i and C_i is a constant independent of ε. Let $\{u_i^\varepsilon\}_{i=1}^\infty$ be the corresponding sequence of eigenfunctions which are assumed to be an orthonormal basis of the space $\mathbf{V}^\varepsilon$, where $\mathbf{V}^\varepsilon$ is the completion of $\{u \in \mathcal{D}(\bar{\Omega}) \, / \, u = 0 \text{ on } \Gamma_\Omega \cup \bigcup T^\varepsilon\}$ in the topology of $H^1(\Omega)$.

Certain convergence results for *the low frequencies*, the eigenvalues of order $O(\varepsilon^{m-2})$ of (31.2), can be found in [1]–[4]. Also, the limit behavior of some sequences of eigenvalues of order $O(1)$, the so-called *high frequencies*, is in [1]–[3] and [5]. As in the case of one single concentrated mass, in general, the low frequencies are associated with *the local vibrations* of the concentrated masses, each one independent of the others. We have found only one exception: for $n = 3$ and $\alpha > 0$, these frequencies also give rise to *global vibrations* affecting the whole structure (cf. Remark 2). Apart from this exception, the low frequencies and the corresponding eigenfunctions are asymptotically described, in a certain way, by the so-called *local eigenvalue problem* (31.3).

The local problem is an eigenvalue problem posed in an unbounded domain:

$$-\Delta_y U = \lambda U \text{ in } B,$$

$$-\Delta_y U = 0 \text{ in } \mathbb{R}^{n-} - \bar{B},$$

$$[U] = \left[\frac{\partial U}{\partial n_y}\right] = 0 \text{ on } \Gamma,$$

$$U = 0 \text{ on } T, \quad \frac{\partial U}{\partial y_n} = 0 \text{ on } \{y_n = 0\} - \bar{T}, \tag{31.3}$$

$$U(y) \to c \text{ as } |y| \to \infty, \ y_n < 0 \text{ when } n = 2,$$

$$U(y) \to 0 \text{ as } |y| \to \infty, \ y_n < 0 \text{ when } n = 3,$$

where the brackets denote the jump across Γ, $\bar{n}_y$ is the unit outward normal to Γ, and c is some unknown constant. $\mathbb{R}^{n-}$ is the half-plane $\{(y_1, y_2) \, / \, y_2 < 0\}$ for $n = 2$ and the half-space $\{(y_1, y_2, y_3) \, / \, y_3 < 0\}$ for $n = 3$. The variable y is a *local variable* defined by

$$y = \frac{x - \tilde{x}_k}{\varepsilon}, \tag{31.4}$$

which dilates the neighborhood of each point $\tilde{x}_k$ and transforms B_k^ε into B. As is known (31.3) can be written as a standard eigenvalue problem with a discrete spectrum in the space $\tilde{\mathcal{V}}$, where $\tilde{\mathcal{V}}$ is the completion of $\{U \in \mathcal{D}(\overline{\mathbb{R}^{n-}}) \, / \, U = 0 \text{ on } T\}$ for the Dirichlet norm $\|\nabla_y U\|_{L^2(\mathbb{R}^{2-})}$ (see [1] and [3]).

In Section 2 we give results on the multiplicity of the low frequencies, $\lambda^\varepsilon = O(\varepsilon^{m-2})$, depending on the multiplicity of the low frequencies of the local problem (31.3) and on the number of concentrated masses. We also give results on the structure of the corresponding eigenfunctions u^ε. It should be mentioned that a first study of the low frequency vibrations is performed in [4]. The results in this paper improve and complement those in [4] (see Remark 2).

We refer to [1]–[3] and [5] for an extensive study of the high frequencies and the global vibrations. We refer to [8] and [9] for the connection of the local problem (31.3) with the high frequency vibrations in the case of one single concentrated mass inside Ω.

31.2 The Low Frequencies and the Local Vibrations

The main results in the paper are stated in Theorems 1–3 below. For the sake of brevity, here we only outline their proofs. Theorem 1 characterizes the sequences of eigenvalues of (31.2) giving rise to local vibrations. Roughly speaking, Theorem 2 allows us to assert that there are at least $l_0 N(\varepsilon)$ values $\lambda^\varepsilon_{i(\varepsilon)}/\varepsilon^{m-2}$ converging towards each eigenvalue λ^0 of (31.3), l_0 being the multiplicity of λ^0. The corresponding eigenfunctions U^ε (cf. (31.5)) are approached in the space $\tilde{\mathbf{V}}^\varepsilon$ by the eigenfunctions of (31.3) associated with λ^0, concentrating their support asymptotically in neighborhoods of the concentrated masses as stated in Theorem 2. Theorem 3 shows that the limit of any converging subsequence of $\lambda^\varepsilon_1/\varepsilon^{m-2}$ is bounded by the first eigenvalue λ^0_1 of the local problem (31.3). Besides, in the case where $\lambda^\varepsilon_1/\varepsilon^{m-2} \to \lambda^0_1$, as $\varepsilon \to 0$, Theorem 2 ensures the convergence $\lambda^\varepsilon_i/\varepsilon^{m-2} \to \lambda^0_1$, as $\varepsilon \to 0$, for any fixed $i = 1, 2, \ldots$ (cf. Remark 1). See [6]–[8] to compare these results with the stronger results in the case of one single concentrated mass.

Let us change the variable in (31.2) by setting $y = x/\varepsilon$. We obtain

$$\int_{\Omega_\varepsilon} \nabla_y U^\varepsilon . \nabla_y V^\varepsilon \, dy = \gamma^\varepsilon \int_{\Omega_\varepsilon} \beta^\varepsilon(y) U^\varepsilon V^\varepsilon \, dy \quad \forall V^\varepsilon \in \tilde{\mathbf{V}}^\varepsilon; \qquad (31.5)$$

here Ω_ε is the domain $\{\, y \, / \, \varepsilon y \in \Omega \,\}$, $\gamma^\varepsilon = \lambda^\varepsilon/\varepsilon^{m-2}$, and λ^ε are the eigenvalues of (31.2). Also, $\beta^\varepsilon(y)$ in (31.5) is defined by $\beta^\varepsilon(y) = 1$ if $y \in \bigcup \tau_y B^\varepsilon$, and $\beta^\varepsilon(y) = \varepsilon^m$ if $y \in \Omega_\varepsilon - \overline{\bigcup \tau_y B^\varepsilon}$, where $\tau_y B^\varepsilon$ denote the transformed domains of the regions B^ε contained in Ω to the y variable. $\tilde{\mathbf{V}}^\varepsilon$ is the functional space $\{\tilde U = U(y) \, / \, U(\varepsilon y) \in \mathbf{V}^\varepsilon\}$. We assume that $\|U^\varepsilon\|_{\tilde{\mathbf{V}}_\varepsilon} = 1$.

Let us introduce an operator $\mathcal{A}^\varepsilon$ on $\tilde{\mathbf{V}}^\varepsilon$, $\mathcal{A}^\varepsilon$ by means of the equality

$$\langle \mathcal{A}^\varepsilon U, V \rangle_{\tilde{\mathbf{V}}_\varepsilon} = \int_{\bigcup \tau_y B^\varepsilon} UV \, dy + \varepsilon^m \int_{\Omega_\varepsilon - \overline{\bigcup \tau_y B^\varepsilon}} UV \, dy \quad \forall U, V \in \tilde{\mathbf{V}}^\varepsilon. \quad (31.6)$$

It is evident that $\mathcal{A}^\varepsilon$ is a selfadjoint, positive, and compact operator on $\tilde{\mathbf{V}}^\varepsilon$ whose eigenvalues are $1/\gamma_i^\varepsilon$, γ_i^ε being the eigenvalues of (31.5).

Let us consider λ^0 an eigenvalue of (31.3) of multiplicity l_0 and let U_1^0, $U_2^0, \cdots, U_{l_0}^0$ be the corresponding eigenfunctions, orthogonal in $\tilde{\mathcal{V}}$, satisfying $\|\nabla_y U_i^0\|_{L^2(\mathbb{R}^{n-})} = 1$.

Let us introduce $\tilde{\varphi}^\varepsilon(y)$ a function defined depending on the value of n. For $n = 2$, we consider $R_\varepsilon = \sqrt{\frac{\varepsilon + \eta/4}{\varepsilon}}$, and we define: $\tilde{\varphi}^\varepsilon(y) = 0$ if $|y| \geq R_\varepsilon^2$,

$$\tilde{\varphi}^\varepsilon(y) = 1 \text{ if } |y| \leq R_\varepsilon \, , \ \tilde{\varphi}^\varepsilon(y) = 1 - \frac{\ln|y| - \ln R_\varepsilon}{\ln R_\varepsilon} \text{ if } R_\varepsilon \leq |y| \leq R_\varepsilon^2 \, . \tag{31.7}$$

For $n = 3$, we consider $\tilde{\varphi}^\varepsilon$ as a smooth function which takes the value 1 in the semi-ball of radius $((\varepsilon + \eta/8)/\varepsilon)$, $B((\varepsilon + \eta/8)/\varepsilon)$, and is zero outside the semi-ball of radius $((\varepsilon + \eta/4)/\varepsilon)$, $B((\varepsilon + \eta/4)/\varepsilon)$:

$$\tilde{\varphi}^\varepsilon(y) = \varphi(2\frac{|\varepsilon y| - \varepsilon}{\eta}), \tag{31.8}$$

where $\varphi \in C^\infty[0,1]$, $0 \leq \varphi \leq 1$, $\varphi = 1$ in $[0, 1/4]$ and $Supp(\varphi) \subset [0, 1/2]$.

Obviously, the elements of $\tilde{\mathbf{V}}^\varepsilon$ extended by zero in $\mathbb{R}^{n-} - \overline{\Omega_\varepsilon}$ are elements of $\tilde{\mathcal{V}}$ (cf. Section 1). Moreover, it has been shown in [3] when $n = 2$ (in [1] when $n = 3$) that $U_p^0 \tilde{\varphi}^\varepsilon \in \tilde{\mathbf{V}}^\varepsilon$, and,

$$U_p^0 \tilde{\varphi}^\varepsilon \longrightarrow U_p^0 \text{ in } \tilde{\mathcal{V}}, \text{ as } \varepsilon \to 0. \tag{31.9}$$

For each $k = 1, 2, \ldots, N(\varepsilon)$, $p = 1, 2, \ldots, l_0$, we introduce the function

$$Z_{k,p}^\varepsilon(y) = \frac{U_p^0(y - \frac{\tilde{x}_k}{\varepsilon})\tilde{\varphi}^\varepsilon(y - \frac{\tilde{x}_k}{\varepsilon})}{\|\nabla_y(U_p^0 \tilde{\varphi}^\varepsilon)\|_{L^2(\mathbb{R}^{n-})}}. \tag{31.10}$$

We use the technique in [4] along with the orthogonality properties of the eigenfunctions U_p^0 in $\tilde{\mathcal{V}}$ to obtain the estimates:

$$|\langle \mathcal{A}^\varepsilon Z_{k,p}^\varepsilon - \frac{1}{\lambda^0} Z_{k,p}^\varepsilon, V\rangle_{\tilde{\mathbf{V}}^\varepsilon}| \leq o_\varepsilon^1 \|V\|_{\tilde{\mathbf{V}}^\varepsilon} \, , \ \forall V \in \tilde{\mathbf{V}}^\varepsilon, \quad \forall k, p, \quad \text{and} \tag{13.11}$$

$$\langle Z_{k_1,p}^\varepsilon, Z_{k_2,q}^\varepsilon\rangle_{\tilde{\mathbf{V}}^\varepsilon} = 0 \text{ for } k_1 \neq k_2 \, , \ |\langle Z_{k,p}^\varepsilon, Z_{k,q}^\varepsilon\rangle_{\tilde{\mathbf{V}}^\varepsilon} - \delta_{p,q}| = o_\varepsilon^2 \, , \ \forall k, p, q, \tag{31.12}$$

where o_ε^1 and o_ε^2 do not depend on k and p, and tend to 0 as $\varepsilon \to 0$:

$$o_\varepsilon^1 = C_1\left(\ln\frac{\varepsilon + \eta/4}{\varepsilon}\right)^{-\frac{1}{2}} \text{ and } o_\varepsilon^2 = C_2\left(\ln\frac{\varepsilon + \eta/4}{\varepsilon}\right)^{-\frac{1}{2}} \text{ when } n = 2, \tag{31.13}$$

$$o_\varepsilon^1 = C_1 \max\{(\frac{\varepsilon}{\eta})^{\frac{1}{2}}, \varepsilon^{m-2}\} \text{ and } o_\varepsilon^2 = C_2\frac{\varepsilon}{\eta} \text{ when } n = 3, \tag{31.14}$$

with constants C_1 and C_2 independent of ε and $\alpha \geq 0$ (see Remark 1).

Theorem 1. *Each eigenvalue λ^0 of the local problem (31.3) is an accumulation point of values $\lambda^\varepsilon_{i(\varepsilon)}/\varepsilon^{m-2}$. In addition, if $\lambda^\varepsilon_{i(\varepsilon)}/\varepsilon^\beta \to \lambda^*$ and the corresponding eigenfunctions $U^\varepsilon_{i(\varepsilon)} \to U^*$ weakly in $\tilde{\mathcal{V}}$, as $\varepsilon \to 0$, with $U^* \neq 0$ and $\lambda^* \neq 0$, then $\beta = m - 2$ and (λ^*, U^*) is an eigenelement of (31.3).*

Proof. The first assertion in the theorem has been proved in [1]–[3]. The second assertion is easily proved by taking limits in (31.5) for suitable V^ε.

Theorem 2. *Let us consider λ^0 an eigenvalue of (31.3) of multiplicity l_0 and let $U^0_1, U^0_2, \cdots, U^0_{l_0}$ be the corresponding eigenfunctions which are assumed to be orthonormal in $\mathcal{V}$. For any $K > 0$ there is $\varepsilon^*(K)$ such that, for $\varepsilon < \varepsilon^*(K)$, $K < l_0 N(\varepsilon)$ and the interval $[\lambda^0 - d^\varepsilon, \lambda^0 + d^\varepsilon]$ contains eigenvalues of (31.5), $\lambda^\varepsilon_{i(\varepsilon)}/\varepsilon^{m-2}$, with total multiplicity greater than or equal to K; d^ε is a certain sequence, $d^\varepsilon \to 0$ as $\varepsilon \to 0$ and the interval $[\lambda^0 - d^\varepsilon, \lambda^0 + d^\varepsilon]$ does not contain eigenvalues of (31.3) different from λ^0.*

In addition, there are $l_0 N(\varepsilon)$ functions, $\{U^\varepsilon_{k,p}\}^{p=1,l_0}_{k=1,N(\varepsilon)}$, $U^\varepsilon_{k,p} \in \tilde{\mathbf{V}}^\varepsilon$, such that $\|U^\varepsilon_{k,p}\|_{\tilde{\mathbf{V}}^\varepsilon} = 1$, $U^\varepsilon_{k,p}$ belongs to the eigenspace associated with all the eigenvalues in $[\lambda^0 - d^\varepsilon, \lambda^0 + d^\varepsilon]$, and

$$\|U^\varepsilon_{k,p} - Z^\varepsilon_{k,p}\|_{\tilde{\mathbf{V}}^\varepsilon} \leq 2(o^1_\varepsilon)^{1-\beta}. \tag{31.15}$$

In (31.15), β is a constant $0 < \beta < 1$, $o^1_\varepsilon(1) \to 0$ is given by (31.13) when $n = 2$ ((31.14) when $n = 3$), $Z^\varepsilon_{k,p}$ is defined by (31.10) and $\tilde{\varphi}^\varepsilon(y)$ is defined by (31.7) when $n = 2$ ((31.8) when $n = 3$). These functions, $\{U^\varepsilon_{k,p}\}^{p=1,l_0}_{k=1,N(\varepsilon)}$, satisfy that for any extracted subset of K functions $\{U^\varepsilon_{j_1}, U^\varepsilon_{j_2}, \ldots, U^\varepsilon_{j_K}\}$, they are linearly independent functions.

Proof. On account of (31.11)–(31.14), we prove that

$$\|\mathcal{A}^\varepsilon Z^\varepsilon_{k,p} - \frac{1}{\lambda^0} Z^\varepsilon_{k,p}\|_{\tilde{\mathbf{V}}^\varepsilon} \leq o^1_\varepsilon \quad \text{for } k = 1, 2, \ldots, N(\varepsilon), \quad p = 1, 2, \ldots, \lambda_0.$$

and, for $(i,p) \neq (j,q)$,

$$|\langle U^\varepsilon_{i,p}, U^\varepsilon_{j,q} \rangle_{\tilde{\mathbf{V}}^\varepsilon}| \leq C o^3_\varepsilon, \quad i, j = 1, 2, \ldots, N(\varepsilon), \quad p, q = 1, 2, \ldots, \lambda_0,$$

where C is a constant independent of i, j, p, q and ε, and, $o^3_\varepsilon \to 0$ as $\varepsilon \to 0$.

Taking the operator $\mathcal{A}^\varepsilon$ and the Hilbert space $\tilde{\mathbf{V}}^\varepsilon$ defined in (31.6), we apply Lemma 1.1 in Section III.1 of [6] to obtain (31.15) for a certain $d^\varepsilon = C(o^1_\varepsilon)^\beta$. Finally, we use the technique in [4] to prove the results on the linear independence of the eigenfunctions and the total multiplicity of the eigenvalues in the statement of the theorem.

Theorem 3. *Let λ^ε_1 and λ^0_1 be the first eigenvalues of (31.2) and (31.3) respectively. Then, there exist a constant $\lambda^* \leq \lambda^0_1$ and a sequence $o^\varepsilon_4 \to 0$, as $\varepsilon \to 0$, such that $\lambda^* \leq \frac{\lambda^\varepsilon_1}{\varepsilon^{m-2}} \leq \lambda^0_1 + o^4_\varepsilon$.*

Proof. We use the minimax principle (cf. [1]–[3] for the technique), the change of variables (31.4), the convergence (31.9) for U_1^0 an eigenfunction associated with λ_1^0 and different Poincaré inequalities for variables x and y, to obtain the result of the theorem for $\alpha \geq 0$ (see (31.1) and Remark 1).

Remark 1. Note that (31.13) and (31.14) are slightly different when $\alpha = +\infty$ (see (31.1)). In the case where $n = 3$ and $\alpha = +\infty$ certain restrictions on ε, η and m should be imposed in order to get the results in Theorems 2 and 3: for example, it suffices that $\eta = O(\varepsilon^{1/3})$ or $\eta^3 \varepsilon^{m-3} = O(1)$.

Remark 2. We refer to [4] for the result of Theorem 2 in the case where $l_0 = 1$ and for complementary results based on asymptotic expansions of the eigenfunctions u^ε associated with $\frac{\lambda^\varepsilon}{\varepsilon^{m-2}} \approx \lambda^0$. We also observe that results in Theorems 1-3 are in good agreement with those in previous papers [1]–[5]: as a matter of fact, in [1] we prove that for $n = 3$ and $\alpha > 0$ there are other different accumulation points of the sequences $\lambda_{i(\varepsilon)}^\varepsilon / \varepsilon^{m-2}$.

References

1. M. Lobo and E. Pérez, On vibrations of a body with many concentrated masses near the boundary, *Math. Models Methods Appl. Sci.* **3** (1993), 249–273.

2. M. Lobo and E. Pérez, Vibrations of a body with many concentrated masses near the boundary: high frequency vibrations, in *Spectral Analysis of Complex Structures*, Travaux en Cours **49**, E. Sanchez-Palencia ed., Hermann, Paris, 1995, 85–101.

3. M. Lobo and E. Pérez, Vibrations of a membrane with many concentrated masses near the boundary, *Math. Models Methods Appl. Sci.* **5** (1995), 565–585.

4. M. Lobo and E. Pérez, On the local vibrations for systems with many concentrated masses, *C.R. Acad. Sci. Paris Sér. IIb* **324** (1997), 323–329.

5. M. Lobo and E. Pérez, The skin effect in vibrating systems with many concentrated masses, *Math. Methods Appl. Sci.* **24** (2001), 59–80.

6. O.A. Oleinik, A.S. Shamaev, and G.A. Yosifyan, *Mathematical Problems in Elasticity and Homogenization*, North-Holland, Amsterdam, 1992.

7. J. Sanchez-Hubert and E. Sanchez-Palencia, *Vibration and Coupling of Continuous Systems. Asymptotic Methods*, Springer-Verlag, Heidelberg, 1989.

8. D. Gómez, M. Lobo, and E. Pérez, On the eigenfunctions associated with the high frequencies in systems with a concentrated mass, *J. Math. Pures Appl.* **78** (1999), 841–865.

9. E. Pérez, On the whispering gallery modes on the interfaces of membranes composed of two materials with very different densities, *Math. Models Methods. Appl. Sci.* **13** (2003), 75–98.

32 Terminal-Edge Algorithms: an Integrated Approach for Mesh Generation

Maria-Cecilia Rivara and Nancy Hitschfeld-Kahler

32.1 Introduction

In the adaptive finite element context, several algorithms for the refinement and/or derefinement of quality unstructured triangulations, based on the bisection of triangles by its longest edge, have been discussed and used in the last 20 years (see [1]–[5]). In two dimensions, they guarantee the construction of refined, nested and irregular triangulations of analogous quality as the input triangulation.

However, the use of two new and related mathematical concepts (the longest-edge propagation path of a triangle and its associated terminal edge), has allowed the development of new longest-edge algorithms for dealing with more general aspects of the mesh generation problem: (1) triangulation refinement problem, (2) triangulation improvement problem, (3) automatic quality triangulation problem, and (4) quality less-obtuse triangulation problem. Either for improving or refining a mesh, the algorithms use a terminal-edge point selection criteria as follows.

For any target element to be improved or refined, the midpoint of an associated terminal edge is selected for point insertion. Each terminal edge is a special edge in the mesh which is the common longest edge of all the elements (triangles or tetrahedra) that share this terminal edge in the mesh. Once the point is selected, this is inserted in the mesh.

In the case of the refinement algorithm, this is done by longest-edge bisection of all the elements that share the terminal edge, which is a very local operation. In the case of the improvement algorithm, the point insertion on the terminal edge is performed by using a constrained Delaunay algorithm. The process is repeatedly performed until the target element is refined/improved in the mesh.

In this paper, different aspects of these mesh generation problems in two dimensions and some of the algorithms proposed to deal with them, are reviewed and illustrated. In particular, the following specific applications are discussed: quality triangulations as needed for finite element methods, and quality less-obtuse triangulations as needed for finite volume methods.

This work was partially supported by Fondecyt Project No. 1030672.

32.2 Mesh-Generation Related Problems

The polygon triangulation problem, an important issue for finite element applications, can be formulated as follows.

Definition 1. Polygon Triangulation Problem: given N representative points of a polygonal region, join them by nonintersecting straight line segments so that every region internal to the polygon is a triangle. The resulting triangulation is a conforming triangulation (the intersection of adjacent triangles is either a common vertex or a common side).

Many criteria have been proposed as to what constitutes a "good" triangulation for numerical purposes, some of which involve maximizing the smallest angle and some that make the maximum angles less obtuse. The Delaunay algorithm, which constructs triangulations satisfying the first criterion, has been routinely used in engineering applications, followed by a postprocess step that relates to the boundary of the polygon.

In the adaptive finite element context, the triangulation refinement problem is also critical. To state this problem, some requirements and criteria about how to define the set of triangles to be refined and how to obtain the desired resolution need to be specified. To simplify we shall introduce a subregion R to define the refinement area, and a condition over the diameter (longest edge) of the triangles (given by a resolution parameter ε) to fix the desired resolution.

Definition 2. Triangulation Refinement Problem: given an acceptable triangulation of a polygonal region Ω, construct a locally refined triangulation such that the diameters of the triangles that intersect the refinement area R are less than ε, and such that the smallest (or the largest) angle in the mesh is bounded.

In the case when we dispose of a bad-quality triangulation of the polygonal geometry (having a non-adequate distribution of vertices) the triangulation improvement problem has to be considered. To state this problem, a triangle quality indicator function $q(t)$, a minimum angle tolerance parameter ε, and a local triangle improvement criterion need to be specified.

Definition 3. Triangulation Improvement Problem: given a non-quality triangulation τ_0 of a polygonal region Ω (having triangles such that its quality indicator q(t) $< \varepsilon$), construct an improved triangulation τ such that each triangle t satisfies q(t) $\geq \varepsilon$.

Note that if an initial coarse triangulation of the boundary polygonal vertices is considered, the more general (automatic) quality polygon triangulation problem can be stated.

Definition 4. Quality Triangulation Problem: Given an initial (boundary) triangulation τ_0 of the boundary vertices which define the polygonal geometry, construct a geometry-adapted triangulation τ such that for each triangle t of τ, $q(t) \geq \varepsilon$.

For finite volume applications, a less-obtuse triangulation problem can also be stated. To this end a quality indicator $\tilde{q}(t)$ that measures the obtuse (largest) angles is considered. In this context, an obtuse angle tolerance parameter $120 \leq \tilde{\varepsilon} < 180°$ is used.

Definition 5. Less-Obtuse Triangulation Problem: given a nonobtuse boundary Delaunay triangulation τ_0 (triangulation without obtuse angles opposite to boundary edges) of a polygonal region Ω having triangles such that its quality indicator $q(\tilde{t}) > \tilde{\varepsilon}$, construct an improved less-obtuse triangulation τ such that for each triangle t of $\tau, \tilde{q}(t) \leq \tilde{\varepsilon}$.

At this point, a few remarks are in order.

(1) The triangulation problems stated in Definitions 2 to 5 are essentially different than the classical triangulation problem in the sense that instead of having a fixed set of points to be triangulated, one has the freedom to choose the points to be added in order to construct a mesh either with a desired resolution or with a given mesh-quality. The construction of the mesh is dynamically performed. Furthermore it is possible to exploit the existence of the reference triangulation (constructed for instance by means of the Delaunay algorithm) in order to reduce the computational cost to construct the output mesh.

(2) To cope with the triangulation refinement problem, the longest-edge refinement algorithms guarantee the construction of good quality irregular triangulations. This is due in part to their natural refinement propagation strategy farther than the (refinement) area of interest R. Furthermore, asymptotically, the number N of points inserted in R to obtain triangles of prescribed size is optimal, and in spite of the unavoidable propagation outside the refinement region R, the time cost of the algorithm is linear in N, independent of the size of the triangulation [6].

In the rest of this paper, terminal-edge algorithms for the triangulation problems of Definitions 2 to 5 will be discussed.

32.3 Lepp and Terminal Edges

The longest-edge propagation path associated to a triangle is defined as follows:

Definition 6. For any triangle t_0 of any conforming triangulation τ, the longest-edge propagation path of t_0 will be the ordered list of all the triangles $t_0, t_1, t_2, \ldots, t_n$, such that t_i is the neighbor triangle of t_{i-1} by the longest edge of t_{i-1}, for $i = 1, 2, \ldots, n$. In addition, we shall refer to it as the Lepp(t_0).

Proposition 1. *For any triangle t_0 of any conforming triangulation of any bounded 2-dimensional geometry Ω, the following properties hold: (a) for any t, the Lepp(t) is always finite; (b) the triangles $t_0, t_1, \ldots, t_{n-1}$ have strictly increasing longest edge (if $n > 1$); (c) for the triangle t_n of the longest-edge propagation path of any triangle t_0, it holds that either (i) t_n has its longest edge along the boundary, which is greater than the* *longest edge of* t_{n-1}, *or (ii)* t_n *and* t_{n-1} *share the same common longest edge.*

The more general concept of terminal edge is defined as follows.

Definition 7. Any edge l in τ will be called a terminal edge if l is the common longest edge of every triangle that shares the edge l in τ. The triangles that share the edge l (one or two triangles in 2-dimensions) will be called terminal triangles.

Proposition 2. *For any triangle* t_0 *and its associated Lepp(t_0), the longest edge of the last triangle in Lepp(t_0) is a terminal edge in the mesh.*

Note that the Lepp(t_0) corresponds to an associated polygon, which in a certain sense measures the local quality of the current point distribution induced by t. In addition, the Lepp(t) allows us to find a terminal edge associated with t. To illustrate these ideas, see Fig. 1(a), where the Lepp(t_0) corresponds to the ordered list of triangles (t_0, t_1, t_2, t_3), while the edge shared by the triangles t_2 and t_3 is the associated terminal edge.

In what follows we shall see that the terminal edge associated with t is the best place for point insertion in order to refine/improve the triangle t.

32.4 Lepp-Bisection Algorithm for the Refinement of Quality Triangulations

By using the Lepp(t) and the terminal-edge concepts, an improved refinement algorithm [7] for non-Delaunay triangulations can be formulated, where the refinement of a target triangle t_0 (see Fig. 1) essentially means the repetitive longest-edge partition of pairs of terminal triangles sharing the terminal edge associated with the current Lepp(t_0), until the triangle t_0 itself is partitioned.

Figure 1 illustrates the refinement of the triangle t_0 over the initial triangulation of Fig. 1(a) with associated Lepp(t_0)=$\{t_0, t_1, t_2, t_3\}$. The triangulations (b) and (c) illustrate the first two steps of the Lepp-bisection procedure and their respective current Lepp(t_0), while that triangulation (d) is the final mesh obtained. Note that the new vertices have been enumerated in the order they were created.

The Lepp-bisection procedure is a nonrecursive algorithm essentially based on repeatedly refining pairs of triangles (in the general case of interior triangles) sharing the current terminal edge, which is found by using the current Lepp associated to the target triangle t, until the triangle t itself is bisected.

```
Lepp-bisection (t, τ)
while (t remains without being bisected) do
   Find the Lepp(t) and associated terminal-edge l
   Bisect the (one or two) terminal triangles that share l
endwhile
```

Since the Lepp-bisection algorithm is an improved version of the previous longest-edge algorithm, the following theorem, based on the properties of the longest-edge bisection, holds.

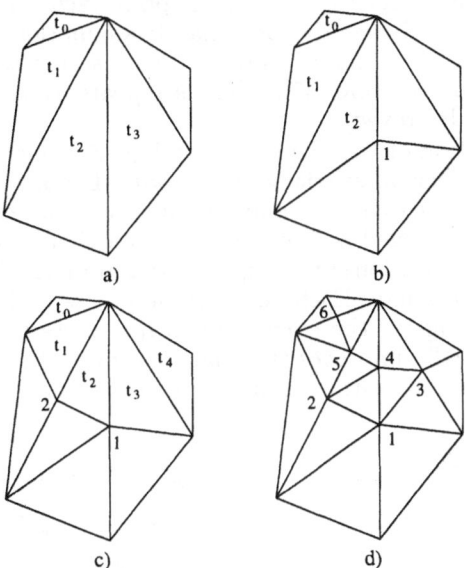

Fig. 1. Lepp-bisection of triangle t_0. (a) Initial triangulation. (b) First step of the process. (c) Second step of the process. (d) Final triangulation.

Theorem 1. *The repetitive use of the Lepp-bisection algorithms produce triangulations such that: (a) the smallest angle α_t of any triangle t obtained throughout this process, satisfies that $\alpha_t \geq \alpha_0/2$, where α_0 is the smallest angle of the initial triangulation; (b) for any conforming triangulation τ, the global iterative application of the algorithm covers, in a monotonically increasing form, the area of t with quasi-equilateral triangles.*

Theorem 1 guarantees the construction of good-quality irregular and nested triangulations. Next Theorem 2 assures in exchange that the Lepp-bisection algorithm solves the triangulation refinement problem with linear time complexity [6], provided that an initial good-quality triangulation is used and the current Lepp(t) is updated, rather than computed from scratch. Note that, the new Lepp bisection algorithm produces the same triangulation as the previous recursive algorithm, in a simpler, cleaner, easy-to-implement and more direct way.

Theorem 2. *Let τ be any conforming triangulation of any bounded polygonal region Ω. Then, for any circular refinement subregion C of radius r, the use of the Lepp-bisection algorithm to produce triangles of size ε inside C, asymptotically introduces N_i points inside C and N_o points outside C, where*

$$N_i = O(n^2), \ N_o = O(n \log n), \ and \ n = \frac{2r}{\varepsilon}.$$

32.5 Lepp–Delaunay Algorithm for the Improvement of Triangulations

The Lepp–Delaunay algorithm uses the Lepp of each target triangle (to be improved in the mesh) over the Delaunay triangulation of the vertices, in order to decide which is the best point to be inserted to produce a good-quality distribution of points [7]. This is repeatedly performed until the target triangle is destroyed.

For an illustration of the algorithm see Fig. 2, where the triangulation (a) is the initial Delaunay triangulation with Lepp(t_0) = $\{t_0, t_1, t_2, t_3\}$, and the triangulations (b), (c) and (d) illustrate the complete sequence of point insertions needed to improve t_0. Note that in this example, the improvement (modification) of t_0 implies the automatic Delaunay insertion of three additional points. Each one of these points is the midpoint of the current terminal edge in the current Lepp(t_0). Note that each Delaunay point insertion locally improves the triangulation in the current Lepp(t_0), and in this way this algorithm improves the triangulations obtained.

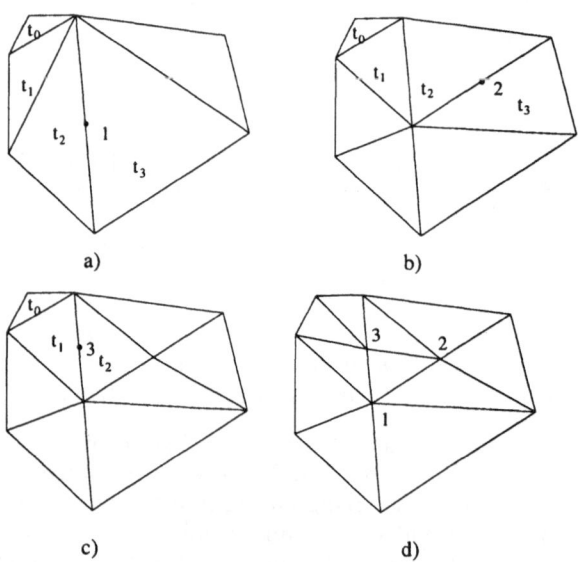

Fig. 2. Lepp–Delaunay improvement of triangle t_0.

Note also that we have used the word improvement instead of bisection or refinement. This is to make explicit the fact that one step of the procedure does not necessarily produce a smaller triangle. More important however, is the fact that the procedure improves the triangle t in the sense of Theorem 2 of section 6.

32.6 Automatic Quality Triangulations

By combining the basic Lepp-procedure over constrained Delaunay triangulations with adequate boundary considerations, a simple 2-dimensional automatic quality-triangulation algorithm can be formulated (see Algorithm 3), where δ is a threshold parameter less than or equal to 30° that can be easily adjusted. For this algorithm the following theorem holds:

```
Quality-Polygon-Triangulation (P, δ)
Input:  A general polygon P (defined by a set of vertices
and edges) and an angle tolerance parameter (δ < 30°)
Construct τ, a coarse constrained Delaunay triangulation
of P.
Find S, the set of the worst triangles t of τ (of smallest
angle < δ)
for( each t in S )
   Lepp-Delaunay-Improvement (τ, t)
   Update the set S (by adding the new small-angled triangles
   and eliminating those destroyed throughout the process)
endfor
Lepp-Delaunay-Improvement (τ, t)
while( t remains without being modified ) do
     Find the Lepp(t), the associated terminal-edge l
     and terminal triangles (t_{n-1}, t_n)
     if((t_n has a boundary edge l_b and
         l_b is not the smallest edge of t_n) or
         (t_{n-1} has a boundary edge l_b, and
         l_b is not the smallest edge of t_{n-1})) then
            l = l_b
     endif
     Perform the Delaunay insertion of p, the midpoint of l
endwhile
```

Theorem 3. *For any Delaunay triangulation τ, the repetitive use of the Lepp–Delaunay improvement algorithm over the worst triangles of the mesh (with smallest angle $\alpha < 30°$) produces a quality triangulation of smallest angles greater than or equal to 30°.*

A few remarks are now in order.

(1) Even when Theorem 3 guarantees the construction of quality triangulations, it says nothing about the size of these triangulations. More mathematical results in this sense are certainly needed. However, in practice, the 2-dimensional triangulations obtained are size-optimal. In fact, they are of analogous quality as those obtained with the circumcenter point insertion strategy [8].

(2) The triangulation of Fig. 3 illustrates the practical behavior of the algorithm. Note that the input data was the polygon with the minimum number of vertices to describe the geometry.

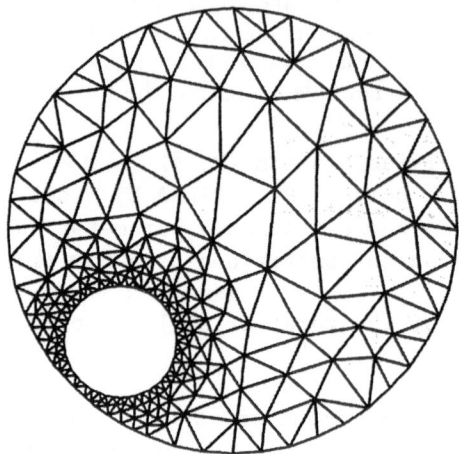

Fig. 3. Automatic triangulation obtained (smallest angles greater than 30°).

32.7 Quality Less-Obtuse Triangulations

This section discusses an algorithm to solve the quality nonobtuse boundary problem of Definition 5, such as needed for control volume discretization methods used in semiconductor device applications. The control volume-method tolerates very well long triangles with longest edges along the flow and shortest edges perpendicular to the flow. Small angles do not produce convergence problems but it is recommended that big obtuse angles and nodes with too high vertex-edge connectivity (number of edges that converges to a vertex) be avoided. The Delaunay triangulation and its dual, the Voronoi diagram fit very well with the Box-method because the Voronoi cells act as the control volumes, which are in turn used to compute the numerical integration around each mesh point [9].

The generation of quality nonobtuse boundary Delaunay triangulations consists of two steps: (a) the construction of a nonobtuse boundary Delaunay triangulation of an input geometry [9], and (b) a postprocess procedure such that, given a maximum angle parameter γ and a maximum vertex-edge connectivity parameter c, generates an output nonobtuse boundary/interface mesh where the maximum angle of every triangle is less than or equal to γ, and no mesh point has vertex-edge connectivity greater than or equal to c [10]. Both to destroy an interior too-obtuse triangle t and a vertex with high vertex-edge connectivity, a Lepp–Delaunay algorithm [11] is used. This algorithm repeatedly selects the midpoint P of a terminal edge for point insertion and performs Delaunay point insertion of P until either the target obtuse triangle or the target vertex-edge connectivity is destroyed. The terminal-edge is found as the longest-edge of the greatest last triangle in the Lepp of the target triangle (either an obtuse triangle t or the triangle with smallest angle that shares the vertex having high vertex-edge connectivity).

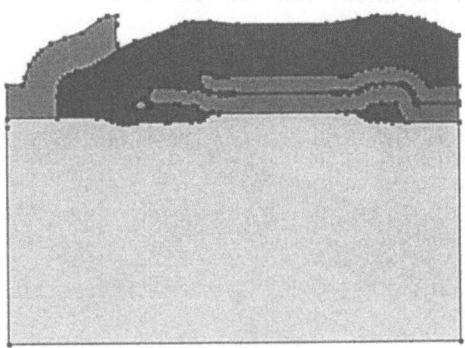

Fig. 4. Input geometry (220 points).

The improvement algorithm has been applied to model several semiconductor devices. Fig. 4 shows the geometry of a device and Fig. 5 shows the input triangulation of a part of the device. In the improved mesh (Fig. 6), the largest angles of each triangle are less than or equal to 120° and the vertex-edge connectivity or each vertex is less than 10, while the initial triangulation had 129 triangles with largest angle greater than 120° .

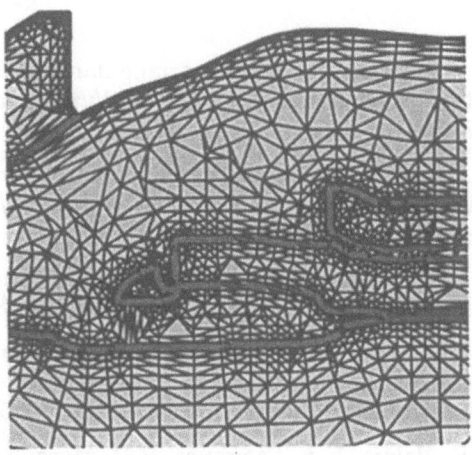

Fig. 5. Input triangulation (3342 points).

32.8 Conclusions

Finally note that the generalization of the algorithms presented in this paper define an integrated approach for mesh generation also in 3-dimensions. In particular the computational experimentation with a 3-dimensional terminal-edge Delaunay algorithm reported in [12] has shown analogous practical improvement behavior to the 2-dimensional algorithm of this paper.

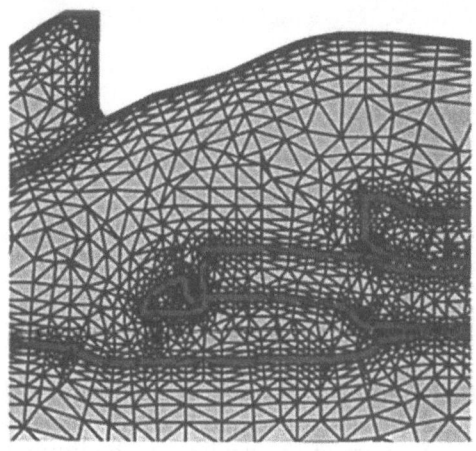

Fig. 6. Improved triangulation (4196 points).

References

1. M.C. Rivara, Algorithms for refining triangular grids suitable for adaptive and multigrid techniques, *Internat. J. Numer. Methods Engrg.* **20** (1984), 745–756.

2. M.C. Rivara, Selective refinement/derefinement algorithms for sequences of nested triangulations. *Internat. J. Numer. Methods Engrg.* **28** (1989), 2889–2906.

3. M.C. Rivara and C. Levin, A 3d refinement algorithm for adaptive and multigrid techniques, *Comm. Appl. Numer. Methods* 8 (1992), 281–290.

4. S.N. Muthukrishnan, P.S. Shiakolos, R.V. Nambiar, and K.L. Lawrence, Simple algorithm for adaptive refinement of three-dimensional finite element tetrahedral meshes, *AIAA J.* **33** (1995), 928–932.

5. N. Nambiar, R. Valera, K.L. Lawrence, R.B. Morgan, and D. Amil, An algorithm for adaptive refinement of triangular finite element meshes, *Internat. J. Numer. Methods Engrg.* **36** (1993), 499–509.

6. M.C. Rivara and M. Venere, Cost analysis of the longest-side (triangle bisection) refinement algorithms for triangulations, *Engineering with Computers* **12** (1996), 224–234.

7. M.C. Rivara, New longest-edge algorithms for the refinement and/ or improvement of unstructured triangulations, *Internat. J. Numer. Methods Engrg.* **40** (1997), 3313–3324.

8. J. Ruppert, A Delaunay refinement algorithm for quality 2-dimensional mesh generation, *J. Algorithms* **18** (1995), 548–585.

9. N. Hitschfeld and M.C. Rivara, Automatic construction of non-obtuse boundary and/or interface Delaunay triangulations for control volume methods, *Internat. J. Numer. Methods Engrg.* **55** (2002), 803–816.

10. N. Hitschfeld, L. Villablanca, J. Krause, and M.C. Rivara, Improving the quality of meshes for the simulation of semiconductor devices using Lepp-based algorithms, *Internat. J. Numer. Methods Engrg.* **58** (2003), 333–347.

11. M.C. Rivara, N. Hitschfeld, and R.B. Simpson, Terminal edges Delaunay (small angle based) algorithm for the quality triangulation problem, *Computer-Aided Design* **33** (2001), 263–277.

12. M.C. Rivara and M. Palma, New LEPP algorithms for quality polygon and volume triangulation: implementation issues and practical behavior, in *Trends in Unstructured Mesh Generation*, S.A. Cannan (ed.), Saigal, AMD **220** (1997), 1–8.

33 On the Identifiability of a Geometric Inverse Problem of Parabolic Type

Keijo Ruotsalainen

33.1 Introduction

We shall consider the moving boundary value problem

$$\frac{\partial u}{\partial t} - \Delta u = 0 \quad \text{in Q,}$$

$$u(x,t) = g(t) \quad \text{on } \Sigma_0,$$

$$u(x,t) = f(x,t) \quad \text{on } \Sigma_1, \tag{33.1}$$

$$u(x,0) = 0.$$

Here the set Q is a non-cylindrical space-time domain with a sufficiently smooth boundary. The internal boundary Σ_0 is time-dependent, i.e. it moves in time. On the contrary the external boundary Σ_1 is fixed for all time. The problem consists of the determination of the unknown moving boundary Σ_0 using the additional boundary flux measurement on an open part of the external boundary. The nonlinear problem can be formally written as solving the unknown domain from the equation

$$F(\Sigma_0) = \left.\frac{\partial u}{\partial \nu}\right|_\Gamma = q,$$

where Γ is a open and connected part of the external boundary Σ_1.

Using the mathematical tools developed for the shape optimization techniques (the domain derivatives [1], [2]), and the maximum principle [3], [4] we are able to prove the uniqueness and the local stability of the problem. In the proof we closely follow the approach utilized in [5],[6], [7] and [8] for the geometric inverse problems of elliptic type. To do this, we assume that

$$g(t) \text{ is strictly increasing and } g(t) > f(x,t) \ \forall t > 0.$$

This work was supported by the Academy of Lille while the author was visiting the University of Valenciennes in France.

33.2 Variational Formulation

The geometrical setting of the problem is as follows. Let Ω be a bounded domain in $\mathbb{R}^d$ with a C^1-boundary $\partial\Omega$. In the interior of the space-time domain $\Omega \times (0, T)$ there is another domain W with a moving boundary Σ_0. Furthermore, we assume that for all time there are three concentric balls B_0, B_1, B_2 such that

$$\overline{B}_0 \subset B_1 \subset \overline{B}_1 \subset B_2 \subset \Omega,$$
$$B_0 \times (0, T) \subset W \subset B_1 \times (0, T).$$

The boundaries Σ_0 and Σ_1 of the domain Q have the C^1-regularity. Besides this we assume that for all time the normal vector is not parallel to the time-axis. The cross-section of Q with the plane $t = \tau$ is denoted by $\Omega(\tau)$. By the assumptions made above the cross-section $\Omega(\tau)$ is topologically equivalent with an annular region for all $0 < \tau < T$. The Sobolev spaces $H^m(\Omega(t))$ and $H_0^m(\Omega(t))$ on $\Omega(t)$ as well as the functional spaces

$$L^2(I; H^1(\Omega(t))), \quad W = W(I; H^1(\Omega(t)))$$

for the variational treatment of the parabolic problem are defined in the usual manner, cf. [3]. The space with homogeneous boundary condition is denoted by $W_0 = W_0(I; H_0^1(\Omega(t)))$ and the anisotropic Sobolev space by $H^{r,\frac{r}{2}}(Q)$. All the function spaces above satisfy the initial condition $u(x, 0) = 0$.

We shall consider the variational problem that consists in finding $u = u_0 + w$ such that $u_0 \in W_0$, $w \in W_{\text{aff}} = \{w \in W \mid w|_{\Sigma_0} = g, \; w|_{\Sigma_1} = f\}$ for which

$$\int_0^T \int_{\Omega(t)} \frac{\partial}{\partial t}(u_0 + w)\phi - \Delta(u_0 + w)\phi \, dx dt = 0$$

for every $\phi \in \{H^{2,1}(Q) \mid \phi|_{\Sigma_0 \cup \Sigma_1} = 0, \; \phi(x, T) = 0\}$.

By the standard techniques of the variational methods [3] one finds that the variational problem admits a unique solution, which satisfies the Green formula

$$\int_0^T \int_{\Omega(t)} \frac{\partial}{\partial t}(u_0 + w)\phi + \nabla(u_0 + w) \cdot \nabla\phi \, dx \, dt = 0$$

for all the test functions $\phi \in \{H^{2,1}(Q) \mid \phi|_{\Sigma_0 \cup \Sigma_1} = 0, \; \phi(x, T) = 0\}$.

33.3 Identifiability and Local Stability

In this section we consider the identifiability of the domain problem for the heat equation with overspecified lateral boundary data.

To begin with, we shall recall the following theorem from [9], whose proof is based on the maximum principle and the Holmgren's theorem.

Theorem 1. *Assume that $u_i(x,t)$, $i = 1,2$, are solutions of the boundary value problems*

$$\frac{\partial u}{\partial t} - \Delta u = 0 \quad \text{in } Q,$$

$$u(x,t) = g(t) \quad \text{on } \Sigma_{0,i},$$

$$u(x,t) = f(x,t) \quad \text{on } \Sigma_1,$$

$$u(x,0) = 0$$

with the same boundary data. If the normal derivatives satisfy

$$\frac{\partial u_1}{\partial \nu}\big|_\Gamma = \frac{\partial u_2}{\partial \nu}\big|_\Gamma$$

on an open part Γ of the boundary Σ_1, then $\Sigma_{0,1} = \Sigma_{0,2}$.

We want to identify the domain Σ_0 from the domain to data map

$$F : U_{add} \to H^{-\frac{1}{2},-\frac{1}{4}}(\Sigma_1) : \ \Sigma_0 \to \frac{\partial u}{\partial \nu}\big|_\Gamma = q,$$

where u is the solution of (33.1). By the previous theorem this mapping is injective. In general the surjectivity of this mapping is an open problem. To prove the local stability of the mapping we will show that it is Fréchet differentiable and that the Fréchet derivative $F'(\Sigma_0)$ is injective "at the point Σ_0".

Let Q_h be a domain such that for every $(x',t) \in Q_h$ there exists $(x,t) \in Q$, $0 < t < T$ such that $(x',t) = (x + h(x,t),t)$, where $h(x,t)$ is a C^1-vector field that vanishes in the open neighborhood of Σ_1. By our regularity assumption there exists a diffeomorphism $\varphi = I + h$ between the domains Q and Q_h. Let us denote by ψ its inverse.

By the change of variables to the original domain the variational problem can be written in the form

$$\int_0^T \int_{\Omega(t)} \det(J_\varphi) \frac{\partial}{\partial t}(\tilde{u}_h \circ \varphi + \tilde{w} \circ \varphi)\phi$$

$$+ \det(J_\varphi)J_\psi J_\psi^* \nabla(\tilde{u}_h \circ \varphi + \tilde{w} \circ \varphi) \cdot \nabla\phi \, dx \, dt = 0$$

$$\forall \phi \in \{H^{2,1}(Q) \mid \phi|_{\Sigma_0 \cup \Sigma_1} \doteq 0, \ \phi(\cdot,T) = 0\}.$$

The function $\tilde{w}$ can be chosen such that $\tilde{w} \circ \varphi = w$, where w is as in the variational problem for $u_0 + w$ in section 2. We write $u_h = \tilde{u}_h \circ \varphi$ for convenience. In the previous equations we have denoted by J_φ, J_ψ, J_ψ^*, etc. the Jacobian matrices and their adjoints for the vector fields in question.

The solution u^1 of the boundary value problem

$$\partial_t u^1 - \Delta u^1 = -\nabla \cdot [(J_h + J_h^* - \operatorname{div}(h)I)\nabla u] - \operatorname{div}(h)\partial_t u \text{ in } Q, \tag{33.2}$$

$$u^1 = 0 \text{ on } \Sigma_0 \cup \Sigma_1, \quad u^1(\cdot,0) = 0 \text{ on } \Omega(0)$$

is the directional derivative with respect to the variation of the domain since the function $v = u_h - u - u^1$ satisfies the following property [9]:

$$\lim_{\|h\|_{C^1} \to 0} \frac{\|v\|_{W_0(I, H_0^1(\Omega(t)))}}{\|h\|_{C^1}} = 0.$$

This follows from the a priori estimates for the parabolic equations [2],[8]. Using the Poincaré inequality we obtain the estimate

$$\frac{\|\frac{\partial v}{\partial \nu}\|_{H^{-1/2, -1/4}(\Sigma_1)}}{\|h\|_{C^1}} \to 0, \text{ as } h \to 0$$

for the normal derivative on the boundary. But this means that $F(\Sigma_0)$ is Fréchet-differentiable and that

$$\left. \frac{\partial u^1}{\partial \nu} \right|_{\Sigma_1} = F'(\Sigma_0)h.$$

Next, we want to prove that the Fréchet derivative is injective. In the proof of the statement we make use of the following assertion [9].

Lemma 1. *Let X_0 be the space of functions $\phi \in H^{2,1}(Q)$ satisfying*

$$\partial_t \phi + \Delta \phi = 0,$$
$$\phi|_{\Sigma_0} = 0,$$
$$\phi|_{\Sigma_1} \in C_{00}^\infty(\Sigma_1),$$
$$\phi(\cdot, T) = 0.$$

Then the linear space $X = \{\chi \mid \chi = \frac{\partial \phi}{\partial \nu}|_{\Sigma_0}, \ \phi \in X_0\}$ is dense in $L^2(\Sigma_0)$.

Theorem 2. *$F'(\Sigma_0)$ is injective.*

Proof. Assume that for some transversal vector field $h(x, t)$,

$$F'(\Sigma_0)h = \left. \frac{\partial u^1}{\partial \nu} \right|_{\Sigma_1} = 0. \tag{33.3}$$

Multiplying the equation (33.2) by the test function $\phi \in X_0$, integrating over Q and using the Green formula together with the assumption (33.3) we get the identity

$$0 = \int_0^T \int_{\Omega(t)} (J_h + J_h^* - \text{div}(h)I) \nabla u \cdot \nabla \phi - \text{div}(h)(\partial_t u)\phi \, dx dt. \tag{33.4}$$

Since on the inner boundary Σ_0 of the domain the boundary conditions are independent of the space variable, the tangential derivative of the solution is zero; hence,

$$h \cdot \nabla u|_{\Sigma_0} = h_\nu \frac{\partial u}{\partial \nu},$$

where h_ν is the normal component of the vector field h. The function $\omega = h \cdot \nabla u$ satisfies the equation

$$\frac{\partial \omega}{\partial t} - \Delta \omega = \partial_t h \cdot \nabla u + h \cdot \nabla(\partial_t u) - \Delta[h \cdot \nabla u].$$

As above, multiplying by the test function $\phi \in X_0$ and integrating by parts, we get the identity

$$\int_{\Sigma_0} h_\nu \frac{\partial u}{\partial \nu} \frac{\partial \phi}{\partial \nu} ds_x dt - \int_Q \partial_t h \cdot (\phi \nabla u) dx dt = 0.$$

On the other hand, the shape derivative $w^1 = u^1 - h \cdot \nabla u$ satisfies the boundary value problem

$$\frac{\partial w^1}{\partial t} - \Delta w^1 = -\nabla \cdot [(J_h + J_h^* - \operatorname{div}(h)I)\nabla u] - \operatorname{div}(h)\partial_t u$$

$$- \frac{\partial}{\partial t}(h \cdot \nabla u) + \Delta(h \cdot \nabla u),$$

$$w^1|_{\Sigma_0} = -h_\nu \frac{\partial u}{\partial \nu}|_{\Sigma_0}, \quad w^1|_{\Sigma_1} = 0, \quad w^1(\cdot, 0) = 0.$$

Multiplying the right-hand side by the test function ϕ and integrating by parts, as in the case of the function u^1 we get

$$- \int_Q (\partial_t h \cdot \nabla u)\phi \, dx dt = 0.$$

Combining this result with the identity (33.4), we obtain the orthogonality relation

$$\int_{\Sigma_0} h_\nu \frac{\partial u}{\partial \nu} \frac{\partial \phi}{\partial \nu} ds_x dt = 0$$

for all test functions $\phi \in X_0$. By Lemma 1, we deduce that

$$h_\nu \frac{\partial u}{\partial \nu}|_{\Sigma_0} = 0.$$

By our asssumption, the vector field is nonzero on an open subset $\tilde{\Sigma}_0 \subset \Sigma_0$, implying that on some open subset the normal derivative is zero.

On the other hand, the function u solves the boundary value problem

$$\partial_t u - \Delta u = 0,$$
$$u|_{\Sigma_0} = g,$$
$$u|_{\Sigma_1} = f,$$
$$u(\cdot, 0) = 0.$$

Since the function g is strictly increasing and $g(t) > f(x,t)$ for all $t > 0$, the function $u(x,t)$ attains its positive maximum on Σ_0. But then, by the maximum principle, $\frac{\partial u}{\partial \nu} \neq 0$, ([10], Theorem 3.3.5, p. 173), which contradicts our assumption.

References

1. F. Murat and J. Simon, *Quelques résultats sur le contrôle par un domaine géometrique*, preprint, Université de Paris VI, 1974.

2. J. Sokolowski and J.-P. Zolesio, *Introduction to Shape Optimization*, Springer-Verlag, 1992.

3. J.-L. Lions and E. Magenes, *Nonhomogeneous Boundary Value Problems and Applications*, vols. 1, 2, Springer-Verlag, Berlin, 1972.

4. A. Friedman, *Partial Differential Equations of Parabolic Type*, Robert E. Krieger Publishing Company, Malabar, Florida, 1983.

5. S. Andrieux, A. Ben Abda, and M. Jaoua, Identifiabilité de frontière inaccessible par des mesures de surface, *C.R. Acad. Sci. Paris Sér. I* **316** (1993), 429–434.

6. S. Andrieux, A. Ben Abda, and M. Jaoua, On the inverse emerging plane crack problem, *INRIA Rapport de Recherche* **3012** (1996).

7. S. Andrieux, A. Ben Abda, and M. Jaoua, On a non-linear geometrical inverse problem of Signorini type: identifiability and stability, *INRIA, Rapport de Recherche (Theme 4)* **3175** (1997).

8. S. Nicaise and O. Zair, Identifiability and stability results of one emerging crack in heteregeneous media by one boundary measurements, *Preprint LIMAV University of Valenciennes* **98-4** (1998).

9. S. Nicaise, L. Paquet, and K. Ruotsalainen, On the detection of the moving internal boundary by a single boundary flux measurement on the fixed external boundary (to appear).

10. M.H. Protter and H.F. Weinberger, *Maximum Principles in Differential Equations*, Prentice-Hall, Englewood Cliffs, NJ, 1984.

34 Multiple Scattering Theory and Integral Equations

Bernard Rutily

34.1 Introduction

Multiple scattering theory describes the transport of particles interacting with a host medium through the processes of scattering, absorption, and emission, the first-mentioned being the main one. It is based on the transport equation, a kinetic equation satisfied by the distribution function of the traveling particles. This is a mixed equation: the unknown function is differentiated with respect to some variables and integrated with respect to other variables. It can be transformed into an integral equation, whose kernel depends on the nature of the scattering process we are considering.

In the present paper, one is interested mainly in the integral version of the transport equation. The physics will be simplified to such an extent that the resulting equation is simple but still mathematically relevant. The main lines of two approaches to solve it will be described. An example of a complete problem involving this equation, namely the problem of light propagation in a stellar atmosphere, is in [1].

34.2 The Basic Physics

A multiple scattering experiment involves two groups of particles, denoted by p and q. The p-particles are the ones that propagate, while suffering repeated scatterings by the q-particles. They all are identical: good candidates are photons, neutrons, neutral molecules of the same chemical species.... On the contrary, the scattering centers q may be members of different types of atoms, molecules, grains, etc... depending on the problem. They are not necessarily at rest, in most problems they are in thermal motion.

It is supposed that the system $\{p, q\}$ is so dilute that there is little chance a p-particle interacts with more than one particle q at the same time. Interactions have to be seen as brief and short-range collisions, so that the transport of the p-particles is dominated by their streaming between well-defined scattering collisions. This is the general framework of the kinetic theory of gases, based on the "molecular chaos" assumption.

I am indebted to M. Ahues, A. Largillier, and O. Titaud for many helpful discussions on this topic and comments on this paper.

Particles p can be scattered or absorbed when colliding with particles q. During a scattering event, a p-particle is not destroyed, just changes its velocity. Sometimes, it can give rise to the emission of some new p-particles by the scattering center. It disappears when being absorbed. A third (non-collisional) process is introduced—emission—during which a p-particle is spontaneously created by a q-particle. Induced emission is ignored, since it can be treated as negative absorption.

The two following assumptions are valid in most multiple scattering experiments:

(i) the p-particles do not interact among themselves;

(ii) they move freely apart from their collisions with the q-particles.

These assumptions are really suited to photons, which are massless particles. They are also appropriate to uncharged, massive particles with number density far lower than the density of the scattering centers, and with average potential energy far lower than their average kinetic energy. Note that the second assumption completes the comment we made about the short-rangeness of the interactions between the particles p and q. It means that between two successive scatterings, the p-particles do not interact with the q-particles, nor with an external field. Their trajectories are thus broken lines joining the scattering centers.

The ability of a medium to scatter some type of particles in a repetitive way can be described by two dimensionless parameters: the albedo ϖ and the thickness τ^*. The albedo ϖ is the average number of secondary particles emitted during a collision event. For pure absorption, $\varpi = 0$, for pure scattering $\varpi = 1$. Most of time, $0 \leq \varpi \leq 1$ and the albedo is interpreted as the conditional probability of being scattered, assuming a collision holds. $1 - \varpi$ is the probability of being absorbed. In a multiplying medium, each collision is followed, on the average, by the emission of more than one particle, and $\varpi > 1$. The thickness $\tau^* > 0$ gives information on the difficulty in going through the medium: it is the ratio of its size to the mean free path of the p-particles. A highly scattering medium is a medium with $\varpi \sim 1$ and $\tau^* \gg 1$.

A familiar example is the atmosphere of the earth irradiated by sunlight. Here, the p-particles are photons and the q-particles are air molecules. The molecular scattering of light is a pure scattering, its albedo ϖ is 1. Despite this fact, the earth atmosphere is a weakly scattering medium since its optical thickness τ^* is low, about 0.1 for visible light. That's why the sun can be seen as a sharp edged disk when the sky is clear. Suppose the weather is cloudy: the sun is no longer visible since a cloud is a very optically thick medium. For the scattering of light on the water droplets it consists of, its optical thickness is as large as, say, $\tau^* = 50$. The solar image is reduced in intensity by a factor of $\exp(-\tau^*) \sim 2 \times 10^{-22}$, which explains why it disappears. A cloud is a highly scattering medium, since its albedo is close to 1 (for instance $\varpi \sim 0.999$ or 0.9999 in visible light) and its thickness is large ($\tau^* = 50$).

Another multiple scattering experiment takes place in a nuclear reactor, which produces energy thanks to the fission of heavy atomic nuclei (the q-particles) under the impact of neutrons (the p-particles). The repeated scattering of neutrons by atomic nuclei is made easier by the electric neutrality of neutrons. The number densities of neutrons and atomic nuclei are typically in the ratio of 1 to 10^{13}, which justifies the above assumption (i). This scattering process is multiplied in a fissile material ($\varpi > 1$), and the possibility of a chain reaction arises.

Photons and neutrons are the best candidates for the multiple scattering process, but they are not the only ones. Neutral molecules, sometimes charged particles (electrons, light ions), acoustic or electromagnetic waves, etc., can also be repeatedly scattered in very different physical situations, all leading to the same mathematical problem we introduce in Section 3. A good synthesis book on the subject is that by Duderstadt and Martin [2]. It provides, among other things, an overview of the many applications of multiple scattering theory in disciplines as different as astrophysics, external geophysics, nuclear physics, rarefied gas dynamics, chemical technology, biology and medicine, ... Other monographs on transport theory are that by Pomraning [3] or Ozisik [4] as regards the transport of photons, and by Williams [5] when neutrons or molecules are involved. The main mathematical aspects of the theory are explained in Chapter 21 of Dautray and Lions [6].

34.3 The Transport Equation

The ultimate goal of multiple scattering theory is to determine, in a given geometrical configuration, the distribution of the p-particles in terms of the physical properties of the host medium. Since this statistical problem involves the velocities of the colliding particles, we seek the mean distribution of the p-particles in space (variable $\vec{r} \in \mathbb{R}^3$), velocity (variable $\vec{v} \in \mathbb{R}^3$), and time (variable $t \geq 0$). The appropriate unknown function is the velocity distribution function $f(\vec{r}, \vec{v}, t)$ of the p-particles, which is such that $f(\vec{r}, \vec{v}, t) d\vec{r} d\vec{v}$ is the average number of particles in the volume element $(\vec{r}, d\vec{r})$ with velocities in the range $(\vec{v}, d\vec{v})$ at time t. This function contains all the information that is usually required to describe transport processes. In particular, it yields the average number density of particles $n(\vec{r}, t)$ by integration with respect to velocity: $n(\vec{r}, t) := \int f(\vec{r}, \vec{v}, t) d\vec{v}$.

The distribution function satisfies a kinetic equation of the form [2]

$$\frac{\partial f}{\partial t} + \vec{v} \cdot \vec{\nabla} f + \vec{\gamma} \cdot \vec{\nabla}_{\vec{v}} f = \left(\frac{\partial f}{\partial t}\right)_{\text{coll}} + E. \tag{34.1}$$

Here $\vec{\nabla}$ is the gradient with respect to $\vec{r}$: $\vec{\nabla} := (\partial/\partial x, \partial/\partial y, \partial/\partial z)$, $\vec{\nabla}_{\vec{v}}$ the gradient with respect to $\vec{v}$: $\vec{\nabla}_{\vec{v}} := (\partial/\partial v_x, \partial/\partial v_y, \partial/\partial v_z)$, and $\vec{\gamma}$ is the mean acceleration of a p-particle due to external forces, a known function of $\vec{r}, \vec{v}, t$. The source term on the right-hand side describes the contribution of collisions (first term) and emission (function E).

In this general framework of the kinetic theory of gases, we have to take into account the specific features of the multiple scattering process as described in the preceding section. The collision term is the sum of two terms, describing the processes of scattering and absorption we have introduced:

$$\left(\frac{\partial f}{\partial t}\right)_{\text{coll}} = \left(\frac{\partial f}{\partial t}\right)_{\text{sca}} + \left(\frac{\partial f}{\partial t}\right)_{\text{abs}}. \qquad (34.2)$$

For binary collisions, the scattering term has the form of the Boltzmann collision term of the kinetic theory of gases.[2] It is quadratic in the unknown f if the particles p collide with each other. Actually, the assumption (i) in Section 2 allows one to write this term in the form

$$\left(\frac{\partial f}{\partial t}\right)_{\text{sca}}(\vec{r}, \vec{v}, t) = -|\vec{v}| f(\vec{r}, \vec{v}, t) \int \sigma(\vec{r}, \vec{v}, \vec{v}', t) d\vec{v}'$$

$$+ \int \sigma(\vec{r}, \vec{v}', \vec{v}, t) |\vec{v}'| f(\vec{r}, \vec{v}', t) d\vec{v}',$$

which is linear for the unknown f ($|\cdot|$ = euclidian norm in $\mathbb{R}^3$). The first term in the right-hand side describes the loss of p-particles by scattering events changing their velocity from $\vec{v}$ to any velocity $\vec{v}'$, the second term describing the gain by scattering events with emergent velocity $\vec{v}$. The rate at which collisions of the form $p[\vec{v}] + q \rightarrow p[\vec{v}'] + q$ occur in a unit volume is proportional to the incident flux $|\vec{v}| f(\vec{r}, \vec{v}, t)$, which represents the average number of particles crossing a unit surface perpendicular to $\vec{v}$ in a unit time. The factor of proportionality is the differential scattering coefficient $\sigma(\vec{r}, \vec{v}, \vec{v}', t)$ of the host medium, whose physical meaning is clear.

For similar reasons, the absorption term in the right-hand side of (34.2) can be written as

$$\left(\frac{\partial f}{\partial t}\right)_{\text{abs}}(\vec{r}, \vec{v}, t) = -\kappa(\vec{r}, \vec{v}, t) |\vec{v}| f(\vec{r}, \vec{v}, t),$$

which defines the absorption coefficient $\kappa(\vec{r}, \vec{v}, t)$. Finally, the output of p-particles is described by the emission coefficient $E(\vec{r}, \vec{v}, t)$ in the right-hand side of (34.1): it is supposed independent of f.

Another simplification is possible in (34.1), due to the second assumption (ii) we adopted in Section 2: the acceleration term on the left-hand side is insignificant when compared to the source term, which allows one to remove it.

Introducing the (integrated) scattering coefficient

$$\sigma(\vec{r}, \vec{v}, t) := \int \sigma(\vec{r}, \vec{v}, \vec{v}', t) d\vec{v}$$

and the extinction coefficient

$$\chi(\vec{r}, \vec{v}, t) := \kappa(\vec{r}, \vec{v}, t) + \sigma(\vec{r}, \vec{v}, t),$$

one obtains the transport equation in the final form

$$\frac{\partial f}{\partial t}(\vec{r}, \vec{v}, t) + \vec{v}.\vec{\nabla} f(\vec{r}, \vec{v}, t) = -\chi(\vec{r}, \vec{v}, t)|\vec{v}|f(\vec{r}, \vec{v}, t) + E(\vec{r}, \vec{v}, t)$$

$$+ \int \sigma(\vec{r}, \vec{v}', \vec{v}, t)|\vec{v}'|f(\vec{r}, \vec{v}', t)d\vec{v}'. \quad (34.3)$$

This equation determines the evolution in time of the distribution of the p-particles, taking account of their motion (via the advection term $\vec{v}\cdot\vec{\nabla}f$) and their interaction with the background medium (via the right-hand side). The latter means a loss of particles by extinction (= absorption plus scattering) and a gain by emission and scattering. Losses and gains are expressed in terms of three independent coefficients: σ, κ and E. These coefficients are macroscopic, depending on the physics of the interactions between the particles p and q, and on the state and the distribution of the particles q. They are calculated by first ascertaining the laws governing the scattering, the absorption and the emission of a single particle p by a single particle q, then by solving the statistical problem of determining the result of a large number of such interactions governed by these laws. This step is generally ignored in transport theory, since it is supposed that the coefficients of the transport equation are given. Actually, it is not the case with concrete problems, and the state and the distribution of the q-particles is dependent on their interaction with the p-particles. To describe the coupling between the dynamics of both sets of particles, we need some more equations accounting for the transformation of the background medium resulting from the migration of the p-particles. The problem becomes much more complicated, it may contain a large number of non-linear, strongly coupled equations. The description of a complete transport problem is not possible in the general context of this section, since it is based on the physical aspects of the problem. An example taken from astrophysics is given in [1].

For given coefficients of interaction, the transport equation (34.3) is linear. This is a substantial simplification when deriving this equation from the far more general (34.1). Another simplification comes from the lack of an acceleration term in the left-hand side. The linearity of (34.3) allows one to undertake its resolution with some confidence, provided that additional information is given: the domain of the variables $\vec{r}, \vec{v}, t$, boundary and initial conditions, functional spaces.... This is the subject of the next section.

34.4 A Simplified Model

The transport equation poses a mixed problem, in the sense that the unknown function f, defined at any time in a six-dimensional phase space, is differentiated with respect to the space variables and integrated with respect to the velocity variables. Solving this equation is a formidable task, in spite of its linear character! To clarify its mathematical structure, we concentrate on a "model problem" arising in systems with highly idealized physical characteristics.

34.4.1 The One-Speed Transport Equation for Isotropic Scattering

The three coefficients of (34.3) simplifies by assuming that
(iii) the host medium is isotropic with regard to its interactions with the
p-particles;
   (iv) scattering is isotropic and leaves the speed of the p-particles unchanged (on the average).
   Assumption (iii) means that the background medium, when irradiated
by a beam of p-particles, behaves independently of the direction of the
beam. As a result, the absorption coefficient and the differential scattering
coefficient verify $\kappa = \kappa(\vec{r}, |\vec{v}|, t)$ and $\sigma = \sigma(\vec{r}, |\vec{v}|, |\vec{v}'|, \theta, t)$, where θ is the
angle between $\vec{v}$ and $\vec{v}'$. It follows that the integrated scattering coefficient
does not depend on the direction: $\sigma = \sigma(\vec{r}, |\vec{v}|, t)$. The emission of p-
particles is also isotropic: $E = E(\vec{r}, |\vec{v}|, t)$. Assumption (iv) specifies that
the differential scattering coefficient is in fact independent of θ and contains
$\delta(|\vec{v}| - |\vec{v}'|)$ as a factor, where δ is the Dirac distribution at zero. The
scattering kernel is then

$$\sigma(\vec{r}, \vec{v}, \vec{v}', t) = \frac{1}{4\pi |\vec{v}|^2} \sigma(\vec{r}, |\vec{v}|, t) \delta(|\vec{v}'| - |\vec{v}|).$$

   Inserting this expression in the scattering term of the transport equation
(34.3) [with $d\vec{v}' = |\vec{v}'|^2 d|\vec{v}'| d\Omega'$], we note that the speed $|\vec{v}|$ of the p-
particles is unchanged during their migration. The "one-speed transport
equation" is thus attached to a given $|\vec{v}| > 0$. Putting $c := |\vec{v}|$ and $\vec{s} =:$
$\vec{v}/|\vec{v}|$ = the direction in which the particle moves, we define $\kappa_c(\vec{r}, t) :=$
$\kappa(\vec{r}, c, t), \sigma_c(\vec{r}, t) := \sigma(\vec{r}, c, t), \chi_c(\vec{r}, t) := \chi(\vec{r}, c, t), E_c(\vec{r}, t) := E(\vec{r}, c, t)$, and
we introduce the new unknown function $g_c(\vec{r}, \vec{s}, t) := cf(\vec{r}, c\vec{s}, t)$, which is
an angular (scalar) flux. The aim of this change of notation is to make
$c = |\vec{v}|$ appear as a parameter. The one-speed transport equation for
isotropic scattering now reads

$$\frac{1}{c}\frac{\partial g_c}{\partial t}(\vec{r}, \vec{s}, t) + \vec{s}.\vec{\nabla}g_c(\vec{r}, \vec{s}, t) = -\chi_c(\vec{r}, t)g_c(\vec{r}, \vec{s}, t) + E_c(\vec{r}, t)$$

$$+ \sigma_c(\vec{r}, t)\frac{1}{4\pi}\int g_c(\vec{r}, \vec{s}', t)d\Omega'. \quad (34.4)$$

   The scattering term is much simpler than the one of (34.3), since it
involves the angular average of the unknown function g_c: the integral is
taken over all directions around $\vec{s}' \in S^2$ (unit sphere of $\mathbb{R}^3$). Nevertheless,
the one-speed transport equation remains difficult to solve, because its
solution still depends on six variables. We have to introduce a new group
of assumptions in view of reducing the number of variables from 6 to 3.

34.4.2 Steady State Transport in Slab Geometry

From now on, our aim is to solve the steady state problem associated to
(34.4) in the simplest one-dimensional geometry, the slab geometry:
  (v) we seek the steady state solution to the transport equation;
  (vi) the host medium is a slab with plane-parallel symmetry.
The solution to the steady state transport equation is derived by removing the derivative with respect to time in the left-hand side of the
transport equation (34.4). Time dependence of any function appearing in
this equation can be ignored, since the t-variable is no longer transformed.
It becomes a parameter, which we omit to simplify the notation.

In slab geometry, the host medium is assumed to vary in only one dimension, taken to be the vertical direction (denoted by z). We thus have
$\kappa_c = \kappa_c(z), \sigma_c = \sigma_c(z), \chi_c = \chi_c(z)$, and $E_c = E_c(z)$. The depth variable z
covers the range $[0, z^*]$, where $z^* > 0$ is the thickness of the slab. If we fur-
ther assume that the boundary conditions are applied uniformly across the
slab boundaries, the unknown function does not depend on the horizontal
position either: $g_c = g_c(z, \vec{s})$. The direction $\vec{s}$ is specified by a polar angle
$\theta \in [0, \pi]$ and by an azimuthal angle $\varphi \in [0, 2\pi]$. θ is the inclination angle
of the vector $\vec{s}$ measured from the positive z-axis (Fig. 1). The choice of
the orientation is the usual one in astrophysics, it may be opposite in other
disciplines.

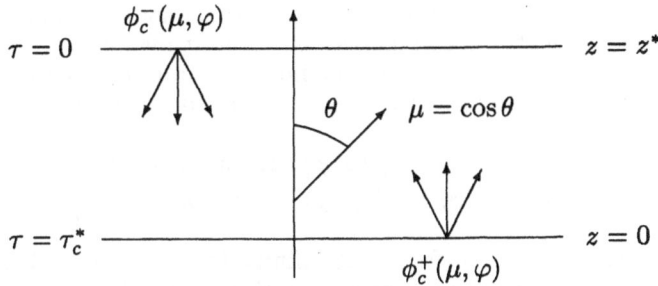

Fig. 1. Slab geometry.

It is convenient to replace the θ-variable by $\mu = \cos\theta \in [-1, +1]$. We
thus have $d\Omega' = \sin\theta' d\theta' d\varphi' = d\mu' d\varphi'$ in the integral term of (34.4). Since
the mapping $\zeta : \vec{s} \in S^2 \rightarrow (\mu, \varphi) \in [-1, +1] \times [0, 2\pi]$ is a bijection, the function $\tilde{g}_c(z, \mu, \varphi) := g_c[z, \zeta^{-1}(\mu, \varphi)]$ satisfies the following transport equation:

$$\mu\frac{\partial \tilde{g}_c}{\partial z}(z, \mu, \varphi) = -\chi_c(z)\tilde{g}_c(z, \mu, \varphi) + E_c(z)$$

$$+\sigma_c(z)\frac{1}{4\pi}\int_{-1}^{+1}\int_0^{2\pi}\tilde{g}_c(z, \mu', \varphi')d\mu'd\varphi'. \qquad (34.5)$$

A new change of variable is required to simplify this equation. The
extinction coefficient χ_c can be supposed strictly positive and continuous

on the interval $[0, z^*]$, so that the function

$$\tau_c(z) := \int_z^{z^*} \chi_c(z') dz'$$

is a bijection from $[0, z^*]$ to $[0, \tau_c^*]$, where $\tau_c^* := \tau_c(0) > 0$. It defines a new scale of depth in the slab, which could be called c-depth since it depends on the parameter c (τ_c^* is then the c-thickness of the slab).

Adopting $\tau \in [0, \tau_c^*]$ as the new spatial variable, we introduce $\hat{\kappa}_c(\tau) :=$ $\kappa_c[\tau_c^{-1}(\tau)]$, $\hat{\sigma}_c(\tau) := \sigma_c[\tau_c^{-1}(\tau)]$, $\hat{\chi}_c(\tau) := \chi_c[\tau_c^{-1}(\tau)]$, $\hat{E}_c(\tau) := E_c[\tau_c^{-1}(\tau)]$ and $\phi_c(\tau, \mu, \varphi) := \tilde{g}_c[\tau_c^{-1}(\tau), \mu, \varphi] = g_c[\tau_c^{-1}(\tau), \zeta^{-1}(\mu, \varphi)]$. After division of both members of (34.5) by $-\chi_c(z)$, the transport equation in slab geometry reads finally

$$\mu \frac{\partial \phi_c}{\partial \tau}(\tau, \mu, \varphi) = \phi_c(\tau, \mu, \varphi) - S_c^*(\tau)$$

$$-\varpi_c(\tau) \frac{1}{4\pi} \int_{-1}^{+1} \int_0^{2\pi} \phi_c(\tau, \mu', \varphi') d\mu' d\varphi', \qquad (34.6)$$

where $S_c^*(\tau) := \hat{E}_c(\tau)/\hat{\chi}_c(\tau)$ and $\varpi_c(\tau) := \hat{\sigma}_c(\tau)/\hat{\chi}_c(\tau)$ are the primary source function and albedo respectively. The primary source function describes the emission of p-particles by the internal sources, and the albedo characterizes the scattering properties of the host medium.

We shall adopt inhomogeneous boundary conditions on the boundary planes $\tau = 0$ (top surface) and $\tau = \tau_c^*$ (bottom surface), viz.

$$\phi_c(0, \mu, \varphi) = \phi_c^-(\mu, \varphi) : \text{given for } \mu < 0, \qquad (34.7)$$
$$\phi_c(\tau_c^*, \mu, \varphi) = \phi_c^+(\mu, \varphi) : \text{given for } \mu > 0. \qquad (34.8)$$

Possibly, the bottom surface is at infinity ($\tau_c^* = +\infty$), and the second boundary condition is replaced by a condition at infinity.

34.4.3 The Integral Version of the Transport Equation

This equation is readly derived as follows: introduce the angular average scalar flux

$$\Phi_c(\tau) := \frac{1}{4\pi} \int_{-1}^{+1} \int_0^{2\pi} \phi_c(\tau, \mu, \varphi) \, d\mu \, d\varphi$$

and the source function

$$S_c(\tau) := S_c^*(\tau) + \varpi_c(\tau) \Phi_c(\tau), \qquad (34.9)$$

and solve the transport equation

$$\mu \frac{\partial \phi_c}{\partial \tau}(\tau, \mu, \varphi) = \phi_c(\tau, \mu, \varphi) - S_c(\tau)$$

for a given source term. The unique solution satisfying the boundary conditions (34.7)–(34.8) is, for $\mu < 0$,

$$\phi_c(\tau, \mu, \varphi) = \phi_c^-(\mu, \varphi) \exp(\tau/\mu) - \frac{1}{\mu} \int_0^\tau S_c(\tau') \exp[(\tau - \tau')/\mu] d\tau', \quad (34.10)$$

for $\mu > 0$,

$$\phi_c(\tau, \mu, \varphi) = \phi_c^+(\mu, \varphi) \exp[-(\tau_c^* - \tau)/\mu)]$$
$$+ \frac{1}{\mu} \int_\tau^{\tau_c^*} S_c(\tau') \exp[-(\tau' - \tau)/\mu] d\tau', \quad (34.11)$$

and for $\mu = 0$, $\phi_c(\tau, 0, \varphi) = S_c(\tau)$. Integrating these expressions with respect to μ and φ yields the average scalar flux

$$\Phi_c(\tau) = \Phi_c^{\text{ext}}(\tau) + \frac{1}{2} \int_0^{\tau_c^*} E_1(|\tau - \tau'|) S_c(\tau') d\tau'. \quad (34.12)$$

Here, Φ_c^{ext} is the average flux of the direct (i.e., unscattered) field coming from the external sources of p-particles

$$\Phi_c^{\text{ext}}(\tau) := \frac{1}{4\pi} \int_0^{2\pi} \int_{-1}^0 \phi_c^-(\mu, \varphi) \exp(\tau/\mu) d\mu d\varphi$$
$$+ \frac{1}{4\pi} \int_0^{2\pi} \int_0^{+1} \phi_c^+(\mu, \varphi) \exp[-(\tau_c^* - \tau)/\mu] d\mu d\varphi, \quad (34.13)$$

and E_1 is the first exponential integral function:

$$E_1(\tau) := \int_0^1 \exp(-\tau/\mu) \frac{d\mu}{\mu} \quad (\tau > 0).$$

Entering the expression (34.12) for $\Phi_c(\tau)$ into the definition (34.9) for the source function leads to the following integral equation for S_c:

$$S_c(\tau) = F_c(\tau) + \varpi_c(\tau) \frac{1}{2} \int_0^{\tau_c^*} E_1(|\tau - \tau'|) S_c(\tau') d\tau', \quad (34.14)$$

with free term
$$F_c(\tau) := S_c^*(\tau) + \varpi_c(\tau) \Phi_c^{\text{ext}}(\tau). \quad (34.15)$$

Once this equation is solved, its solution $S_c(\tau)$ is substituted into Eqs. (34.10)–(34.11) for the calculation of $\phi_c(\tau, \mu, \varphi)$.

The physical meaning of this integral equation is clear: the free term describes the output of p-particles at level τ by the internal and external sources, and the integral term describes the multiple scattering process (hence the presence of the albedo as a factor). It prevails in problems that are dominated by scattering ($\varpi_c \sim 1$ and $\tau_c^* \gg 1$).

Equation (34.14) is a Fredholm integral equation of the second kind. Its kernel is weakly singular, since the E_1-function has a logarithmic behavior in a neighborhood of 0: $E_1(\tau) \sim -\ln\tau$ as $\tau \to 0^+$. This function is integrable over $\mathbb{R}^+$, and we have

$$\int_0^{\tau_c^*} E_1(\tau)d\tau = 1 - E_2(\tau_c^*) \leq 1$$

(equality holds for $\tau_c^* = +\infty$). $E_2(\tau) := \int_0^1 \exp(-\tau/\mu)d\mu$ is the second exponential integral function. Finally, $E_1(\tau)$ is rapidly decreasing as τ increases, and

$$E_1(\tau) \sim (1/\tau)\exp(-\tau) \quad \text{as } \tau \to +\infty.$$

It follows that the integral on the right-hand side of (34.14) is close to $S_c(\tau)$, specially when τ_c^* is large, which makes the iterative solution of (34.14) difficult when $\varpi_c(\tau)$ is close to 1: this is precisely what happens in a highly scattering medium!

34.4.4 Functional Spaces

This important question has been left aside, since it depends partially on the physics of the problem we are interested in. Actually, some minimum requirements can be deduced from the equations of the preceding section. For instance, the unknown function ϕ_c is a positive function of (τ, μ, φ) which must be differentiable with respect to τ on $]0, \tau_c^*[$, and integrable with respect to μ and φ on $[-1, +1] \times [0, 2\pi]$. The incoming boundary conditions must be such that (34.13) makes sense (distributions are possible).

To clarify the dependence with τ of the functions appearing in (34.14), let us introduce three functional spaces, denoted as C^0, L^1 and M^1. C^0 and L^1 have their usual meaning regarding functions from $[0, \tau_c^*]$ to $\mathbb{R}^+$ and

$$M^1 := \{f : [0, \tau_c^*] \to \mathbb{R}^+ \text{ such that } \forall \mu > 0\ \tau \to f(\tau)\exp(-\tau/\mu) \text{ is in } L^1\}.$$

If $\tau_c^* < +\infty$, $C^0 \subset L^1 \subset M^1$, the last inclusion being still valid when $\tau_c^* = +\infty$.

In most (not all) transport problems, it can be supposed that the functions ϖ and S_c^* are in C^0; we have seen that E_1 is in L^1, and it follows from (34.10)–(34.11) that S_c is in M^1. The function Φ_c^{ext} is currently in C^0 or L^1. It is not necessarily defined at $\tau = 0$ and $\tau = \tau_c^*$, as can be seen by entering $\phi_c^\pm(\mu, \varphi) = 1/|\mu|$ into (34.13) (a realistic example). From Eqs. (34.15), (34.12), and (34.9), this remark still applies to the functions F_c, Φ_c and S_c.

We are not aware of any topological structure on the space M^1; the main thing is that it contains L^1, which allows one to solve the integral equation (34.14) in this space (Section 5).

34.4.5 Concluding Remarks

The model we have introduced (one-speed, one-dimensional transport with isotropic scattering) is a basic one in transport theory. It has been the subject of an abundant literature since the 1930's (see [2]–[5] and references therein). Of course the single-speed approximation is unrealistic for most scattering processes; even so, this model is useful for a lot of reasons. The other assumptions (slab geometry, isotropic scattering) are realistic in several concrete problems, arising for instance in stellar atmospheres modeling (Section 6). The greatest merit of this idealized model is that it is simple enough to generate analytic solutions, against which numerical codes can be tested: this chance is not so frequent in physics!

34.5 Analytic/Numerical Solution

We are only concerned with the simplified problem of the previous section, formulated as the mixed equation (34.6) (plus boundary conditions (34.7)–(34.8)) or, equivalently, as the integral equation (34.14), with expressions (34.10)–(34.11) for the angular flux. Both formulations have been widely studied using analytic or numerical approaches. The basic methods are explained in [2]–[5] and synthesized in [7], which contains a great number of references. Instead of reviewing these methods, we would like to summarize the researches jointly carried out in the "Transfer" team of the Centre de Recherche Astronomique de Lyon and the "Spectral Analysis" team of the Université Jean Monnet of Saint-Etienne (France). Both groups have focused on the integral version (34.14) of the transport equation.

In this section, the albedo is supposed independent of τ and less than 1, i.e., $\varpi(\tau) = \varpi \in [0,1]$. An unvarying albedo is essential for solving analytically the transport equation (Section 5.1). Although this assumption is no longer necessary when using the numerical approach of Section 5.2, it will be maintained to make the presentation easier.

Omitting the c-subscript everywhere in (34.14), this equation reads

$$S = F + TS, \qquad (34.16)$$

where

$$TS(\tau) := \frac{\varpi}{2} \int_0^{\tau^*} E_1(|\tau - \tau'|)S(\tau')d\tau'. \qquad (34.17)$$

Note that this operator depends on two parameters: $\varpi \in [0,1]$ and $\tau^* \in]0,+\infty]$. It has been proved in [8] that it leaves C^0 and L^1 invariant, and that it is bounded from each space into itself, with norm

$$\|T\|_\infty = \|T\|_1 = \varpi \int_0^{\tau^*/2} E_1(\tau)d\tau = \varpi[1 - E_2(\tau^*/2)] \le 1.$$

This norm is close to 1 in a highly scattering medium; it is even just 1 in a conservative, semi-infinite medium ($\varpi = 1$ and $\tau^* = +\infty$). Except (possibly) in this case, the Fredholm integral equation (34.16) is uniquely solvable for any F in C^0 or L^1.

34.5.1 An Example of Analytic Method: the FLT Method

In the following, "analytic" method (solution, expression,...) stands for a method which don't discretize the variables of the problem it is supposed to solve. The first analytic solution to (34.16) goes back to the work of Wiener and Hopf in the 1930s. They treated the case $\varpi = 1, \tau^* = +\infty$ using the Fourier transform.[7] Integral (Fourier or Laplace) transform methods were extensively developed in the fifties and the sixties, as may be seen in the list of references of [9]. In this paper, three different methods for solving the integral equation (34.16) in L^1 are reviewed. They all use the finite Laplace transform (FLT), as defined for $f \in L^1$ by

$$\overline{f}(z) = \int_0^{\tau^*} f(\tau) \exp(-\tau z) d\tau \quad (z \in D).$$

The domain $D \subset \mathbb{C}$ of this function is the half-plane $\mathrm{Re}(z) > c$ when $\tau^* = +\infty$, with $c < 0$ depending on the behavior of the function f at infinity. When $\tau^* < +\infty$, $D = \mathbb{C}$. An inversion formula exists at any $\tau \in (0, \tau^*)$ where f is continuous, viz

$$f(\tau) = \frac{1}{2i\pi} \int_{c-i\infty}^{c+i\infty} \overline{f}(z) \exp(\tau z) dz \quad (c \in D \cap \mathbb{R})$$

(the integral is a principal value at infinity).

We shall not elaborate on the use of the FLT method for solving problem (34.16), since details can be found in [9]. We stress the central role played by the theory of Cauchy integral equations in this approach. In a semi-infinite medium ($\tau^* = +\infty$), the solution is expressed analytically in terms of a single auxiliary function, the famous "H-function" of transport theory. This function can be defined *explicitly*, i.e., by an integral. In a finite slab ($\tau^* < +\infty$), the solution is expressed analytically in terms of two auxiliary functions, denoted as ρ_+ and ρ_- in [10]. These functions are *implicitly* defined as the solution to a Fredholm integral equation of the second kind with a regular kernel. These integral equations can be solved accurately, and their solutions are smooth everywhere (contrary to H). Anyway, the existing solution is conceptually less satisfactory than the one reached for $\tau^* = +\infty$, since it is not expressed in closed form.

To illustrate the above comment, let us write the complete definition of the basic auxiliary functions H, ρ_+, and ρ_-.

The semi-infinite case ($\tau^* = +\infty$, $\varpi < 1$).

A possible analytic expression of $H = H(\varpi, z)$ is

$$H(\varpi, z) = \frac{1+z}{1+zk(\varpi)} \exp\left[z \int_0^1 \theta(\varpi, u) \frac{du}{u(u+z)} \right] \quad (z \in \mathbb{C} \setminus [-1, 0[),$$

where

$$k(\varpi) := \sqrt{1 - \varpi} \exp\left[\int_0^1 \theta(\varpi, u) \frac{du}{u}\right],$$

$$\theta(\varpi, u) := \frac{1}{\pi} \arctan\left[\frac{\pi(\varpi/2)u}{T(\varpi, u)}\right] \quad (0 \le u < 1),$$

$$T(\varpi, u) := 1 - \frac{\varpi}{2} u \ln\left[\frac{1 + u}{1 - u}\right] \quad (0 \le u < +1).$$

Continuous values on $[0, \pi]$ of the arctan function are used in the expression of $\theta(\varpi, u)$, i.e., the branch is not the principal one. With this choice, the function $u \to \theta(\varpi, u)$ is continuous from $[0, 1[$ to $[0, 1[$, although $T(\varpi, u)$ vanishes once in the interval $[0, 1[$.

The finite case $(\tau^* < +\infty)$

The functions $\rho_\pm = \rho_\pm(\varpi, \tau^*, z)$ satisfy the equations $(z \in \mathbb{C})$

$$\rho_\pm(\varpi, \tau^*, z) = 1 \pm \frac{\varpi}{2} z \int_0^1 \frac{g(\varpi, u)}{H^2(\varpi, u)} \exp(-\tau^*/u) \rho_\pm(\varpi, \tau^*, u) \frac{du}{u + z},$$

where

$$g(\varpi, u) := \frac{1}{T^2(\varpi, u) + (\pi\varpi u/2)^2} \quad (0 \le u < +1).$$

They are Fredholm integral equations over $[0, 1]$, the solution of which allows one to calculate the functions $\rho_\pm$ outside this interval (the integral is a Cauchy principal value when $z \in\,]-1, 0[$). Since the functions $\rho_\pm$ are in $C^1([0, 1])$, they can be very accurately computed over $[0, 1]$, then outside this interval.

To conclude, we think, contrary to the commonly accepted opinion (see, e.g., [2], p. 100), that analytic methods in transport theory are necessary from a theoretical and *practical* standpoint. We have ascertained that the analytic solution to (34.16) is numerically tractable, in spite of its apparent complexity. It can act its role of benchmark solution for numerical methods, which are of vital importance for solving realistic problems.

34.5.2 An Example of Numerical Method: the FRA Method

Numerical techniques for solving the transport equation were initially developed in connection with the multiple scattering of neutrons during the second world war. They were enjoying a new boom in the 1970's, thanks to the increase in computing power that enables more complex problems to be solved. It is not possible to give a brief survey of the many methods worked out since Chandrasekhar's discrete ordinates method. They may be quite different, depending on the physics of each model.[7]

The finite rank approximation (FRA) methods involve an approximation of finite rank of the T-operator as defined by (34.17). The main lines of this approach are briefly summarized below, regarding the problem (34.16)–(34.17) posed in the Banach space $X = C^0$ or L^1. Details on these methods can be found in [11], and in [12] for an application to the particular problem (34.16)–(34.17). In this subsection, it is supposed that $\tau^* < +\infty$.

The main idea is to replace T by an operator T_n of finite rank $n \in \mathbb{N}^*$ in (34.16). The sequence $(T_n)_{n \geq 1}$ must be built in such a way that it tends to T—in a certain manner that depends particularly on X—as n tends to infinity. Moreover, for n large enough, the corresponding approximate equation

$$(T_n - I)S_n = -F \tag{34.18}$$

must have a unique solution. In this case S_n is determined by

$$S_n = -R_n\, F\,,$$

where $R_n := (T_n - I)^{-1}$ is the resolvent operator of T_n at 1. The resolution of (34.18) leads to an n-dimensional linear system

$$(A_n - I_n)x_n = b_n\,, \tag{34.19}$$

where A_n is a matrix of rank n and I_n denotes the identity matrix of order n. Once this system is solved, the approximate solution S_n is obtained in terms of x_n. The main point is that this expression can be used to evaluate S_n in the whole interval $[0, \tau^*]$, i.e., without any additional interpolation. Details on some particular finite rank approximations in X can be found in [8] or [13].

There is no doubt that the dimension of the system (34.19) increases when the expected precision on the approximate solution S_n tends to zero. In our case, as we need to consider large values of τ^*, this dimension may be prohibitively large from a computational point of view. It is important to consider some way of reducing the computational cost—in time and memory—of the resolution.

First, A_n is generally a full matrix containing a lot of entries which are close to zero. The possibility of zeroing these entries is studied in [12]. Secondly, refinement schemes allow one to attain iteratively a given precision on the exact solution of a large scale linear system by means of the resolution of a sequence of linear system of moderated *fixed* size. A detailed description of such schemes is given in [11]. See also [8], [13] and [14] for applications to the problem (34.16)–(34.17). Each scheme is based on an approximation G_n of the resolvent operator $R = (T - I)^{-1}$ of the original operator T. Their common structure is the following:

$$\begin{cases} S_n^{(0)} := -G_n F, \\ S_n^{(k+1)} := S_n^{(0)} - G_n(T S_n^{(k)} - S_n^{(k)} + F) & (k \geq 0)\,. \end{cases}$$

We emphasize that the convergence of these schemes is performed *with respect to k*, not with respect to n which is definitively fixed. These schemes need the evaluation of T. In pratice T is replaced by an operator T_m, with

$m \gg n$. The corresponding scheme will converge to the solution S_m of the—large scaled—approximate problem $(T_m - I)S_m = -F$. A good way to stop these schemes is to require the relative residual to be less than a given tolerance:

$$r^{(k)} = \|(T_m - I)S_n^{(k)} + F\|/\|F\| < \text{TOL},$$

where $\|\cdot\|$ denotes the norm in X. The number of iterations needed by a scheme to attain a given precision—its efficiency—depends firstly on the choice of the finite rank approximation and on the choice of the approximate resolvent G_n. Secondly, it depends on the ratio m/n and on the norm of T_m that is, in our problem, on ϖ and τ^*.

Let us take an example from [13]. Three refinement schemes, denoted by (A), (B), and (C), are introduced there; they correspond to $G_n = R_n$, $G_n = R_n T - I$, and $G_n = T R_n - I$, respectively. (A) is Atkinson's iterative scheme, (B) is Brakhage's scheme, and (C) is a new one. For $n = 100$, the solution S_m with $m = 1000$ is reached after ten iterations using the scheme (A), five iterations using the schemes (B) and (C). Of course, the number of iterations increases with decreasing n: for $n = 100$, 16 iterations are necessary with scheme (A) and nine with schemes (B) and (C); for $n = 20$, the number of iterations is 43 using (A) and 22 using (B) or (C). These calculations have been performed with the following input parameters: $\tau^* = 100$, $\varpi = 0.7$, $F(\tau) = 1 - \varpi$, using uniform grids. The tolerance TOL is 10^{-10}. Figure 2 shows the first three relative residuals $r^{(k)}$ ($k = 0, 1, 2$) for $m = 900$ and $n = 60$.

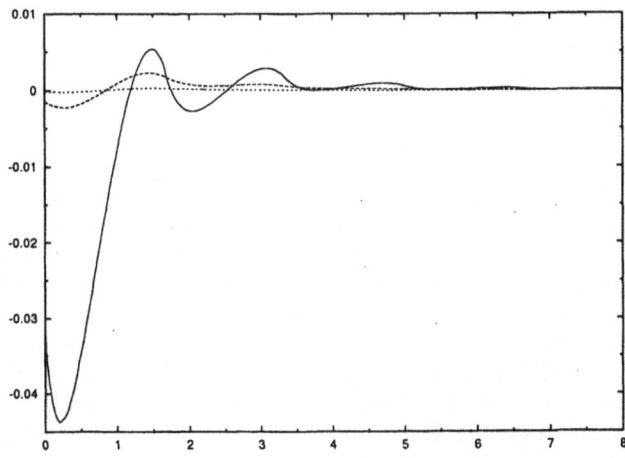

Fig 2. First three residuals $r^{(k)}$ obtained with the (C) refinement scheme for $m = 900$ and $n = 60$: $k = 0$ (solid), $k = 1$ (long dashed), $k = 2$ (short dashed).

The FRA methods provide a fine course of action for tackling singular integral equations such as (34.16). They rest on solid foundations and include many tracks. Good numerical results have already been achieved for not too large values of τ^*. When tackling the problem (34.16) on a very large domain $[0, \tau^*]$, these methods are still too demanding of memory and computational time, in spite of the use of the refinement schemes. Three tracks are currently explored to overcome this difficulty: kernel truncation [12], parallel computations [14], asymptotic decomposition of the domain $[0, \tau^*]$ (see [15] and [16]).

34.6 A Concrete Example from Astrophysics

The problem (34.16)–(34.17) is easily solvable in C^0 for "reasonable" values of the input parameters ϖ and τ^*. Things are far less evident when $F \in L^1$ diverges at $\tau = 0$, ϖ is close to 1 ($\varpi = 0.999999$), τ^* is very large ($\tau^* = 10^6$) and $S(\tau)$ has to be computed accurately for τ close to 0 ($\tau = 10^{-3}$). Since these values are realistic in at least one application we know, it is appropriate to linger over the "schedule of conditions" imposed by physical considerations. The example we choose is that of stellar atmospheres, since it is a wide domain of application of the multiple scattering theory, the first one from the historical standpoint.

34.6.1 The Radiative Transfer Equation

We are concerned with the transport of photons through the boundary layers of a star, which constitute its atmosphere. These photons interact with the particles of the stellar plasma (ions, atoms, free electrons) through the three processes involved in transport theory: scattering, absorption, emission.

The description of Sections 2–4 adapts to the transport of photons, provided the velocity $\vec{v}$ of a particle is denoted by two quantities: the magnitude $|\vec{v}|$ and the direction $\vec{s} = \vec{v}/|\vec{v}|$. A quantity equivalent to the speed $|\vec{v}|$ is the kinetic energy

$$E = (1/2)m|\vec{v}|^2,$$

where m is the mass of the particle. We could thus rewrite the developments of Sections 2–4 with the variables $(E, \vec{s})$ in place of $\vec{v}$. The resulting equations would be suitable for photons, so long as the variable E is replaced by the photon energy $h\nu$, where h is the Planck constant and ν the photon frequency. We conclude that transport theory applies to photons, provided the velocity variable $\vec{v}$ is replaced by the "frequency-direction" variable $(\nu, \vec{s})$, with $\nu \in]0, +\infty[$ and $\vec{s} \in S^2$. This change of variable is necessary, since the velocity $\vec{v}$ makes no sense for photons ($|\vec{v}| = $ constant).

A change of unknown function is also usual when tackling the transport of photons. The physical quantity of interest is generally the energy transported by the radiation field, rather than the number flux of photons. The unknown function is then the product of the angular flux of photon $cf(\vec{r}, \vec{s}, \nu, t)$ by the photon energy $h\nu$, viz. $I(\vec{r}, \vec{s}, \nu, t) = ch\nu f(\vec{r}, \vec{s}, \nu, t)$. The terminology is the following: I is the "specific intensity" of the radiation field, the equation it satisfies is the "radiative transfer equation", and the theory of propagation of energy by radiation is the "radiative transfer theory". In what follows, RTE is an abbreviation for "radiative transfer equation".

34.6.2 Validity of Assumptions (i)–(vi) in a Stellar Atmosphere

The assumptions (i)–(vi) of Sections 2–4 are suitable for the transport of photons in most stellar atmospheres, except (iv) at the beginning of Section 4.1. The scattering of photons on the components of a stellar plasma is nearly isotropic, but not monochromatic, which means that it is generally accompanied with a change of the photon frequency. It follows that a "one-frequency RTE", corresponding to the one-speed transport equation of Section 4.1, is not acceptable. The same is true for the integral equation (34.14), whose kernel depends on the assumption (iv). Actually, in an important class of realistic (in stellar atmospheres) scattering processes, the integral version of the transport equation can be written in the form (34.14), just replacing the E_1-function by a function K as defined by a Laplace-type integral

$$K(\tau) = \int_0^{+\infty} k(s) \exp(-\tau s) ds \quad (\tau > 0).$$

The explicit expression of $k(s)$ depends on the physics of the scattering process. It has possibly (but not necessarily) a compact support in $\mathbb{R}^+$. The E_1-function corresponds to $k(s) = 0$ for $0 < s < 1$ and $k(s) = 1/s$ for $s \geq 1$. Solving the integral equation (34.16) for an operator T defined by a K-function instead of the E_1-function is not really constraining, it can even be easier since the K-function is generally algebrically decreasing at infinity, while the E_1-function is exponentially decreasing at infinity. The E_1-function thus provides a simple, but numerically relevant model for the scattering of photons in a stellar atmosphere.

A last comment about the assumptions (i)–(vi) concerns the last one: (vi). It may seem surprising to adopt the slab geometry in a spherical star. Actually, a stellar radius is generally so great compared to the mean free path of photons that the plane layer approximation is justified in most (not all) stellar atmospheres.

34.6.3 How Many Integral Equations?

The first thing to do when solving the RTE in a stellar atmosphere is to restore the subscript c we have dropped in Section 5. Since $c = |\vec{v}|$ is replaced by the frequency when photons are concerned, we have to solve a single RTE for each frequency ν in $]0, +\infty[$, assuming monochromatic scattering as we did in Section 5. In a concrete problem, the range $]0, +\infty[$ is replaced by $[\nu_{min}, \nu_{max}]$, and a grid is defined on this interval in order to integrate the solution of the RTE with respect to the frequency. Such an integration is necessary for solving the equations coupled to the RTE, for instance the energy equation.[1] The number of frequencies of the grid depends on the variation with frequency of the coefficients of the RTE. A representative coefficient is the opacity of the stellar material, which is nothing else than the extinction coefficient χ_c (here χ_ν) appearing in (34.5). The variation of the function $\nu \to \chi_\nu(z)$ at a given z is strongly connected to the chemical composition of the atmosphere. Suppose we are faced with a pure hydrogen atmosphere: the only components are hydrogen atoms (with, say, 10 possible energy states), protons and free electrons. The opacity curve at a given z is shown in Fig. 3 (in fact $\log \chi_\nu$ in terms of $u = h\nu/kT$, where h, k and T are constant).

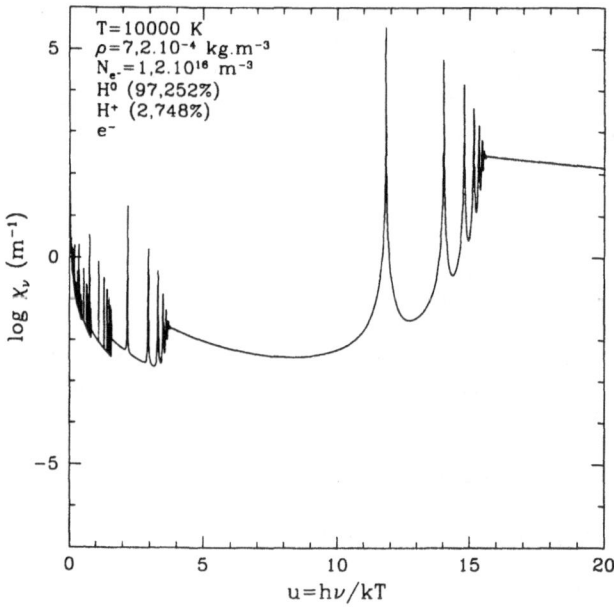

Fig. 3. Opacity curve in a pure hydrogen stellar atmosphere (from J.F. Gonzalez, Centre de Recherche Astronomique de Lyon).

It turns out that opacity is smooth in some parts of the closed interval $[\nu_{min}, \nu_{max}]$—the continuum—and that it rises suddenly in narrow regions—the spectral lines. Considering that (a) the variation of the opacity in a spectral line is correctly described by, say, 10 frequency points on the average, (b) there are 10–20 significant lines (among 55) in Fig. 3, and (c) some additional frequency points must be chosen in the continuum, we conclude that about two hundred frequency points are required to reproduce the variation with frequency of the opacity in a pure hydrogen atmosphere. In a realistic atmosphere, there are many chemical elements far more complex than hydrogen, and the number of spectral lines is very high: from some thousands to some millions or tens of millions, depending on the degree of evolution of the star! Since the number of frequency points is about ten times the number of lines, the problem is untractable.

Fortunately, some statistical processing allows one to reduce the number of frequencies by a factor of about 10^4.

On the other hand, we have seen that a single RTE was to be solved for each frequency of the grid, supposing that the scattering of photons is monochromatic. As mentioned at the beginning of Section 6.2, that is not really the case, and we have to take account of the frequency redistribution of photons during each scattering event. The RTE needs then to be solved several times for each frequency point.

Including in the discussion the great number of spectral lines, the statistical processing and the frequency redistribution of photons, we conclude as follows: the RTE has to be solved some thousands of times in a "standard" stellar atmosphere model, and some tens of thousands of times in the most complex models currently developed by astrophysicists.

34.6.4 Some Difficulties when Solving a Single Radiative Transfer Equation

Let us add a few comments on the difficulties met when solving the RTE at a given frequency.

1) The range for the parameters ϖ_ν and τ_ν^* goes from 0 to 0.999999... for ϖ_ν (supposing the albedo is constant) and from some units to $10^6 - 10^8$ for τ_ν^*. Large values of τ_ν^* are reached at those frequencies where the opacity is large (since $\tau_\nu^* = \int_0^{z^*} \chi_\nu(z)dz$), that is at the central frequency of strong spectral lines. Since the albedo is close to 1 at these frequencies, we have *simultaneously* $\varpi_\nu \sim 1$ and $\tau_\nu^* \to +\infty$ in the core of strong spectral lines.

2) As a function of τ, the solution of the RTE is smooth in a large part of the interval $[0, \tau_\nu^*]$; it varies suddenly when approaching the surface of the atmosphere $\tau = 0$. This effect is due to the escape of photons in the interstellar vacuum. The sudden change occurs at the boundary of the optically thick and the optically thin parts of the atmosphere. There is some analogy with the transitional flows in aerodynamics. As a result, the discretization of the τ-variable, which is conceptually straightforward, is difficult in a stellar atmosphere model. Uniform grids are quite out of the question, logarithmic grids are acceptable, adaptative grids are the best.

3) Accurate computations are easy in the deep layers of an atmosphere, but far more difficult close to the surface. Now, astrophysicists are primarily interested in the description of the radiation field in those superficial layers where the solution is difficult to reach. It is so because the main interesting features of stellar spectra form at very small τ.

34.6.5 The Overall Problem

Solving the RTE is a step in the iterative resolution of a system of equations that describes the structure of the atmosphere.[1] We emphasize the strong coupling between all equations, which form a nonlinear system. We have seen that the radiative transfer of energy depends on the optical properties of the atmosphere via the coefficients of the RTE. Conversely, these coefficients depend strongly on the radiation field by means of the structural equations of the atmosphere. It follows that the RTE with explicit coefficients is a nonlinear equation. It is not possible to give briefly a general survey on the resolution of the "overall problem" posed by stellar atmospheres modeling. Here, our sole aim is to put forward the huge amount of equations and data involved in a model.

As mentioned in [1], the main unknown quantities of a model are the specific intensity of the radiation field $I(z, \mu, \nu)$, the number densities $n_i(z)$ of each chemical species in all possible ionisation and energy states, the electron number density $n_e(z)$ and the temperature $T(z)$. In a realistic model, there are some tens of chemical species, with possibly tens, hundreds, thousands, tens of thousands of energy states for each ion of each species (depending on the electronic structure of the ion). It yields theoretically a huge number of unknowns n_i. Once again, statistical processing gives help to astrophysicists, and current models include typically some hundreds of number densities n_i (only).

The variable z (or τ) is discretized using a grid of typically one hundred points. The grid for the angular variable μ contains say, some tens or hundreds of points (depending on z) and we have seen that some thousands or tens of thousands of frequencies were required in a typical model.

34.6.6 How Do They Do?

The standard algorithms used in astrophysics for solving the above problem are the ALI methods [17]. ALI means accelerated (or approximate) lambda iteration, the lambda operator being the T-operator of Section 5 divided by ϖ. These methods solve the integral version (34.16) of the RTE. The T-operator is replaced by a finite rank operator T_n defined by an approximate product integration in (34.17). This approximate integration uses a parabolic interpolation of the S-function. The associated n-dimensional linear system (34.19) is solved by remarking that the main entries of the A_n matrix lie on its diagonal or close to it. Classical iterative methods (Jacobi, Gauss–Seidel,...) for solving large linear systems with band matrix are applied [18]. This is equivalent to the Atkinson scheme starting from an approximate version of the resolvent operator R_n. This crude starting point is rapidly improved by accelerating the iterative scheme, using for instance the Ng-acceleration [19].

This brief description reveals the main objective of astrophysicists when tackling the resolution of (34.16): reach the solution as quickly as possible, use as little memory as possible. Of course, this is due to the great number of integral equations to be solved. Another requirement imposed by physics is fully satisfied by the ALI methods: their software performance is not too sensitive to the values of the input parameters ϖ_ν and τ_ν^*, which can be extreme depending on ν. This "robustness" is one of the most surprising features of the ALI methods. Their main drawback is their limitation as far as accurate calculations are concerned. Anyway, a clarification of the mathematical framework of the ALI methods would be an apposite work.

34.7 Conclusion

Multiple scattering theory provides a wide field of application to mathematicians working on integral equations. The basic equation of the theory is really attractive, since it is linear. It has already generated some beautiful developments revealing a rich mathematical structure, which can be studied for its own intrinsic interest or in view of some application.

Solving the transport equation poses a difficult but within reach problem. In the simplified model of Section 4, the integral version of this equation is a weakly singular integral equation with an apparently simple kernel. This straightforwardness is deceptive: in the concrete example we have described in Section 6, everything works towards making the solution difficult to reach. Nature is sometimes mischievous!

References

1. L. Chevallier, Stellar atmospheres modeling, this volume, 37–40.

2. J. J. Duderstadt and W.R. Martin, *Transport Theory*, John Wiley and Sons, New York, 1979.

3. G.C. Pomraning, *The Equations of Radiation Hydrodynamics*, Pergamon Press, Oxford, 1973.

4. M.N. Ozisik, *Radiative Transfer*, John Wiley and Sons, New York, 1973.

5. M.M.R. Williams, *Mathematical Methods in Particle Transport Theory*, Butterworths, London, 1971.

6. R. Dautray and J.-L. Lions, *Analyse Mathématique et Calcul Numérique pour les Sciences et les Techniques*, Masson, Paris, 1984.

7. J. Lenoble (ed.), *Radiative Transfer in Scattering and Absorbing Atmospheres: Standard Computational Procedures*, A. Deepak, Hampton, VA, 1985.

8. M. Ahues, A. Largillier, and O. Titaud, The roles of weak singularity and the grid uniformity in relative error bounds, *Numer. Functional Anal. Optimization* **22** (2002), 789–814.

9. B. Rutily and J. Bergeat, The solution of the Schwarzschild-Milne integral equation in a homogeneous isotropically scattering plane-parallel medium, *J. Quantitative Spectr. Radiative Transfer* **51** (1994), 823–847.

10. B. Rutily and L. Chevallier, A study of two basic auxiliary functions in radiative transfer theory (in preparation).

11. M. Ahues, A. Largillier, and B.V. Limaye, *Spectral Computations for Bounded Operators*, Chapman and Hall/CRC, Boca Raton, 2001.

12. O. Titaud, Reduction of computation in the numerical resolution of a second kind weakly singular Fredholm equation, this volume, 255–260.

13. O. Titaud, *Analyse et résolution numérique de l'équation de transfert*, PhD Thesis, Université Jean Monnet, Saint-Etienne (France), available at http://www.univ-st-etienne.fr/anum/annuaire/otitaud, 2001.

14. P.B. Vasconcelos and F. d'Almeida, A parallel code for integral equations on a cluster of computers, this volume, 261–266.

15. G.P. Panasenko, Method of asymptotic partial decomposition of domain, *Math. Models Methods Appl. Sci.* **8** (1998), 139–156.

16. G. Panasenko, B. Rutily, and O. Titaud, Asymptotic analysis of integral equations for great interval and its application to stellar radiative transfer, *C.R. Acad. Sci. Paris Sér. IIb* **330** (2002), 735–740.

17. F. Paletou, Transfert de rayonnement: méthodes itératives, *C.R. Acad. Sci. Paris Sér. IV* **2** (2001), 885–898.

18. R.S. Varga, *Matrix Iterative Analysis*, Prentice-Hall, Englewood Cliffs, NJ, 1962.

19. L.H. Auer, Acceleration of convergence, in *Stellar Atmospheres: Beyond Classical Models*, L. Crivellari et al. (eds.), NATO ASI Ser., Kluwer, Holland, 1991, 9–17.

35 A Resonance Problem for a Second-Order Vector Differential Equation

Seppo Seikkala and Markku Hihnala

35.1 Introduction

We shall consider the boundary value problem (BVP)

$$
\begin{cases}
\begin{pmatrix} x_1'' \\ x_2'' \end{pmatrix} + A \begin{pmatrix} x_1 \\ x_2 \end{pmatrix} = \begin{pmatrix} f_1(ax_1 + bx_2) \\ f_2(cx_1 + dx_2) \end{pmatrix} + \begin{pmatrix} b_1(t) \\ b_2(t) \end{pmatrix}, \\
x_1(0) = x_2(0) = x_1(\pi) = x_2(\pi) = 0.
\end{cases}
\tag{35.1}
$$

Here f_1, f_2, b_1 and b_2 are continuous and bounded and a, b, c and d are real numbers. The matrix A is diagonalizable and such that the corresponding homogeneous problem has nontrivial solutions, i.e., we have a BVP at resonance. Problems of this type arise in mechanics (coupled oscillators) or in coupled circuits theory [1].

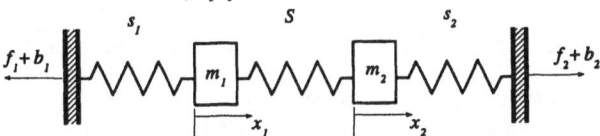

Fig. 1. A coupled spring-mass system.

For example, for the spring-mass system in Fig. 1 (m_1 and m_2 are the masses and s_1, s_2, and S the stiffnesses) we have

$$
A = \begin{pmatrix} a_1 & -\alpha \\ -\beta & a_2 \end{pmatrix},
\tag{35.2}
$$

where $a_1 = \frac{s_1}{m_1} + \frac{S}{m_1}, a_2 = \frac{s_2}{m_2} + \frac{S}{m_2}, \alpha = \frac{S}{m_1}$ and $\beta = \frac{S}{m_2}$. In the special case $\frac{s_1}{m_1} = \frac{s_2}{m_2} = 1$, which was studied in ([2]), we have

$$
A = \begin{pmatrix} \alpha + 1 & -\alpha \\ -\beta & \beta + 1 \end{pmatrix}
$$

and the problem (35.1) is a resonance problem. If furthermore $s_1 = s_2 = m_1 = m_2$, then

$$
A = \begin{pmatrix} \alpha + 1 & -\alpha \\ -\alpha & \alpha + 1 \end{pmatrix}.
$$

The problem (35.1) with this A was studied in ([3]).We shall consider here more general problems (35.1) at resonance, special cases of which are, for example,

$$\frac{S}{m_1} = \frac{5}{2}, \frac{S}{m_2} = \frac{18}{5}, \frac{s_1}{m_1} = \frac{3}{2}, \frac{s_2}{m_2} = \frac{2}{5} \tag{35.3}$$

and

$$m_1 = \frac{64}{81}, m_2 = 1, s_1 = \frac{32}{81}, s_2 = \frac{13}{9}, S = \frac{32}{9}. \tag{35.4}$$

By change of variables $z = T^{-1}x$, where T is a matrix diagonalizing A, having the eigenvectors of A as columns, problem (35.1) is transformed into a system

$$\begin{cases} \begin{pmatrix} z_1'' + d_1 z_1 \\ z_2'' + d_2 z_2 \end{pmatrix} = T^{-1} \begin{pmatrix} f_1(ax_1 + bx_2) \\ f_2(cx_1 + dx_2) \end{pmatrix} + T^{-1} \begin{pmatrix} b_1(t) \\ b_2(t) \end{pmatrix}, \\ z_1(0) = z_2(0) = z_1(\pi) = z_2(\pi) = 0, \end{cases} \tag{35.5}$$

where $x = \begin{pmatrix} x_1 \\ x_2 \end{pmatrix} = Tz$ and d_1, d_2 are the eigenvalues of A. For the matrix A in (35.2) we may choose

$$T = \begin{pmatrix} \alpha & \gamma \\ \gamma & -\beta \end{pmatrix}, \quad T^{-1} = \frac{1}{\alpha\beta + \gamma^2} \begin{pmatrix} \beta & \gamma \\ \gamma & -\alpha \end{pmatrix}, \tag{35.6}$$

where either $\gamma = \frac{a_1-a_2}{2} + \frac{\sqrt{(a_1-a_2)^2+4\alpha\beta}}{2}$ or $\gamma = \frac{a_1-a_2}{2} - \frac{\sqrt{(a_1-a_2)^2+4\alpha\beta}}{2}$.

If either $d_1 = k^2$ or $d_2 = l^2$, where k and l are positive integers, then we have a resonance case and the null space of the differential operator E in the coupled system (35.1),

$$Ex = \begin{pmatrix} x_1'' \\ x_2'' \end{pmatrix} + A \begin{pmatrix} x_1 \\ x_2 \end{pmatrix},$$

with the given Dirichlet boundary conditions, is spanned by either

$$\Phi(t) = T \begin{pmatrix} 1 \\ 0 \end{pmatrix} \sin kt$$

or by

$$\eta(t) = T \begin{pmatrix} 0 \\ 1 \end{pmatrix} \sin lt.$$

In this paper we shall choose the former notation and study the existence and multiplicity of solutions of (35.1) depending on the parameter $\bar{b}$ in the decomposition

$$b(t) = \begin{pmatrix} b_1(t) \\ b_2(t) \end{pmatrix} = \bar{b}\Psi(t) + \tilde{b}(t), \tag{35.7}$$

where $\tilde{b}$ is orthogonal to $\Psi(t) = \sqrt{\frac{2}{\pi}}\sqrt{\frac{1}{\beta^2+\gamma^2}}\begin{pmatrix}\beta\\\gamma\end{pmatrix}\sin kt$ and $\bar{b} = (b|\Psi)$, the inner product $(x|y)$ in $X = C[0,\pi] \times C[0,\pi]$ being defined by

$$(x|y) = \int_0^\pi x(t)y(t)dt = \int_0^\pi [x_1(t)y_1(t) + x_2(t)y_2(t)]dt.$$

The motivation for decomposition (35.7) is that the BVP

$$Ex = g, \ x(0) = x(\pi) = 0$$

has a solution if and only if g is orthogonal to Ψ. We shall generalize results of [2], [3], and [4].

In the more complicated case when $d_1 = k^2$ and $d_2 = l^2$ (e.g. (35.4)), which was mentioned in [3] but has not been studied in the literature (as far as we know), and hence when the null space of E is spanned by $\{\Phi(t), \eta(t)\}$, we shall present numerical results for existence of solutions of (35.1) in a forthcoming paper.

35.2 Results

Suppose that A is of the form (35.2), $d_1 = k^2$, $d_2 \neq l^2$ and that T is given by (35.6). For a fixed $\lambda \in \mathbb{R}$ consider the integral equation system

$$z = \lambda\psi + KNz, \tag{35.8}$$

where

$$Nz = N\begin{pmatrix}z_1\\z_2\end{pmatrix} = T^{-1}\begin{pmatrix}f_1(ax_1 + bx_2) + b_1\\f_2(cx_1 + dx_2) + b_2\end{pmatrix}, \quad \psi = \begin{pmatrix}1\\0\end{pmatrix}\phi,$$

$x = Tz$, $\phi(t) = \sin kt$, the linear operator $K : X \to X$ is defined by

$$Kz = \begin{pmatrix}\int_0^\pi k_1(t,s)z_1(s)ds\\\int_0^\pi k_2(t,s)z_2(s)ds\end{pmatrix},$$

k_1 defined by

$$\begin{cases} Lk_1(t,s) = \delta(t-s) - \phi(t)\phi(s),\\ k_1(0,s) = k_1(\pi,s) = 0,\\ \int_0^\pi k_1(t,s)\phi(t)dt = 0, \end{cases}$$

and k_2 is the ordinary Green's function for problem $z_2'' + d_2 z_2 = 0$, $z_2(0) = z_2(\pi) = 0$.

Since f_1 and f_2 are continuous and bounded, then for a fixed $\lambda \in \mathbb{R}$ the system (35.8) has at least one solution. For any such solution z^λ we denote $x^\lambda = Tz^\lambda$,

$$\delta(\lambda) = \int_0^\pi [f(x^\lambda(t)) + b(t)]\Psi(t)dt$$

and

$$\tilde{\delta}(\lambda) = \int_0^\pi f(x^\lambda(t))\Psi(t)dt.$$

Thus, we have $\delta(\lambda) = \tilde{\delta}(\lambda) + \bar{b}$.

We now easily deduce that the BVP (35.5) is equivalent to the pair of equations

$$\begin{cases} z = \lambda\psi + KNz, \\ \delta(\lambda) = 0, \end{cases}$$

and from this it follows that the BVP (35.1) is equivalent to the pair

$$\begin{cases} x = \lambda\Phi + Fx, \\ \delta(\lambda) = 0, \end{cases} \tag{35.9}$$

where $F = TKNT^{-1}$. Note that $\tilde{\delta}(\lambda)$ depends on $\bar{b}$, however, $\tilde{\delta}(\lambda)$ is independent of $\bar{b}$ because x^λ is independent of $\bar{b}$. Denote

$$a = \inf\{\tilde{\delta}(\lambda) : \lambda \in \mathbb{R}, \ x^\lambda \text{ is a solution of (35.9)}\},$$
$$b = \sup\{\tilde{\delta}(\lambda) : \lambda \in \mathbb{R}, \ x^\lambda \text{ is a solution of (35.9)}\}.$$

Theorem 1. *The BVP (35.1) has (i) at least one solution if $-\bar{b} \in (a,b)$ and (ii) no solution if $-\bar{b} \notin [a,b]$. If $c \in (a,b)$ is a limit point of both $\{\tilde{\delta}(\lambda) : \lambda \in (-\infty, d]\}$ and $\{\tilde{\delta}(\lambda) : \lambda \in [d, \infty)\}$ for a $d \in \mathbb{R}$ and if $-\bar{b} \in (a,b)\backslash\{c\}$, then problem (35.1) has at least two solutions.*

Proof. (cf.[2],[5]) Since f_1 and f_2 are continuous and bounded, then for any closed bounded interval $I = [\alpha, \beta]$ there exists a closed bounded convex subset B of $C[0,\pi] \times C[0,\pi]$ such that the mapping G,

$$G(x,\lambda) = \lambda\Phi + Fx, \ \lambda \in I, x \in B,$$

is a compact continuous mapping from $B \times I$ into B. Then, by ([6],Fixed Point Theorem, p.341), there exists a connected set $S \subset B \times I$ of fixed points of T, and S meets both $B \times \{\alpha\}$ and $B \times \{\beta\}$. Now, if $a < -\bar{b} < b$, then there exist λ_1 and λ_2 such that $\lambda_1 < \lambda_2$ and $\tilde{\delta}(\lambda_1) < -\bar{b} < \tilde{\delta}(\lambda_2)$. Hence, as a continuous real-valued function on a connected set S_1 associated with the interval $I_1 = [\lambda_1, \lambda_2]$, $(x, \lambda) \to \tilde{\delta}(\lambda) + \bar{b}$ assumes the value 0 on S_1, i.e. problem (35.1) has a solution. This proves (i), and (ii) follows immediately. If (iii) holds, then for any $-\bar{b} \in (a,b)\backslash\{c\}$ we can find $\tilde{\delta}(\lambda_1), \tilde{\delta}(\lambda_2)$ and $\tilde{\delta}(\lambda_3)$

such that $\lambda_1 < \lambda_2 < \lambda_3$ and $\tilde{\delta}(\lambda_1) < -\bar{b} < \tilde{\delta}(\lambda_2)$, $\tilde{\delta}(\lambda_3) < -\bar{b} < \tilde{\delta}(\lambda_2)$. Hence, on a connected set S_1 associated with the interval $I_1 = [\lambda_1, \lambda_2]$, $(x, \lambda) \rightarrow \tilde{\delta}(\lambda) + \bar{b}$ assumes the value 0 on S_1, i.e. problem (35.1) has a solution. The same conclusion holds true for the interval $I_2 = [\lambda_2, \lambda_3]$, i.e problem (35.1) has at least two solutions.

Example 1. The case (35.3) above leads, for example, to the problem

$$\begin{cases} x_1'' + 4x_1 - \frac{5}{2}x_2 = \tan^{-1}(x_1 + 2x_2) + b_1(t), \\ x_2'' - \frac{18}{5}x_1 + 4x_2 = \tan^{-1}(2x_1 - x_2) + b_2(t), \ . \\ x_1(0) = x_2(0) = x_1(\pi) = x_2(\pi) = 0 \end{cases} \qquad (35.10)$$

Now we have $\alpha = \frac{5}{2}$, $\beta = \frac{18}{5}$, $\gamma = 3$, $d_1 = 1$, $d_2 = 7$, and

$$\tilde{\delta}(\lambda) = \eta \int_0^\pi \left\{ \frac{18}{5}\tan^{-1}[x_1{}^\lambda(t) + 2x_2{}^\lambda(t)] \right.$$

$$\left. + 3\tan^{-1}[2x_1{}^\lambda(t) - x_2{}^\lambda(t)] \right\} \sin t \, dt,$$

where $\eta = \sqrt{\frac{2}{\pi}}\sqrt{\frac{1}{(\frac{18}{5})^2 + 9}}$, from which we deduce that $\lim_{\lambda \to \infty} \tilde{\delta}(\lambda) = -\eta\frac{3\pi}{5}$ and $\lim_{\lambda \to -\infty} \tilde{\delta}(\lambda) = \eta\frac{3\pi}{5}$. Hence, for problem (35.1) we have $(-\eta\frac{3\pi}{5}, \eta\frac{3\pi}{5}) \subseteq (a, b)$. Thus, problem (35.10) has at least one solution if $\bar{b} \in (-\eta\frac{3\pi}{5}, \eta\frac{3\pi}{5}) \approx (-3.53, 3.53)$. Numerically we have found the curve $\tilde{\delta}(\lambda)$ shown in Fig. 2 when $\bar{b} \equiv 0$. It indicates that problem (35.10), in case $\bar{b} \equiv 0$, has a unique solution when $\bar{b} \in (-3.53, 3.53)$ and no solution otherwise.

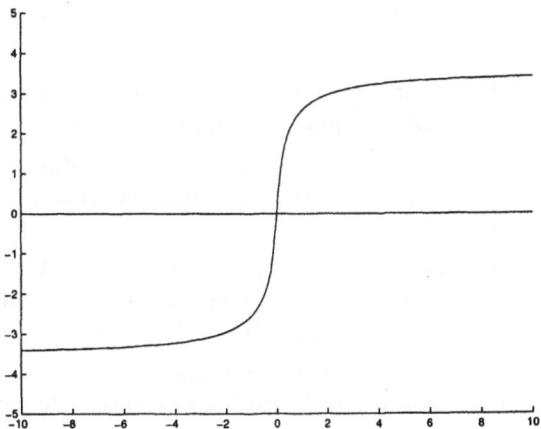

Fig. 2. The $\tilde{\delta}(\lambda)$ curve for problem (35.10).

Example 2. [2] It can be shown that the problem

$$\begin{cases} x_1'' - 4x_1 + 5x_2 = \frac{1}{2}\tan^{-1}(x_1 + 2x_2) + b_1(t), \\ x_2'' - 2x_1 + 3x_2 = \frac{1}{5}\tan^{-1}(2x_1 - x_2) + b_2(t), \\ x_1(0) = x_2(0) = x_1(\pi) = x_2(\pi) = 0, \end{cases} \qquad (35.11)$$

with $\tilde{b} \equiv 0$, for small $|\bar{b}|$ has a solution. Also, $\lim\limits_{\lambda \to \infty} \tilde{\delta}(\lambda) = \lim\limits_{\lambda \to -\infty} \tilde{\delta}(\lambda) = 0$.
Hence, by Theorem 1, problem (35.11) has at least two solutions for small $|\bar{b}|$. Numerically we have found the curve $\tilde{\delta}(\lambda)$ given in Fig. 3 below. It indicates that $(a, b) \approx (-0.206, 0.206)$.

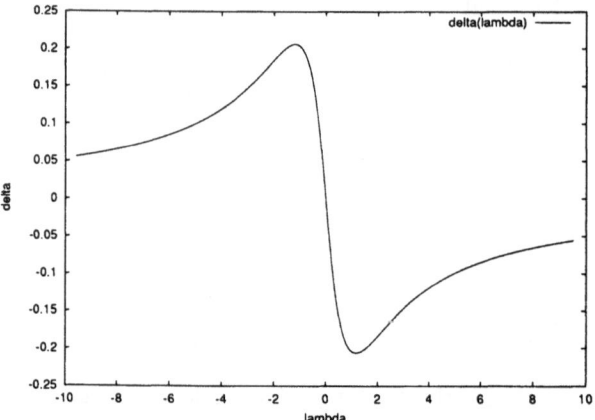

Fig. 3. The $\tilde{\delta}(\lambda)$ curve for problem (35.11).

References

1. I.G. Main, *Vibrations and Waves in Physics*, Cambridge Univ. Press, Cambridge, 1993.

2. S. Seikkala and D. Vorobiev, A Resonance Problem for a System of Second Order Differential Equations (in preparation).

3. A. Canada, Nonlinear ordinary boundary value problems under a combined effect of periodic and attractive nonlinearities, *J. Math. Anal. Appl.* **243** (2001), 174–189.

4. P. Diamond, P.E. Kloeden, A.M. Krasnosel'skii, and A.V. Pokrasovskii, Bifurcation at infinity for equations in spaces of vector-valued functions, *J. Austral. Math. Soc. Ser. A* **63** (1997), 263–280.

5. R. Kannan and S.Seikkala, Existence of solutions to $u'' + u + g(t, u, u')$, $u(0) = u(\pi) = 0$, *J. Math. Anal. Appl.* **263** (2001), 555–564.

6. H. Shaw, A nonlinear elliptic boundary value problem, in *Nonlinear Functional Analysis and Differential Equations*, L. Cesari, R. Kannan, and J.D. Schuur (eds.), Marcel Dekker, 1976, 339–345.

36 Numerical Calculations for a Mullins–Sekerka Problem in 2D

Jianzhong Su and Bao Loc Tran

36.1 Mathematical Model and Formulations

The Mullins–Sekerka equation [1] has been known to be a singular limit of the Cahn–Hillard equation [2] and phase-field equation [3] in the sense that there is a one-to-one relation between their solution sets; knowledge of the solutions of the Mullins–Sekerka equation will help to understand the dendritic solidification phenomena in the phase transition problems [4,5] which is a very important area of study. These problems are also closely related to a classical problem in fluid dynamics studied by Saffman and Taylor [6] and others [4,5,7,8].

We consider the Mullins–Sekerka problem in 2-dimensional space. Let $\Omega = \{(x,y) \in \mathbf{R}^2, -\infty < x < \infty, -1 \le y \le 1\}$, the infinitely long channel with width 2, and Γ_0 be a simple curve in Ω. Consider the free boundary problem of a potential function $u(x,t)$, $x \in \Omega$, $t \ge 0$, and a free boundary $\Gamma_{0,T} = \cup_{0 \le t < T}(\Gamma_t \times \{t\})$ for some $T > 0$ satisfying the Mullins–Sekerka equation

$$\Delta u(x,y,t) = 0 \quad \text{in } \Omega \setminus \Gamma_t,$$

$$\frac{\partial u}{\partial n} = 0 \quad y = \pm 1 \quad \text{or } x \to \infty,$$

$$\nabla u = (U, 0) \quad x \to -\infty \quad \text{and} \quad x \in \Omega \quad \text{and to the left of } \Gamma_t,$$

$$\nabla u = (0, 0) \quad x \to -\infty \quad \text{and} \quad x \in \Omega \quad \text{and to the right of } \Gamma_t, \tag{36.1}$$

$$u = K \quad \text{on } \Gamma_t,$$

$$-\left[\frac{\partial u}{\partial n}\right]_\Gamma = V \quad \text{on } \Gamma_t,$$

$$\Gamma_t|_{t=0} = \Gamma_0.$$

We change Eq. (36.1) into integral formulations as follows.

The first author was partially supported by the Texas ARP grant No. 003656-0009-1999.

Lemma 1. *Let* $x = (\zeta, \chi), y = (\xi, \eta), G(x, y) = \frac{1}{2\pi}(\log|x - y| + h(x, y)),$

$$h(x, y) = \sum_{k=1}^{\infty} \left(\log \frac{|(\zeta, \chi - 4k) - (\xi, \eta)|}{4k} + \log \frac{|(\zeta, \chi + 4k) - (\xi, \eta)|}{4k} \right.$$

$$+ \log \frac{|(\zeta, -\chi - 2(2k - 1)) - (\xi, \eta)|}{2(2k - 1)}$$

$$\left. + \log \frac{|(\zeta, -\chi + 2(2k - 1)) - (\xi, \eta)|}{2(2k - 1)} \right).$$

If $(u, \Gamma_{0,T})$ *is a solution of (36.1), then*

$$K(x) = \frac{1}{2\pi} \int_{\Gamma_t} G(x, y)V(y) \, ds_y + C \quad \text{for} \quad x \in \Gamma_t, \tag{36.2}$$

$$\int_{\Gamma_t} V(y) \, ds_y = UW \tag{36.3}$$

hold for $t \in [0, T]$ *where* $V(x, t)$ *is the normal velocity of* Γ_t, $C = C(t)$, $K(x, t)$ *is the curvature of* Γ_t, *and* W *is the width of* Γ_t *at* $x \to -\infty$.

For details of the Green's function $G(x, y)$ and the derivation of (36.2), see [7,9].

36.2 Discretization and Algorithm

Suppose that Γ_t is discretized by N points, labeled by $z_1, z_2, \ldots, z_N$ counterclockwise. Let z_L, z, and z_R be three counterclockwisely consecutive points on Γ; we approximate Γ near z as a segment of the circle passing through z_L, z and z_R. We estimate the unit tangent τ, outward unit normal n, and the curvature K of the interface Γ by the corresponding unit tangent, outward unit normal and the curvature of the circle passing through z_L, z, and z_R. We denote

$$T_L = \frac{z - z_L}{|z - z_L|}, \quad T_R = \frac{z - z_R}{|z - z_R|},$$

$$N_L = (T_L^y, -T_L^x), \quad N_R = (T_R^y, -T_R^x),$$

$$d_L = |z - z_L|, \quad d_R = |z_R - z|, \quad d_{RL} = |z_R - z_L|.$$

Lemma 2. *Let* V *be the velocity of interface* Γ *at* $z(t)$, *then the new location of the interface at* $z(t + h)$ *is approximated by*

$$z(t + h) = z(t) + h \cdot n(z(t)) \cdot V(t). \tag{36.4}$$

Furthermore, the curvature of the interface at $z(t + h)$ *is determined by*

$$K(t + h) = K(t) + h \cdot B \cdot V(t)$$

where $B = (b_{i,j})$ for $i, j = 1, \ldots, N$ and

$$b_{i,i-1} = -\frac{2(N_L \cdot n_L)(N_R \cdot N_L)}{d_L d_{RL}} + K\frac{(d_R T_R + d_L T_L) \cdot n_L}{d_{RL}^2},$$

$$b_{i,i} = \frac{2(N_R \cdot n)(N_R \cdot N_L)}{d_R d_{RL}} + K\frac{2(N_L \cdot n)(N_R \cdot N_L)}{d_R d_{RL}},$$

$$b_{i,i+1} = -\frac{2(N_R \cdot n_R)(N_R \cdot N_L)}{d_L d_{RL}} - K\frac{(d_R T_R + d_L T_L) \cdot n_R}{d_{RL}^2}, \quad i = 1, \cdots, N,$$

$$b_{i,j} = 0 \quad for\ j \neq i, i+1, i-1;\ i, j = 1, \cdots, N.$$

$$\tag{36.5}$$

For the proof of Lemma 2, see [9].

Let $\Gamma = \bigcup_{j=1}^N \Gamma_j$, where Γ_j is a small segment of Γ that contains z_j and V_j is almost constant on the segment Γ_j. We define

$$a_{ij} = \frac{1}{2\pi d_i} \int_{\Gamma_i} \int_{\Gamma_j} G(z, z')\ ds_{z'}\ ds_z \quad and \quad d_i = \int_{\Gamma_i} d\ s_i = \frac{1}{2}(d_{L[i]} + d_{R[i]}).$$

With the evaluation of K_i at $z_i(t+h)$ in (36.5), (36.2)–(36.3) become

$$\sum_{j=1}^N (a_{ij} - h\, b_{ij})V_j + C = K_i \quad for \quad i = 1, \cdots, N,$$

$$\sum_{j=1}^N d_j V_j = UW.$$

$$\tag{36.6}$$

After solving (36.6) for $(V_1, V_2, \cdots, V_N, C)$ and using $V_1, V_2, \ldots, V_N$ in formula (36.4) to update z_i, we redistribute all $z_i(t+h)$ denoted by z_i according to

$$z_i = z_i + \frac{(d_R - d_L)(T_L - T_R)}{2(1 + T_L \cdot T_R)}.$$

$$\tag{36.7}$$

We have

$$a_{ij} = \frac{1}{2\pi d_i} \int_{\Gamma_i} \int_{\Gamma_j} G(z, z')\ ds_{z'}\ ds_z$$

$$\tag{36.8}$$

$$= \frac{1}{2\pi d_i} \int_{\Gamma_i} \int_{\Gamma_j} \ln|z - z'|\ ds_{z'}\ ds_z + \frac{1}{2\pi d_i} \int_{\Gamma_i} \int_{\Gamma_j} h(z, z')\ d\ s_{z'}\ d\ s_z.$$

The first integration in (36.8) can be evaluated directly, see [9]. The second one can be evaluated by truncating $h(z, z')$ up to $k = 15$:

$$\frac{1}{2\pi d_i} \int_{\Gamma_i} \int_{\Gamma_j} h(z, z')\ ds_{z'}\ ds_z = \frac{1}{2\pi}h(z, z')\ d_j,$$

$$h(x, y) = \sum_{k=1}^{15} \log \frac{|(\zeta, x - 4k) - (\xi, \eta)|}{4k} * \frac{|(\zeta, x + 4k) - (\xi, \eta)|}{4k}$$

$$* \frac{|(\zeta, -x - 2(2k-1)) - (\xi, \eta)|}{2(2k-1)} * \frac{|(\zeta, -x + 2(2k-1)) - (\xi, \eta)|}{2(2k-1)} + R_{15},$$

where
$$R_{15} \leq 0.00819599 * |\zeta - \xi|^2$$

from estimations of the remainder of the infinite series.

In summary, the numerical scheme consists of the following steps:

a. Solve for $(V_1, V_2, \ldots, V_N, C)$ in (36.6).

b. Update z_i by (36.4). The time step h is dynamically updated as

$$h = h_0 * \min\{\frac{1}{V_{max}}, 1\} \quad (h_0 \text{ is the initial time step}),$$

$$V_{max} = \max(|V_i|) \quad \text{for } i = 1, \ldots, N.$$

c. Redistribute z_i by (36.7).

36.3 Numerical Examples

In this section, all the experiments have the initial time step $h_0 = 0.0004$ and $N = 65$. The parameters W and U are as in (36.3). We illustrate below the evolutions of the interface with an initial curve:

$$x(r) = 0.18 \cos\left(r - \frac{\pi}{2}\right),$$

$$y(r) = W \sin\left(r - \frac{\pi}{2}\right), \tag{36.9}$$

$$0 \leq r \leq \pi.$$

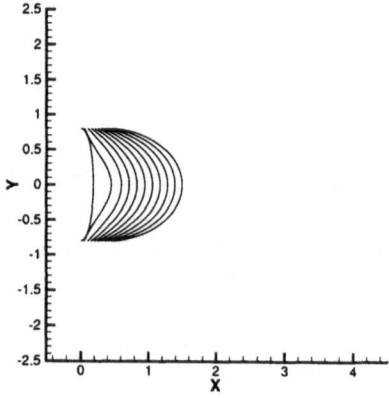

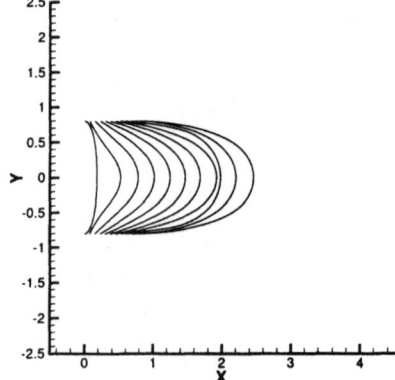

Interface evolution from (36.9) with $W = 0.8$ and $U = 0.5$, from $t = 0.0$ to $t = 1.0$.

Interface evolution from (36.9) with $W = 0.8$ and $U = 0.75$, from $t = 0.0$ to $t = 1.0$.

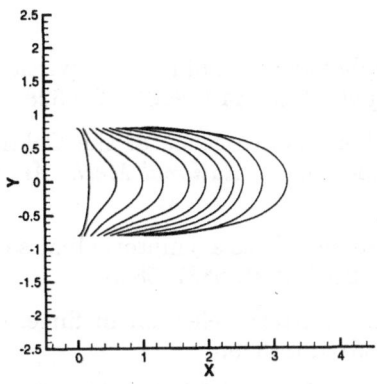

Interface evolution from (36.9) with $W = 0.8$ and $U = 1.0$, from $t = 0.0$ to $t = 1.0$.

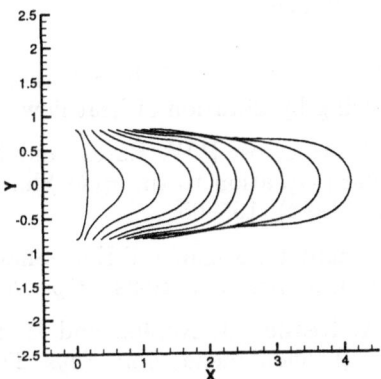

Interface evolution from (36.9) with $W = 0.8$ and $U = 1.25$, from $t = 0.0$ to $t = 1.0$.

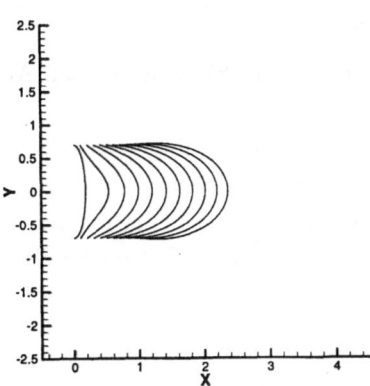

Interface evolution from (36.9) with $W = 0.7$ and $U = 1.0$, from $t = 0.0$ to $t = 1.0$.

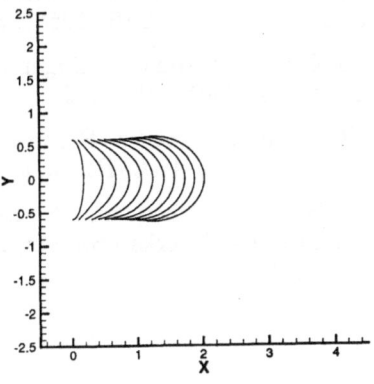

Interface evolution from (36.9) with $W = 0.6$ and $U = 1.0$, from $t = 0.0$ to $t = 1.0$.

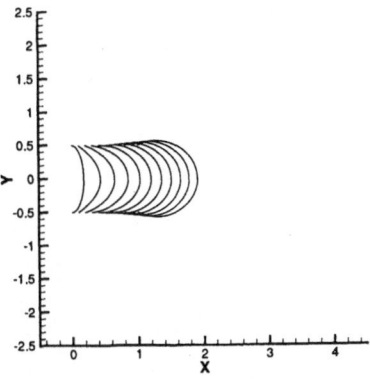

Interface evolution from (36.9) with $W = 0.5$, $U = 1.0$, from $t = 0.0$ to $t = 1.0$.

References

1. W.W. Mullins and R.F. Sekerka, Morphological stability of a particle growing by diffusion of heat flow, *J. Appl. Phys.* **34** (1963), 323–328.

2. N.D. Alikakos, P.W. Bates, and X. Chen, Convergence of the Cahn–Hillard equation to the Hele–Shaw model, *Arch. Rational Mech. Anal.* **128** (1994), 165–05.

3. G. Caginalp, Stefan and Hele–Shaw type models as asymptotic limits of the phase field equations, *Phys. Rev. A* **39** (1989), 5887–5896.

4. D.A. Kessler, J. Koplik, and H. Levine, Pattern selection in fingered growth phenomena, *Adv. Phys.* **37** (1988), 255–329.

5. S. Tanveer, Surprises in viscous fingering *J. Fluid Mech.* **409** (2000), 273–308.

6. P.G. Saffman and G.I. Taylor, The penetration of a fluid into a porous medium or Hele-Shaw cell containing a more viscous liquid, *Proc. R. Soc. London Ser. A* **245** (1958), 312–329.

7. J. Su, On the existence of finger solutions in Hele–Shaw Equation, *Nonlinearity* **14** (2001), 153–166.

8. X. Xie and S. Tanveer, Rigorous results in steady finger selection in viscous fingering *Arch. Rational Mech. Anal.* (in press).

9. P.W. Bates, X. Chen, and X. Deng, A numerical scheme for the two phase Mullins–Sekerka problem, *Electronic J. Differential Equations* **11** (1995), 1–28.

37 On Anisotropic Elliptic Equations in Bounded Domains

Tadie

37.1 Introduction

Because of the anisotropic character of the equations in (Qc) and (Pc) below, in solving those problems, we cannot right away apply the comparison principles. We will pick up some conditions under which the existence of some sub-and-super solutions of the problems leads to the existence theorems for those problems.

In this work, for some $m_0 > 0$ and $c > 0$, we consider the problems

$$Pu \equiv \sum_{i=1}^{n} \mu_i(x)\partial_{ii}u = f(x,u) \quad \text{in } \Omega; \quad u|_{\partial\Omega} = c,$$

$$\mu_i \in C(\overline{\Omega}); \quad \mu_i(x) > m_0 \quad \forall x \in \overline{\Omega}, \; i = 1, 2, \ldots, n, \qquad (Pc)$$

$$\Omega \in C^{2,\alpha} \text{ is open and bounded in } \mathbb{R}^n, \quad \alpha \in (0,\, 1),$$

$$f \in C(\overline{\Omega} \times \mathbb{R}_+; \; \mathbb{R}_+) \text{ is increasing in } u,$$

where $\partial_i = \frac{\partial}{\partial x_i}$, and with the same Ω and f

$$Qu \equiv \sum_{i=1}^{n} \mu_i(x,u)\partial_{ii}u = f(x,u) \quad \text{in } \Omega,$$

$$u|_{\partial\Omega} = c; \quad \mu_i \in C^1(\overline{\Omega} \times \mathbb{R}_+ \; ; \; \mathbb{R}_+) \text{ is decreasing in } u, \qquad (Qc)$$

$$\mu_i(x,u) > m_0 > 0 \quad \forall x \in \overline{\Omega}, \; u > 0, \; i = 1, 2, \ldots, n.$$

The alternative hypothesis on Ω is to be open, bounded, and either $\partial\Omega \in C^2$ or it satisfies the uniform external sphere condition [1]. When the domain satisfies that condition, there is a family of subdomains

(w) $\{\Omega_m\}_{m \in \mathbb{N}}$ such that $\forall m, \; \bar{\Omega}_m \subset \Omega_{m+1} \subset \Omega, \; \bigcup_{m \in \mathbb{N}} \Omega_m = \Omega$ and $\partial\Omega_m$

is a C^∞-submanifold of dimension $n - 1$.

We start with a uniqueness theorem.

The author dedicates this work to his colleagues Dr. L. Kadosh and Dr. M. McIntyre, for the difficult time spent together in Legon.

Theorem 1. *Suppose that $c > 0$ is finite and that $\forall x \in \Omega$ and $i = 1, 2, \ldots, n$ either*
    (a) $\mu_i(x, t)$ is decreasing and $f(x, t)$ increasing for $t > 0$, or
    (b) $\mu_i(x, t)$ is increasing and $f(x, t)$ decreasing for $t > 0$.
Then (Qc) has at most one solution $u \in C^{2,\alpha}(\overline{\Omega})$ which satisfies

$$\partial_{ii} u(x) \geq 0 \quad \forall x \in \Omega, \quad i = 1, 2, \ldots, n.$$

Proof. For such solutions u and v, in Ω and for $\mu_i(w) \equiv \mu_i(x, w)$,

$$\sum_{i=1}^{n} \{\mu_i(u) \partial_{ii}(u - v) + [\mu_i(u) - \mu_i(v)] \partial_{ii} v\}$$
$$+ f(x, v) - f(x, u) = 0. \qquad (37.1)$$

In the case (a) suppose that there is an open and nonempty $A \subset \Omega$ such that $u < v$ in A, and an $x' \in A$ such that $u - v$ reaches its minimum at x'. Then at x', $\partial_{ii}(u - v) > 0$ and the left-hand side of (37.1) is strictly positive; the assumption cannot hold. The case (b) is established in a similar way, with $u > v$ in A.

37.2 Comparison Results for Bounded Solutions

Definition 2. Let $\phi, \psi \in C^1(\overline{\Omega})$. The function ϕ (ψ) will be said to be a *subsolution (supersolution)* of the equation in (Pc) if

$$P\phi - f(x, \phi) \geq 0 \ (\geq P\psi - f(x, \psi)) \quad \text{a.e. in } \Omega.$$

If, in addition, $\phi \leq \psi$ in Ω and $\phi \leq c \leq \psi$ on $\partial\Omega$, the two functions will be said to be *Pc-compatible*.

Theorem 3. *If there are $\phi, \psi \in C^2(\overline{\Omega})$ which are Pc-compatible, then (Pc) has a solution $u \in C^{2,\alpha}(\overline{\Omega})$ such that $\phi \leq u \leq \psi$ in $\overline{\Omega}$.*

Proof. Define on $E \equiv E_{\phi\psi} := \{w \in C(\overline{\Omega}) \mid \phi \leq w \leq \psi \text{ in } \overline{\Omega}\}$ an operator T by

$$Tu = U \iff PU = f(x, u) \quad \text{in } \Omega, \quad U|_{\partial\Omega} = c. \qquad (T)$$

The coefficients of P being bounded with μ_i strictly positive for any u in E, (T) has a solution $U \in C^{2,\alpha}(\overline{\Omega})$ by the classical elliptic theory (see [1] and [2]). The first claim is that

$$TE \subset E \cap C^{2,\alpha}(\overline{\Omega}).$$

In fact, from (T) with $u \in E$,

$$\sum_{i=1}^{n} \mu_i(x) \partial_{ii}(U - \phi) \leq 0 \quad \text{in } \Omega, \quad (U - \phi) \geq 0 \quad \text{on } \partial\Omega. \qquad (37.2)$$

If $U < \phi$ in $\Omega' \subset \Omega$ with $\text{mes}(\Omega') > 0$, let x_0 be the point where $U - \phi$ reaches its minimum; then $\partial_{ii}(U - \phi)(x_0) > 0$ conflicting with (37.2) as each $\mu_i(x_0) > 0$. Thus $U \geq \phi$ in Ω. Similarly $U \leq \psi$ is obtained.

Define for some $u \in E$ the sequence

$$u_1 = Tu, \quad u_{n+1} = Tu_n, \quad n \in \mathbb{N}.$$

By elliptic theory,

$$\forall m \in \mathbb{N}, \ \forall p > 1, \quad ||u_m||_{W_p^2(\Omega)} \leq C(|f(.,\psi)|, \Omega, n),$$

where the constant is independent of m. This uniform bound leads to the existence of a fixed point of T in E which is such a required solution [2].

Theorem 4. *Let ϕ and ψ be those of Theorem 3. Then*
a) if $\mu_i(x) \geq 0 \ \forall x \in \Omega$, $i = 1, 2, \ldots, n$, any classical solution u of (Pc) satisfies

$$\phi \leq u \leq \psi \quad \text{in } \Omega; \tag{37.3}$$

b) if, in addition, $f(x, \cdot)$ is either convex or concave, (Pc) has at most one solution.

Proof. a) For such u,

$$\sum_i \mu_i(x)\partial_{ii}(u - \phi)(x) \leq f(x, u) - f(x, \phi), \quad (u - \phi)|_{\partial\Omega} \geq 0.$$

If there is $A \subset \Omega$ with $\text{mes } A > 0$ such that $u < \phi$ in A, let $x_0 \in A$ be the point where $u - \phi$ has its minimum; then $\sum_i \mu_i(x_0)\partial_{ii}(u - \phi)(x_0) > 0$, conflicting with the fact that $f(x_0, u) - f(x_0, \phi) < 0$ as f is increasing in its second argument. Thus, $u \geq \phi$ in Ω.

Similarly, $u \leq \psi$ in Ω.

b) Under the assumed conditions, if there are two distinct solutions u, v of (Pc), let $w_t = tu + (1 - t)v$, $t \in [0, 1]$. Then

$$Lw_t = \sum_i \mu_i(x)\partial_{ii}w_t$$

$$= tf(x, u) + (1 - t)f(x, v) \begin{cases} \geq f(x, w_t) & \text{if } f \text{ is concave,} \\ \leq f(x, w_t) & \text{if } f \text{ is convex.} \end{cases}$$

Thus w_t is a subsolution when f is concave and is a supersolution when it is convex. We choose t such that on $\partial\Omega$, $\quad \phi \leq c \leq w_t$ for the convex case or $w_t \leq c \leq \psi$ for the concave case. In each of the two cases, (37.3) is violated considering the Pc-compatible $\{\phi, w_t\}$ or $\{w_t, \psi\}$ as both u and v cannot lie between any of the compatible pairs.

Lemma 5. *For (Qc), assume that there are $\phi, \psi \in C^2(\overline{\Omega})$ such that*
1) $\forall x \in \Omega$, $\partial_{ii}\phi(x) \geq 0$ and $\partial_{ii}\psi(x) \geq 0$ in Ω, $i = 1, 2, \ldots, n$, and
2) $Q\phi - f(x, \phi) \geq 0 \geq Q\psi - f(x, \psi)$ in Ω and $\phi \leq c \leq \psi$ on $\partial\Omega$.

Then any solution u of (Qc) satisfies $\phi \le u \le \psi$ in Ω.

Proof. For such u and setting $\mu_i(w)$ for $\mu_i(x, w)$,

$$\sum_i \{(\mu_i(u) - \mu_i(\phi))\partial_{ii}\phi + \mu_i(u)\partial_{ii}(u - \phi) \le f(x, u) - f(x, \phi). \quad (37.4)$$

As before, if we suppose that $u < \phi$ in some $A \subset \Omega$ and x_0 is as before, then at x_0 the right-hand side of (37.4) would be negative, while the left-hand side would be positive. Thus, $u \ge \phi$ in Ω. The inequality $u \le \psi$ is obtained similarly.

Let $\{\Omega_m\}$ be a sequence of subdomains as in (w). Define for any $x \in \Omega$ and $m \in \mathbb{N}$

$$\psi_m := \psi|_{\Omega_m}, \quad \phi_m := \phi|_{\Omega_m},$$

$$\mu_{im}(x, u) := \begin{cases} \mu_i(x, \phi_m) & \text{if } 0 < u < \phi_m, \\ \mu_i(x, u) & \text{if } \phi_m \le u \le \psi_m, \\ \mu_i(x, \psi_m) & \text{if } u > \psi_m, \end{cases}$$

and the problem

$$Q_m u := \sum_{i=1}^n \mu_{im}(x, u)\partial_{ii} u = f(x, u) \quad \text{in } \Omega_m, \tag{Qm}$$

$$u|_{\partial\Omega_m} = c.$$

Lemma 6. *Under the hypotheses of Lemma 5, the problem (Qm) has classical solutions u_m such that $\phi \le u_m \le \psi$ in Ω_m and $\forall m \in \mathbb{N}$,*

$$\|u_{m+k}\|_{W_p^2(\Omega_m)} \le C(|f(., \psi)|_{L^p}) \quad \forall p > 1, \ k \in \mathbb{N}. \tag{37.5}$$

Proof. Because the coefficients μ_{im} are uniformly bounded in Ω_m, (Qm) has a classical solution u_m (see [3] and [4]). By Lemma 5, in Ω_m we have

$$\sum_{i=1}^n \mu_{im}(x, \phi)\partial_{ii}\phi - f(x, \phi) \ge 0 \ge \sum_{i=1}^n \mu_{im}(x, \psi)\partial_{ii}\psi - f(x, \psi),$$

$$\phi \le u_m \le \psi,$$

and estimate (37.5) follows from classical elliptic theory.

Theorem 7. *If there are $\phi, \psi \in C^2(\overline{\Omega})$ satisfying 1) and 2) in Lemma 5, then (Qc) has a classical solution u such that $\phi \le u \le \psi$ in $\overline{\Omega}$.*

Proof. Let $(u_n)_{n \in \mathbb{N}}$ be a sequence of solutions of (Qn) as in Lemma 6. By (37.5), it has a subsequence $(u_m^{(1)})$ that converges in $W_p^2(\Omega_1)$ to U_1, say, which is a weak solution of the problem

$$\sum_{i=1}^{n} \mu_i(x, U)\partial_{ii}U = f(x, U) \quad \text{in } \Omega_1 \quad \text{and} \quad \phi \le U_1 \le \psi. \tag{q1}$$

Taking $p > n$ in (37.5) large enough, the Sobolev imbedding theorem implies that $U_1 \in C^{0,\nu}(\overline{\Omega}_1)$, $\nu \in (0, 1)$. Thus, by elliptic theory, $U_1 \in C^{2,\nu}(\overline{\Omega}_1)$ is a classical solution of $(q1)$.

Also, $(u_m^{(1)})$ has a subsequence $(u_n^{(2)})$ that converges in $W_p^2(\Omega_2)$ to U_2, say, and, as before, $U_2 \in C^{2,\nu}(\overline{\Omega}_2)$ is a classical solution of $(q2)$. Moreover, $U_2|_{\overline{\Omega}_1} = U_1$. So, by induction, we obtain a sequence $(U_k)_{k \in \mathbb{N}}$ such that $\forall m \in \mathbb{N}$,

$$U_m \in C^{2,\nu}(\overline{\Omega}_m) \text{ solves } (Qm), \quad U_{m+i}|_{\overline{\Omega}_m} = U_m \quad \forall i \in \mathbb{N}.$$

This last sequence has an inductive limit [2], which is the required solution of (Qc).

References

1. Z. Zhang, A remark on the existence of explosive solutions for a class of semilinear elliptic equations, *Nonlinear Anal.* **41** (2000), 143–148.

2. Tadie, Weak and classical positive solutions of some semilinear elliptic equations in $\mathbb{R}^n$, $n \ge 3$. Radially symmetric cases, *Quart. J. Math.* **45** (1994), 397–406.

3. O.A. Ladyzhenskaya and N.N. Ural'tseva, *Linear and Quasilinear Elliptic Equations*, Academic Press, New York-London, 1968.

4. D. Gilbarg and N. Trudinger, *Elliptic Partial Differential Equations of Second Order*, 2nd ed., Springer-Verlag, New York, 1983.

38 Uniqueness and Symmetry for Some Singular Ground State Problems in $\mathbf{R}^n$, $n \geq 3$

Tadie

38.1 Introduction

Let $\Omega \subseteq \mathbb{R}^n$, $n \geq 3$, be a domain containing the origin. It is known that if a positive function $u \in C^2(\Omega \setminus \{0\})$ solves (E)1),2) below, then its estimate at 0 is $|x|^{-\alpha}$ (i.e., $\lim_{|x| \searrow 0} |x|^\alpha u(x) = \text{const} > 0$) for some $\alpha \in \{n - 2, 2/(\mu - 2)\}$ (see [1] and [2]); if it solves (E), then its estimate at infinity is $|x|^{2/(1-\mu)}$ and μ has to be in the interval $(1, n/(n-2))$ (see [3] and [4]). In this note we show that the problem (E) has exactly two radially symmetric solutions with the estimate $|x|^{2/(1-\mu)}$ at infinity and one having the estimate $|x|^{2-n}$ and the other $|x|^{2/(1-\mu)}$ at 0.

For $n \geq 3$ and $\mu > 0$, consider positive solutions $u \in C^2(\mathbb{R}^n \setminus \{0\})$ of the problem

$$1) \quad \triangle u = u^\mu, \quad u > 0,$$

$$2) \quad \lim_{|x| \searrow 0} u(x) = \infty, \qquad (E)$$

$$3) \quad \lim_{|x| \nearrow +\infty} u(x) = 0.$$

For $n = 3$ and $\mu = 3/2$, the existence of a unique radial solution with the estimate r^{-1} at 0 has be obtained in [1] as a minimizer of an energy functional in the Thomas–Fermi theory for atoms. We have shown in [3] and [4] that for any $\mu \in (1, n/(n-2))$,

(T1) the problem (E) has a unique radial solution with the estimates r^{2-n} at 0 and $r^{2/(1-\mu)}$ at infinity;

(T2) the problem (E) has a radial solution with the estimate $r^{2/(1-\mu)}$ both at 0 and at infinity (this result did not include uniqueness).

Here we establish the following results for $\mu \in (1, n/(n-2))$.

Theorem 1. *Any solution of (E) with the estimates $|x|^{2-n}$ at 0 and $|x|^{2/(1-\mu)}$ at infinity is radial and unique.*

Theorem 2. *Any solution of (E) with the estimate $|x|^{2/(1-\mu)}$ both at 0 and at infinity is radial and unique.*

38.2 Proof of the Theorems

In the Kelvin transformation $y := x/|x|^2$, let v be the function satisfying

$$|y|^{n-2}v(y) := u(x).$$

With $g(r) := r^{\mu(n-2)-n-2}$, if u is a solution of (E) with the estimates $|x|^{2-n}$ at 0 and $|x|^{2/(1-\mu)}$ at infinity, then for some $k \geq 0$,

$$\triangle v = g(|y|)v^\mu \quad \text{in } \mathbb{R}^n,$$
$$\lim_{|y|\nearrow+\infty} v(y) = k, \quad \lim_{|y|\searrow 0} v(y) = 0, \quad v \in C^2(\mathbb{R}^n \setminus \{0\}). \tag{38.1}$$

If it has the estimate $|x|^{2/(1-\mu)}$ both at 0 and at infinity, then

$$\triangle v = g(|y|)v^\mu \quad \text{in } \mathbb{R}^n,$$
$$\lim_{|y|\nearrow+\infty} v(y) = +\infty, \quad \lim_{|y|\searrow 0} v(y) = 0. \quad v \in C^2(\mathbb{R}^n \setminus \{0\}). \tag{38.2}$$

38.2.1 Proof of Theorem 1

Given (T1), we just need to establish that (E) has at most one solution with the estimate $|x|^{2-n}$ at 0. This follows from the next assertion (to be applied to (38.1)).

Lemma 3. *Let $\Omega \subseteq \mathbb{R}^n$ be a non empty and bounded domain and $f \in C(\Omega \times \mathbb{R})$. If $\forall x \in \Omega$ and either*
  1) $t \mapsto f(x,t)$, or
  2) $t \mapsto \{\partial/\partial t\}f(x,t) > 0$
is increasing in $\mathbb{R}$, then the problem

$$\triangle W = f(x,W) \quad \text{in } \Omega; \quad W|_{\partial\Omega} = 0$$

has at most one bounded classical solution.

For the proof it suffices to note that if there are two solutions U and V, then there is $D \subset \Omega$ simply connected with non zero measure such that $U > V$ in D and $U = V$ on ∂D, say. Suppose that $U - V$ has its maximum at $y_0 \in D$. U and V satisfy the identities

$$i) \quad \triangle(U - V)(y_0) = f(y_0, U) - f(y_0, V),$$
$$ii) \quad \int_{\partial D} U\partial_n(V - U)dS = \int_D UV\{f(x,V)/V - f(x,U)/U\}dx,$$

where ∂_n denotes the derivative along the outward unit normal to D.

If D is bounded, for 1) the both sides of i) have opposite signs and for 2) those for ii) have opposite signs. We then have a contradiction unless $U \equiv V$.

This can be extended to some unbounded domains. If D is unbounded, then the left-hand side of ii) is zero and the right-hand strictly negative, and the same conclusion follows (this applies to our case). If only i) holds and all solutions have the same k, y_0 is an interior point and i) provides a contradiction unless the solutions coincide. If the values of k are distinct, say, k_1 and k_2, then we apply identity i) to the functions $U_1 := U/k_1$ and $V_1 := V/k_2$.

38.2.2 Proof of Theorem 2

Here it is also enough to show that any solution of (38.2) is radial. This follows from Lemma A.1 in [5] since, with $g(r, s) := r^{\mu(n-2)-n-2} s^\mu$, for any $r \geq r_0 > 0$ and $s \geq s_0 > 0$ we have

(g1) $g(r, s)$ and $\{\partial/\partial s\} g(r, s)$ are continuous and positive;

(g2) for $\gamma \in (1, \mu)$ and $d > 1$, as $(r, s) \nearrow (\infty, \infty)$,

$$\liminf(\inf_{v>d}\{g(r, vs)/(v^\gamma g(r, s))\}) = v^{\mu-\gamma} > 1;$$

(g3) since $\mu > 1$, the function

$$r^{2n-2} g(r, s) = r^{n-4+\mu(n-2)} s^\mu$$

is monotonically increasing for $r > 0$;

(g4) finally, since $\mu < n/(n-2)$,

$$\int_{r_0}^\infty rg(r, s) dr < \infty \quad \forall s > s_0 > 0.$$

These four conditions ensure that any solution of (38.2) is radial.

References

1. H. Brezis and L. Veron, Removable singularities of nonlinear elliptic equations, *Arch. Rational Mech. Anal.* **75** (1980), 1–6.

2. L. Veron, Singular solutions of some nonlinear elliptic equations, *Nonlinear Anal. Methods Appl.* **5** (1981), 225–242.

3. Tadie, On singular ground states for $\Delta u = u_+^\gamma$ in $\mathbb{R}^n$, $n > 2$: monotonicity of the atomic radius in the Thomas–Fermi theory, preprint no. 4 (1999), Inst. Math. Sciences, Copenhagen University.

4. Tadie, Monotonicity and boundedness of the atomic radius in the Thomas–Fermi theory : mathematical proof, *Canadian Appl. Math. Quart.* **7** (1999), 301–311.

5. S.D. Taliaferro, Radial symmetry of large solutions of nonlinear elliptic equations, *Proc. Amer. Math. Soc.* **124** (1996), 447–455.

39 Reduction of Computation in the Numerical Resolution of a Second-Kind Weakly Singular Fredholm Equation

Olivier Titaud

39.1 Introduction

Let $X = L^1([0, \tau^*])$ be the Banach space of all (equivalence classes of) Lebesgue integrable functions on $[0, \tau^*]$, where τ^* is a given nonnegative large number. We consider an integral operator $T : X \to X$ defined by

$$x \mapsto Tx : \tau \in [0, \tau^*] \mapsto (Tx)(\tau) = \int_0^{\tau^*} \kappa(\tau, \sigma) x(\sigma) d\sigma.$$

We suppose that the kernel κ is of the form

$$\kappa(\tau, \sigma) := \eta(\tau, \sigma) g(|\tau - \sigma|),$$

where η is a continuous complex-valued function on the square $[0, \tau^*] \times [0, \tau^*]$ and g is a weakly singular function at zero, in the following sense:

(a) $\lim_{\tau \to 0^+} g(\tau) = +\infty$;

(b) $g \in C^0(]0, \tau^*]) \cap L^1([0, \tau^*])$;

(c) g is a positive decreasing function on $]0, \tau^*]$.

It was proved in [1] that T is a compact operator in X. For $z \in \mathrm{re}(T) := \{z \in \mathbb{C} : T - zI \text{ is bijective}\}$ and $f \in X$, there is a unique solution φ to the problem

$$(T - zI)\varphi = f.$$

39.2 Finite Rank Approximations

We study a special class of approximate operators whose range is a finite n-dimensional subspace of X. For all $\ell \in X^*$, the topological adjoint space

---

The author is indebted to Mario Ahues and Alain Largillier for fruitful discussions on this topic.

of X, and all $x \in X$, we define $\langle x, \ell \rangle := \bar{\ell}(x)$. As ℓ is linear conjugate, $\langle x, \ell \rangle$ is linear with respect to x and linear conjugate with respect to ℓ, such as a scalar product of a complex prehilbertian space. We recall that a linear bounded finite rank operator is compact and can be written as (see [2]):

$$T_n := \sum_{j=1}^{n} \langle \, . \, , \ell_{n,j} \rangle e_{n,j},$$

where $n \in \mathbb{N}^*$ and, for $j \in [\![1, n]\!]$, $\ell_{n,j} \in X^*$ and $e_{n,j} \in X$. If $z \in \mathrm{re}(T_n)$, then for all $f \in X$, the approximate equation

$$(T_n - zI)\varphi_n = f \tag{39.1}$$

admits a unique solution. If we set

$$A_n(i,j) := \langle e_{n,j}, \ell_{n,i} \rangle, \quad b_n(i) := \langle f, \ell_{n,i} \rangle, \quad x_n(j) := \langle \varphi_n, \ell_{n,j} \rangle, \tag{39.2}$$

then, by applying each semi-linear functional $\ell_{n,i}$ to each term of equation (39.1), we turn (39.1) into the n-dimensional linear system

$$(A_n - zI_n)x_n = b_n, \tag{39.3}$$

where I_n denotes the identity matrix of order n.

We are interested in approximations obtained with a family of projections on X of finite rank n. Such a projection π_n reads as

$$\pi_n x := \sum_{j=1}^{n} \langle x, \xi_{n,j} \rangle e_{n,j}, \quad x \in X,$$

where $(e_{n,j})_{j=1}^{n}$ is an ordered basis of the range of π_n and $(\xi_{n,j})_{j=1}^{n}$ is an adjoint basis of the former. The projection π_n considered in this paper is built as follows: let $(\tau_{n,i})_{i=1}^{n}$ be a grid on $[0, \tau^*]$ such that

$$0 := \tau_{n,0} < \tau_{n,1} < \cdots < \tau_{n,n-1} < \tau_{n,n} := \tau^*.$$

Set $h_{n,j} := \tau_{n,j} - \tau_{n,j-1}$ for $j \in [\![1, n]\!]$ and define

$$\mu_n := \min\{h_{n,j}, \quad j \in [\![1, n]\!]\}; \quad h_n := \max\{h_{n,j}, \quad j \in [\![1, n]\!]\}.$$

We define for all $x \in X$ and for all $j \in [\![1, n]\!]$,

$$\langle x, \xi_{n,j} \rangle := \frac{1}{h_{n,j}} \int_{\tau_{n,j-1}}^{\tau_{n,j}} x(\sigma)d\sigma,$$

and for all $\tau \in [0, \tau^*]$,

$$e_{n,j}(\tau) := \begin{cases} 0 & \text{if } \tau \in]\tau_{n,j-1}, \tau_{n,j}[, \\ 1 & \text{otherwise.} \end{cases}$$

Notice that if $x \in C^0([0, \tau^*])$, then, for all $j \in [\![1, n]\!]$, $\lim_{h_n \to 0} \langle x, \xi_{n,j} \rangle = x(\tau_{n,j})$. We define the approximate operator T_n by

$$T_n := \pi_n T.$$

It was proved in [1] that $(\pi_n)_{n \geq 1}$ is pointwise convergent to the identity operator in X. Then, as T is compact, the sequence $(T_n)_{n \geq 1}$ is convergent to T in the operator norm. The entries of the matrix A_n and of the second member b_n defined by (39.2) are given for all $(i, j) \in [\![1, n]\!]^2$ by

$$A_n(i, j) = \frac{1}{h_{n,i}} \int_{\tau_{n,i-1}}^{\tau_{n,i}} (Te_{n,j})(\tau) d\tau = \frac{1}{h_{n,i}} \int_{\tau_{n,i-1}}^{\tau_{n,i}} \int_{\tau_{n,j-1}}^{\tau_{n,j}} \kappa(\tau, \sigma) d\sigma d\tau,$$

$$b_n(i) = \frac{1}{h_{n,i}} \int_{\tau_{n,i-1}}^{\tau_{n,i}} \int_0^{\tau^*} \kappa(\tau, \sigma) f(\sigma) d\sigma d\tau. \tag{39.4}$$

39.3 Reduction of Computation

To attain a given precision on the approximate solution φ_n, it may be necessary that the largest grid step h_n be so small that the dimension of the corresponding linear system will be prohibitively large from a computational point of view so as to make the matrix A_n full. Moreover, when g is, for example, an exponentially decreasing function, a lot of the entries of A_n are very close to zero. Here we suggest a consistent way to reset to zero some small entries—in fact, we prove that a truncation on the kernel of T induces the zeroing of some small (in absolute value) entries of A_n.

Consider a sequence $(\varepsilon_n)_{n \geq 0}$ of positive real numbers. We define

$$I_{n,i,j} := [\tau_{n,i-1}, \tau_{n,i}] \times [\tau_{n,j-1}, \tau_{n,j}], \quad (i, j) \in [\![1, n]\!]^2,$$

and

$$\mathcal{E}_n := \left\{ (i, j) \in [\![1, n]\!]^2, \quad \sup_{(t,s) \in I_{n,i,j}} \kappa(t, s) \leq \frac{\varepsilon_n}{h_{n,j}} \right\}.$$

Let κ_n be the function defined for all $(\tau, \sigma) \in [0, \tau^*] \times [0, \tau^*]$ such that $\tau \neq \sigma$, by

$$\kappa_n(\tau, \sigma) := \begin{cases} 0 & \text{if } \exists (i, j) \in \mathcal{E}_n \text{ such that } (\tau, \sigma) \in I_{n,i,j}, \\ \kappa(\tau, \sigma) & \text{otherwise.} \end{cases}$$

Let K_n be the integral operator induced by the kernel κ_n, that is,

$$x \mapsto K_n x: \quad \tau \in [0, \tau^*] \mapsto (K_n x)(\tau) = \int_0^{\tau^*} \kappa_n(\tau, \sigma) x(\sigma) d\sigma,$$

and consider the finite rank approximation

$$\widetilde{T}_n := \pi_n K_n.$$

Let us denote by $\widetilde{A}_n$ the matrix of the linear system (39.3) corresponding to the approximation $\widetilde{T}_n$, i.e.,

$$\widetilde{A}_n(i,j) = \frac{1}{h_{n,i}} \iint_{I_{n,i,j}} \kappa_n(\tau,\sigma) d\tau d\sigma, \quad (i,j) \in [\![1,n]\!]^2. \qquad (39.5)$$

The following theorem shows that the truncation of κ induces the zeroing of some entries of A_n which are less than ε_n in absolute value.

Theorem 1. *Let A_n and $\widetilde{A}_n$ be the matrices defined by (39.4) and (39.5), respectively. Then for all $(i,j) \in \mathcal{E}_n$, $|A_n(i,j)| \leq \varepsilon_n$, and for all $(i,j) \in [\![1,n]\!]^2$,*

$$\widetilde{A}_n(i,j) = \begin{cases} 0 & \text{if } (i,j) \in \mathcal{E}_n, \\ A_n(i,j) & \text{otherwise.} \end{cases}$$

We now justify the use of $\widetilde{T}_n$ instead of T_n in the approximate equation (39.1).

Theorem 2. *If $(\varepsilon_n)_{n \geq 0}$ is such that $\lim\limits_{n \to +\infty} \dfrac{\varepsilon_n}{\mu_n} = 0$, then $\widetilde{T}_n$ converges to T in the operator norm.*

Proof. Let $x \in L^1([0,\tau^*])$. Then

$$\|(T - K_n)x\|_1 \leq \sum_{i=1}^{n} \sum_{j=1}^{n} \iint_{I_{n,i,j}} |\kappa(\tau,\sigma) - \kappa_n(\tau,\sigma)| |x(\sigma)| d\sigma d\tau$$

$$\leq \sum_{(i,j) \in \mathcal{E}_n} \frac{\varepsilon_n}{h_{n,j}} \iint_{I_{n,i,j}} |x(\sigma)| d\sigma d\tau \leq \sum_{(i,j) \in \mathcal{E}_n} \frac{\varepsilon_n h_{n,i}}{h_{n,j}} \int_{\tau_{n,j-1}}^{\tau_{n,j}} |x(\sigma)| d\sigma$$

$$\leq \frac{\tau^* \varepsilon_n}{\mu_n} \|x\|_1. \qquad (39.6)$$

Since $\lim\limits_{n \to +\infty} \dfrac{\varepsilon_n}{\mu_n} = 0$, $\lim\limits_{n \to +\infty} \|T - K_n\|_1 = 0$. Also,

$$\|T - \widetilde{T}_n\|_1 \leq \|T - K_n\|_1 + \|(I - \pi_n)T\|_1 + \|(I - \pi_n)(K_n - T)\|_1,$$

and the conclusion follows.

Remark 1. Note that it is not necessary to compute the entries before their zeroing: the condition which defines $\mathcal{E}_n$ can be used to decide the zeroing of an entry. In the case of a large matrix, this trick allows a gain of time in the construction of the matrix. Finally, note that some entries whose absolute value is less than ε_n may not be zeroed.

39.4 Numerical Example

In this section, we apply the previous reduction of computation to the resolution of an integral equation which appears in a radiative transfer problem. A description of the physical problem is given in [3] and [4]. A parallel code for the resolution of this problem is given in [5]. Let κ and f be defined for all $(\tau, \sigma) \in [0, \tau^*]^2$ by

$$\kappa(\tau, \sigma) := \frac{\varpi_*}{2} E_1(|\tau - \sigma|) = \frac{\varpi_*}{2} \int_0^1 \frac{\exp(-|\tau - \sigma|/\mu)}{\mu} d\mu, \quad \tau \neq \sigma,$$

$$f(\tau) := \begin{cases} \varpi_* - 1 & \text{if } \tau \in [0, \tau^*/2], \\ 0 & \text{otherwise}, \end{cases}$$

where $\varpi_* \in]0, 1[$. This kernel satisfies conditions (a), (b), and (c). The corresponding entries of A_n and b_n are given in [5].

We set $\varpi_* = 0.75$ and used a uniform grid of 100 nodes on $[0, \tau^*]$ to compute the results shown below.

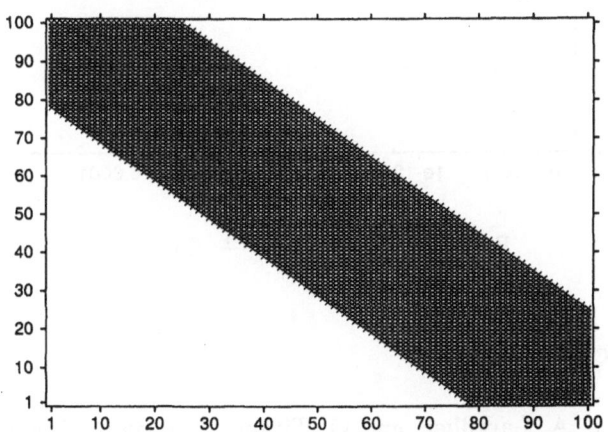

Fig. 1. Profile of the truncated matrix $\tilde{A}_n$ for $\varepsilon_n = 10^{-12}$.

In Fig. 1 we show the profile of the corresponding matrix $\tilde{A}_n$ for $\epsilon_n = 10^{-12}$: in this case, 5853 entries have been zeroed in the matrix A_n, i.e., more than 50%.

Let $\tilde{\varphi}_n$ be the solution of

$$(\tilde{T}_n - zI)\tilde{\varphi}_n = f. \tag{39.7}$$

In Fig. 2 we show the (log-scaled) relative error $\dfrac{\|\varphi_n - \tilde{\varphi}_n\|_1}{\|f\|_1}$ for different values of ε_n. Note that the variation of the relative error is linear and that the numerical results are consistent with respect to the bound (39.6).

Remark 2. We recall that the entries of the second term of the linear system (39.3) corresponding to (39.7) are given for all $i \in [\![1, n]\!]$ by

$$\widetilde{b}_n(i) := \frac{1}{h_{n,i}} \sum_{\substack{j \in [\![1,n]\!] \\ (i,j) \notin \mathcal{E}_n}} \iint_{I_{n,i,j}} \kappa(\tau, \sigma) f(\sigma) d\sigma.$$

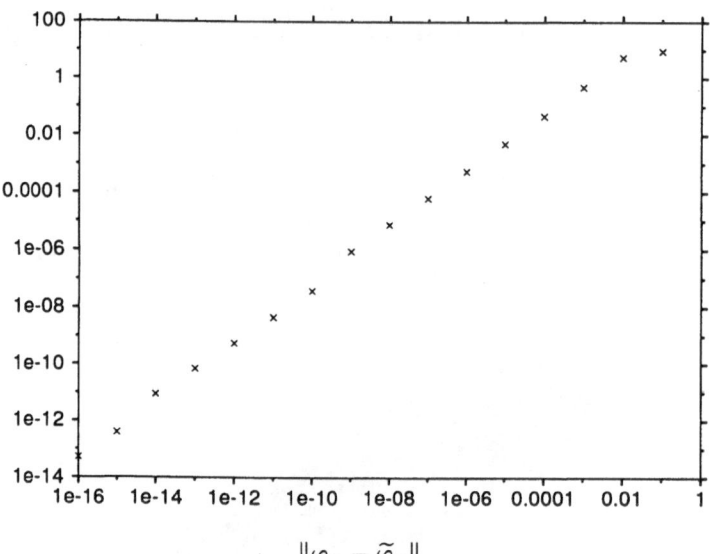

Fig. 2. $\varepsilon_n \longmapsto \dfrac{\|\varphi_n - \widetilde{\varphi}_n\|_1}{\|f\|_1}$ (log scale).

References

1. M. Ahues, A. Largillier, and O. Titaud, *The roles of weak singularity and the grid uniformity in the relative error bounds*, Numer. Functional Anal. Optimization **22** (2002), 789–814.

2. M. Ahues, A. Largillier, and B.V. Limaye, *Spectral Computations for Bounded Operators*, Chapman and Hall/CRC, Boca Raton, 2001.

3. B. Rutily, Multiple scattering theory and integral equations, this volume, 211–232.

4. L. Chevallier, Stellar atmospheres modelling, this volume, 37–40 .

5. P.B. Vasconcelos and F. d'Almeida, A parallel code for integral equations on a cluster of computers, this volume, 261–266.

40 A Parallel Code for Integral Equations on a Cluster of Computers

Paulo B. Vasconcelos and Filomena D. d'Almeida

40.1 Description of the Fredholm Integral Problem

Parallel codes for partial differential equations often use domain decomposition as a technique for parallelization together with discretization methods such as finite differences, finite elements or finite volumes. When building the discretization matrix, in the case of nqnoverlapping domain decomposition, the domain is divided into several subdomains by interfaces of sufficient width to cut off the influence of the nodes of one domain, in the elements corresponding to another subdomain [1]. If the grid nodes are numbered first on the subdomains and then on the interfaces, the resulting matrix is a block bordered diagonal matrix. The discretized problem is first solved in the interfaces by using the Schur complement and then the solution in the subdomains is updated in parallel.

For integral equations the influence of all subintervals of the domain is present at every other subinterval and so domain decomposition is not suited for their parallelization. In the application of interest here, concerning the emission of photons in stellar atmospheres (see [2] in this volume for details on the astrophysics problem), we have a Fredholm integral equation of the second kind

$$T\varphi - z\varphi = f, \quad \varphi \in L^1(I), \quad I = [0, \tau^\star]. \tag{40.1}$$

The integral operator is defined by

$$(T\varphi)(\tau) = \frac{\varpi}{2} \int_0^{\tau^\star} E_1\left(|\tau - \tau'|\right) \varphi\left(\tau'\right) d\tau', \tag{40.2}$$

where z is in the resolvent set of T, $\tau^\star$ is the optical depth of the stellar atmosphere, $\varpi \in \ ]0, 1[$ is the albedo (assumed to be constant) and E_1 is the first function of the family of exponential integrals

This work is part of a joint research with M. Ahues, A. Largillier, and O. Titaud from the Spectral Analysis team of the University of Saint Etienne, France. The application was proposed by B. Rutily of the Observatoire de Lyon, France.

$$E_\nu(\tau) = \int_1^\infty \frac{\exp(-\tau\mu)}{\mu^\nu} d\mu, \quad \nu \geq 1. \tag{40.3}$$

The behavior of E_1 in (40.3) near 0 is logarithmic and T in (40.2) is a weakly singular operator (for details see [3], [4]). The free term f in (40.1) in the test examples will be $f(\tau) = -1$ for $0 \leq \tau \leq \tau^*/2$ and $f(\tau) = 0$ otherwise.

To discretize (40.1) a projection method is used where T is approximated, by a finite rank operator T_n whose restriction to the corresponding finite dimensional subspace X_n is represented by a matrix A_n of size n (more details will be given in Section 2). The parallelization of this stage of the code, which is the heaviest one in terms of computing time, will be discussed in Section 3. The dimension of the matrix necessary to produce a good approximation to φ may be too large. It would be possible to solve the corresponding linear system of equations in a distributed way, but as discussed before (see [3], [5]) it is also possible to use iterative refinement of an initial approximation obtained with a discretization matrix of small size A_n and only a large discretization matrix A_m for matrix-vector products. Section 4 shows the efficiency and scalability of this parallel code in a Beowulf machine with 22 processing elements and MPI as message passing software.

40.2 Sequential Algorithm

To solve problem (40.1) by projection method and iterative refinement schemes the overall algorithm may be summarized as follows:

Algorithm 1
1. Projection method stage: build matrix representations of projection method approximations of T of rank n and m $(m \gg n)$ restricted to subspaces X_n and X_m: $A_n(n \times n)$, $A_m(m \times m)$, $C(n \times m)$, $D(m \times n)$.
2. Solution of the small n-dimensional approximate problem.
3. Iterative refinement stage.

40.2.1 Projection Method Stage

For stage 1 we use the projection method, where equation (40.1) is approximated by

$$T_n \varphi_n - z\varphi_n = f. \tag{40.4}$$

T_n is obtained by setting a nonuniform grid of $n+1$ points on the interval $[0, \tau^*]$: $0 = \tau_{n,0} < \tau_{n,1} < \cdots < \tau_{n,n} = \tau^*$, $j = 1, \ldots, n$, and considering a basis of n piecewise constant functions $e_{n,j}(\tau)$ equal to 0 except for $\tau \in [\tau_{n,j-1}, \tau_{n,j}]$ where they are equal to 1. Let X_n be the subspace spanned by this basis and T_n defined as $T_n = \sum_{j=1}^n \langle \cdot, \ell_{n,j} \rangle e_{n,j}$; $\ell_{n,j}$ are functions of a basis of the adjoint subspace X_n^* and $\langle x, \ell_{n,j} \rangle = \frac{1}{h_{n,j}} \int_{\tau_{n,j-1}}^{\tau_{n,j}} Tx(\tau)d\tau$, for all x in $L^1(I)$, where $h_{n,j} = \tau_{n,j} - \tau_{n,j-1}$. Let $e_{n,j}^*$ be defined by $\langle x, e_{n,j}^* \rangle = \frac{1}{h_{n,j}} \int_{\tau_{n,j-1}}^{\tau_{n,j}} x(\tau)d\tau$. Then $\lim_{n\to\infty} \|T_n - T\| = 0$ and $R_n(z) = (T_n - zI)^{-1}$ exists and it is uniformly bounded in n for n large enough (see [6] and [7]).

Equation (40.4) is solved through a linear system obtained by applying $\ell_{n,i}$, $i = 1, \ldots, n$, to both sides of (40.4): $\langle T_n \varphi_n, \ell_{n,i} \rangle - z \langle \varphi_n, \ell_{n,i} \rangle = \langle f, \ell_{n,i} \rangle$ leading to $A_n x_n - z x_n = b_n$ (see also [9] in this volume). Each column of A_n is obtained by integration of the operator's kernel over $[\tau_{n,i-1}, \tau_{n,i}]$, and we obtain

$$A_n(i,j) = \frac{\varpi}{2h_{n,i}} \int_{\tau_{n,i-1}}^{\tau_{n,i}} \int_0^{\tau^*} E_1(|\tau - \tau'|) e_{n,j}(\tau') d\tau' d\tau,$$

$$b_n(i) = \frac{\varpi}{2h_{n,i}} \int_{\tau_{n,i-1}}^{\tau_{n,i}} \int_0^{\tau^*} E_1(|\tau - \tau'|) f(\tau') d\tau' d\tau. \qquad (40.5)$$

Since $E_\nu' = -E_{\nu-1}$ and $E_\nu(0) = 1/(\nu - 1)$, $\nu > 1$, equations (40.5) may be written in terms of linear combinations of E_3 (see [8]).

40.2.2 Solution of the "Small" System

Once we have built the discretization matrix and the second member corresponding to the coarse grid approximation, we solve the corresponding linear system $A_n x_n - z x_n = b_n$ and we obtain the solution of (40.4) by $\varphi_n = (1/z) \left(\sum_{j=1}^n x(j) e_{n,j} - f \right)$. Since this is a small system, it can be solved by band LU factorization followed by the resolution of two triangular systems. It is also possible to solve this system by nonstationary iterative methods. In the tests reported here we used the first option. For the solution of small systems necessary in the next stage we can use the factors computed here.

40.2.3 Iterative Refinement Stage

Iterative refinement methods may be seen as Newton-type methods to solve $T\varphi - z\varphi - f = 0$ for which the Fréchet derivative of the operator is approximated by $R_n(z)$, $1/z(R_n(z)T - I)$ or $1/z(TR_n(z) - I)$. This leads to three schemes:

$$(A): \quad x^{(0)} = \varphi_n = R_n(z)f,$$
$$x^{(k+1)} = x^{(0)} + R_n(z)(T_n - T)x^{(k)}, \quad k \geq 0, \qquad (40.6)$$

$$(B): \quad x^{(0)} = 1/z(R_n(z)T - I)f,$$
$$\widetilde{x}^{(k+1)} = \widetilde{x}^{(0)} + 1/zR_n(z)(T_n - T)T\widetilde{x}^{(k)}, \quad k \geq 0, \qquad (40.7)$$

$$(C): \quad \widehat{x}^{(0)} = 1/z(TR_n(z) - I)f,$$
$$\widehat{x}^{(k+1)} = \widehat{x}^{(0)} + 1/zTR_n(z)(T_n - T)\widehat{x}^{(k)}, \quad k \geq 0. \qquad (40.8)$$

The first is related to the Atkinson method [10], the second to the Brackhage method [11]. Error bounds for the relative error of the refined approximation can be found in [5].

In equations (40.6)–(40.8) we do not use the operator T to compute the residual, but T_m corresponding to a much finer grid. Then the iterative refinement solution will converge to φ_m, the solution of $T_m\varphi_m - z\varphi_m = f$. We denote the matrix representation of T_m restricted to X_m by A_m and its restriction to the coarse grid subspace by C. The initial approximation T_n will also have a representation matrix, D, when restricted to X_m.

Stage 3 of Algorithm 1 is carried out with one of the three iterative schemes and is detailed for method (A) in Algorithm 2.

Algorithm 2

Given A_n, A_m, C, D, z, $x_n^{(0)}$, and $x_m^{(0)}$,

repeat until convergence
1. $y_n = A_n x_n^{(k)} - C x_m^{(k)}$
2. solve $(A_n - zI)w_n = y_n$
3. $w_m = (Dw_n + A_m x_m^{(k)} - D x_n^{(k)})/z$
4. $x_n^{(k+1)} = x_n^{(0)} + w_n; \quad x_m^{(k+1)} = x_m^{(0)} + w_m$
5. $k = k + 1$

For each function $x^{(0)}$, $x^{(k)}$, and $x^{(k+1)}$ two vectors are kept, one with its projection onto X_n, as in equation (40.6), and the other one with the projection onto X_m, denoted accordingly by a subscript n or m. The same option is taken for two auxiliary functions $y = (T_n - T)x^{(k)}$ and $w = R_n(z)y$. The stopping criteria is that the relative residual of $x_n^{(k)}$ must be less than or equal to some prescribed tolerance denoted by *eps*.

40.3 Parallel Implementation

To build a parallel algorithm out of Algorithm 1, we remark that the most expensive part is the computation of the matrices, mainly of A_m, and this is *embarrassingly parallel*. On the other hand A_m may be so large that it does not fit in one of the available processing nodes. So it is necessary to divide the matrix into blocks and compute each one in a different processor. In order to keep a good load balance and scalability of the computations we need to allocate the blocks of the matrix to the processors in a cyclic way by rows and columns. In practice we have noticed that in this kind of applications the coefficients of matrices A_m and A_n outside a central band of bandwidth bw are smaller than machine precision (see [3]). So a good load balance is obtained with a distribution of the nonzero blocks of matrix A_m cyclic by block diagonals by the processors that will store only the nonzero elements. If matrix A_n fits in the memory it is computed by all processors to avoid the communications required otherwise for the iterative refinement. Matrices C and D are divided into k blocks and computed in parallel.

For the description of the parallelization of Algorithm 2. we will use a SPMD (Single Program Multiple Data) paradigm. In processor number i the block of matrix A_m is denoted by $A_m[P_i]$ and the same with C and D.

Algorithm 3

Given A_n, $A_m[P_i]$, $C[P_i]$, $D[P_i]$, z, $x_n^{(0)}$, and $x_m^{(0)}[P_i]$,

repeat until convergence

1. $y_n[P_i] = -C[P_i]x_m^{(0)}[P_i]$; if $i = 0$, then $y_n[P_0] = A_n x_n^{(k)} + y_n[P_0]$

2. collective sum communication: $y_n = \sum y_n[P_j]$, $j = 0, \dots, p-1$

3. solve $(A_n - zI)w_n = y_n$

4a. $y_m[P_i] = A_m[P_i]x_m^{(k)}[P_i]$ send $y_m[P_i]$

4b. receive $y_m[P_j]$, j in neighborhood, $y_m = \sum y_m[P_j]$

5. $w_m[P_i] = (D[P_i](w_n - x_n^{(k)}) + y_m)/z$

6. $x_n^{(k+1)} = x_n^{(0)} + w_n$; $x_m^{(k+1)}[P_i] = x_m^{(0)}[P_i] + w_m[P_i]$

7. $k = k + 1$

40.4 Numerical Results and Conclusions

These algorithms were tested on a "Beowulf Cluster" machine with Linux operating system from the Faculdade Engenharia Universidade do Porto. This machine is based on 22 PIII processors at 450 MHz with 128 MB RAM connected by a Fast Ethernet switch (100 Mpbs). Data interchange between processors was done with MPI (Message Passing Interface) library.

$m = 4000$	it	$p = 1$	$p = 2$	$p = 4$	$p = 5$	$p = 10$	$p = 20$
$n = 200$	58	428.3	216.7	110.7	87.6	46.1	25.6
S_p	—	1.0	2.0	3.9	4.9	9.3	16.7

Table 1. Elapsed time, number of iterations and *Speedup* for $\tau^* = 4000$, $n = 200$, and $\varpi = 0.750$ and $eps = 10^{-12}$ using several numbers of processors.

In Table 1 we show for $\tau^* = m = 4000$ and $n = 200$ the required number of refinement steps needed to achieve a residual less than 10^{-12} and the elapsed time to solve the overall problem for several numbers of processors p. The performance increases almost linearly with the number of processors used, showing a *Speedup* $S_p = T_1/T_p$ (where T_p represents the time spent in p processors) close to the ideal one ($S_p = p$).

m	n	nnz	it	p	time	p	time	E_p
10000	500	1238190	128	1	2763.2	10	292.8	0.94
20000	1000	2482890	111	1	11158.0	20	602.8	0.93

Table 2. Elapsed time, number of iterations, number of nonzero elements and efficiency for $\tau^* = 10000$, 20000 and $\varpi = 0.999$ and $eps = 10^{-7}$.

In Table 2 we show the number of refinement steps and the total elapsed time to solve the problem using p processors. We also present the number of nonzero elements that need to be stored using a sparse structure, nnz, for $\tau^* = 10000$ and 20000. We conclude that even for larger problems our approach still delivers a high performance with efficiency $E_p = S_p/p \sim 95\%$. In other words, the code reveals scalability since maintaining a constant memory use per node allows efficiency to be maintained.

References

1. B.F. Smith, P. Bjorstad, and W.D. Gropp, *Domain Decomposition*, Cambridge University Press, Cambridge, 1996.

2. B. Rutily, Multiple scattering theory and integral equations, this volume, 211–232.

3. M. Ahues, F. d'Almeida, A. Largillier, O. Titaud, and P. Vasconcelos, An L^1 refined projection approximate solution of the radiation transfer equation in stellar atmospheres, *J. Comput. Appl. Math.* **140** (2002), 13–26.

4. M. Ahues, A. Largillier, and O. Titaud, The roles of a weak singularity and the grid uniformity in the relative error bounds, *Numer. Functional Anal. Optimization* **22** (2001), 789–814.

5. M. Ahues, F. d'Almeida, A. Largillier, O. Titaud, and P. Vasconcelos, Iterative refinement schemes for an ill-conditioned transfer equation in astrophysics, in *Proceedings of Algorithms for Approximation IV*, 2002, 70–77.

6. F. Chatelin, *The Spectral Approximation of Linear Operators*, Academic Press, 1983.

7. M. Ahues, A. Largillier, and B.V. Limaye, *Spectral Computations with Bounded Operators*, Chapman & Hall/CRC, Boca Raton, 2001.

8. M. Abramowitz and I.A. Stegun, *Handbook of Mathematical Functions*, Dover, New York, 1960.

9. O. Titaud, Reduction of computation in the numerical resolution of a second-kind weakly singular Fredholm equation, this volume, 255–260.

10. K. Atkinson, Iterative variants of the Nyström method for the numerical solution of integral equations, *Numer. Math.* **22** (1973), 17–31.

11. H. Brakhage, Über die Numerische Bechandlung von Integralgleichungen nach der Quadraturformelmethod, *Numer. Math.* **2** (1960), 183–196.

41 Analytic Solution of the S_N Equations by Integral Transform Technique

Marco T. Vilhena, Haroldo F. de Campos Velho,
Cynthia F. Segatto, and Glênio A. Gonçalves

41.1 Introduction

The LTS_N method that appeared in the last decades [1, 2], solves the dicrete ordinates equations (S_N equations) by the Laplace transform technique in a slab. The main idea is: (i) application of the Laplace transform to the set of S_N equations, (ii) analytical solution of the resulting linear system depending on the complex parameter s, and (iii) analytic inversion of the transformed angular flux. The convergence of the LTS_N method was proved in the framework of C_0-semigroup theory [3].

Now, it is important to cite that the one-dimensional S_N problem in spherical geometry can be transformed into an S_N problem in a slab [4], and consequently the application of the LTS_N approach is a straightforward task. On the other hand, to solve the one-dimensional S_N problem considering isotropic scattering in cylindrical geometry we proceed as in the LTS_N method, but now applying the Hankel transform. This method can be named as HTS_N approach. Considering the novelty of this approach in what follows, we focus our attention on the issue of presenting the derivation of the HTS_N method for the one-dimensional transport problem in cylindrical geometry.

41.2 The HTS_N Solution

As mentioned before, the application of the Hankel transform is focused on the one-dimensional isotropic problem in cylindrical geometry. Following the work of Mitsis [4], the transport problem for an infinite cylinder reduces to the solution of the problem

$$\left(\frac{\partial^2}{\partial r^2} + \frac{1}{r} \frac{\partial}{\partial r} - \frac{1}{\mu^2} \right) \phi(r, \mu) = -c \int_0^1 \phi(r, \mu) \frac{d\mu}{\mu^2} - (1 - c) \left[Q(r) - G \right],$$

$$(41.1)$$

The authors are indebted to CNPq-Brazil (Conselho Nacional de Desenvolvimento Científico e Tecnológico) for the partial financial support to this work, and to Dr. Ezzat S. Chalhoub for his help with the Tex editing.

subject to the boundary condition

$$K_1 \left(\frac{R}{\mu} \right) \phi(R, \mu) + \mu K_0 \left(\frac{R}{\mu} \right) \frac{\partial \phi(r, \mu)}{\partial r} \bigg|_{r=R} = 0, \qquad (41.2)$$

where K_0 and K_1 are the modified Bessel functions. Here $\phi(r, \mu)$ denotes the pseudo-flux introduced by Mitsis, $Q(r)$ the external source and G the incoming flux at the cylindrical surface of radius R. It is important to remark that the scalar flux of the original problem is written in terms of the pseudo-flux as

$$I(r) = \int_0^1 \phi(r, \mu) \frac{d\mu}{\mu^2} . \qquad (41.3)$$

In order to solve problem (41.1) by the Hankel transform technique, the collocation method is applied to the angular variable μ considering the roots of the Legendre polynomials as collocation roots. After this procedure, the equation (41.1) reads as

$$\left(\frac{\partial^2}{\partial r^2} + \frac{1}{r} \frac{\partial}{\partial r} - \frac{1}{\mu_j^2} \right) \phi_j(r) = -c \sum_{i=1}^{N/2} w_i \frac{\phi_i(r)}{\mu_i^2} - (7 - c) S(r), \qquad (41.4)$$

where μ_i and w_i, $i = 1, \ldots, N$, are, respectively, the roots and weights of the Gaussian quadrature and $S(r) = Q(r) - G$. Applying the Hankel transform of zeroth order to equation (41.4), we arrive at

$$\left(-\xi^2 - \frac{1}{\mu_j^2} \right) \bar{\phi}_j(\xi) = -c \sum_{i=1}^{N/2} w_i \frac{\bar{\phi}_i(\xi)}{\mu_i^2} - (1 - c) \bar{S}(\xi) . \qquad (41.5)$$

Here $\bar{\phi}_j(\xi)$ denotes the Hankel transform [1] of zeroth order of $\phi_j(r)$. Recasting equation (41.5) in matrix form, we write

$$\left(\xi^2 \mathbf{I} + \mathbf{A} \right) \bar{\Phi}(\xi) = (1 - c) \bar{\mathbf{S}}(\xi),$$

---

[1] The Hankel transform of order ν of $f(r)$ is defined by [5]

$$\bar{f}_\nu(\xi) \equiv H_\nu\{f(r)\} = 2\pi \int_0^\infty f(r) J_\nu(2\pi\xi r) r \, dr .$$

If $\nu > -1/2$, then the inverse Hankel transform is

$$f(r) \equiv H_\nu^{-1}\{\bar{f}(\xi)\} = 2\pi \int_0^\infty \bar{f}_\nu(\xi) J_\nu(2\pi\xi r) \xi \, d\xi .$$

where $\mathbf{I}$ is the identity matrix and $\mathbf{A}$ is the LTH_N matrix. Recalling the diagonalization property we are in position to write the pseudo transformed angular flux as

$$\bar{\Phi}(\xi) = \mathbf{U} \left(\xi^2 \mathbf{I} + \mathbf{D}\right)^{-1} \mathbf{V}(1 - c)\, \bar{\mathbf{S}}(\xi),$$

where $\mathbf{U}$ is the eigenvector matrix, $\mathbf{V} = \mathbf{U}^{-1}$, and $\mathbf{D}$ is a diagonal matrix whose entries are the eigenvalues of the matrix $\mathbf{A}$. To this point, it is relevant to mention that a similar procedure was employed in the solution of the transport equation in a slab, regarding the issue of diagonalization, but the Laplace transform was the technique considered. Therefore, the jth component of the vector $\bar{\Phi}$ has the form

$$\bar{\phi}_j(\xi) = (1 - c) \sum_{i,k}^{N/2} \frac{u_{j,k}\, v_{k,i}}{\xi^2 + \lambda_k}\, \bar{S}_i(\xi),$$

where λ_k are the eigenvalues of the matrix $\mathbf{A}$. From the definition of the inverse Hankel transform it follows that

$$H_0^{-1}\left\{\bar{\Phi}(\xi)\right\} = \phi_j(r) = (1-c) \sum_{i,k}^{N/2} u_{j,k}\, v_{k,i} \int_0^\infty \xi\, \frac{\bar{S}_i(\xi)}{\xi^2 + \lambda_k}\, J_0(r\xi)\, d\xi\ . \quad (41.6)$$

Now, using the Parseval relation for Hankel transforms [6] and the equality [7]

$$\int_0^\infty \xi\, \frac{J_0(r\xi)}{\xi^2 + \lambda_k}\, J_0(r'\xi)\, d\xi = \begin{cases} I_0(\alpha_k r')\, K_0(\alpha_k r), & \text{for } 0 < r' < r; \\[2mm] I_0(\alpha_k r)\, K_0(\alpha_k r'), & \text{for } r < r' < \infty. \end{cases}$$

where α_k is the square root of λ_k, we find that (41.6) becomes

$$\phi_j(r) = (1-c) \sum_{i,k}^{N/2} u_{jk}\, v_{ki} \left[K_0(\alpha_k r) \int_0^r r'\, I_0(\alpha_k r')\, S_i(r')\, dr' \right.$$

$$\left. + I_0(\alpha_k r) \int_r^R r'\, K_0(\alpha_k r')\, S_i(r')\, dr' \right],$$

which is a particular solution of (41.4). To obtain the general solution of the corresponding homogeneous equation, we rewrite (41.4) with $S(r) = 0$ in matrix form as

$$\left(\frac{\partial^2}{\partial r^2} + \frac{1}{r}\frac{\partial}{\partial r} \right) \Phi(r) - \mathbf{UDV}\Phi(r) = 0. \quad (41.7)$$

Defining the matrix $\mathbf{H}$ by $\mathbf{H} = \mathbf{V}\boldsymbol{\Phi}$ and recalling that the entries of the matrix $\mathbf{V}$ are real numbers, we see that (41.7) has the form

$$\left(\frac{\partial^2}{\partial r^2} + \frac{1}{r}\frac{\partial}{\partial r} - \mathbf{D}\right)\mathbf{H}(r) = 0.$$

The solution of this equation which is bounded at $r = 0$ is

$$\mathbf{H}(r) = \mathbf{B}\,I_0(\alpha r),$$

where $\mathbf{B}$ is a diagonal matrix whose entries are the integration constants. Therefore, we write the general solution of the homogeneous equation (41.4) as

$$\boldsymbol{\Phi}(r) = \mathbf{U}\,\mathbf{B}\,I_0(\alpha r).$$

The jth component of the vector $\boldsymbol{\Phi}_h(r)$ is

$$\phi_{h,j} = \sum_{k}^{N/2} u_{jk}\,I_0(\alpha_k r)\,b_{kk}.$$

The jth component of the general solution of the full equation (41.4) is now given by

$$\phi_j(r) = \sum_{k}^{N/2} u_{jk}\,I_0(\alpha_k r)\,b_{kk}$$

$$+ (1-c)\sum_{i,k}^{N/2} u_{jk}\,v_{ki}\left[K_0(\alpha_k r)\int_0^r r'\,I_0(\alpha_k r')\,S_j(r')\,dr'\right.$$

$$\left. + I_0(\alpha_k r)\int_r^R r'\,K_0(\alpha_k r')\,S_{(}r')\,dr'\right]. \qquad (41.8)$$

Applying the boundary condition (41.2) to the jth component of $\boldsymbol{\Phi}(r)$, we arrive at the algebraic system

$$\sum_{k=1}^{N/2} u_{jk}\left[\mu_j\,\alpha_k\,K_0\left(\frac{R}{\mu_j}\right)I_1(\alpha_k R) + K_1\left(\frac{R}{\mu_j}\right)I_0(\alpha_k R)\right]b_{kk}$$

$$= (c-1)\sum_{i,j}^{N/2} u_{jk}\,v_{ki}\left[\left(\mu_j\,\alpha_k\,K_0\left(\frac{R}{\mu_j}\right)K_1(\alpha_k R) + K_1\left(\frac{R}{\mu_j}\right)K_0(\alpha_k R)\right)\right.$$

$$\left. \times \int_0^R r'\,I_0(\alpha_k r')\,S_j(r')\,dr'\right],$$

which allows us to determine the unknown coefficients. Therefore, the solution for the one-dimensional problem in cylindrical geometry specialized for isotropic scattering is well determined by replacing (41.8) in equation (41.3).

41.3 Numerical Results and Conclusion

An application of the HTS$_N$ approach is exemplified considering a transport problem with a constant external source $Q(r) = 1/(1-c)$. In Tables 1 and 2, the numerical results encountered by the HTS$_N$ approach are reported for $F(r) = 1 - \phi(r)/Q(r)$, considering $c = 0.3$ and $c = 0.9$, respectively, as well as numerical comparisons with the ones attained by Siewert and Thomas [8]. All the calculations were performed on a PC.

r/R	Hankel	F$_N$
0.0	0.460883(-4)	0.460882(-4)
0.1	0.595325(-4)	0.595325(-4)
0.2	0.112121(-4)	0.112121(-3)
0.3	0.253579(-3)	0.253579(-3)
0.4	0.625300(-3)	0.625301(-3)
0.5	0.162250(-2)	0.162250(-2)
0.6	0.437880(-2)	0.437880(-2)
0.7	0.122887(-1)	0.122887(-1)
0.8	0.362561(-1)	0.362559(-1)
0.9	0.116749	0.116748
1.0	0.558360	0.558361

Table 1. The function $F(r)$ for $R = 10$ cm and $c = 0.3$.

From Tables 1 and 2, the good agreement of the results are promptly realized, at first glance. This fact, reinforced by the small computational effort required in the calculations, show us the aptness of the Hankel transform technique from a computational point of view, to solve the one-dimensional isotropic transport problem in cylindrical geometry. Currently, solution of tha anisotropic problem is under investigation. Finally, bearing in mind the main feature of the nodal approach, which replaces the multidimensional transport equation with a set of one-dimensional equations, and the analytic feature of the LTS$_N$ and HTS$_N$ approaches, which introduce no approximations, we believe that the integral (Laplace and Hankel) transforms constitute an important and promising technique for solving multidimensional problems. This argument is reinforced by the successful application of the LTS$_N$ method to the multidimensional transport problem in Cartesian geometry [9]. We can indeed say that the LTS$_N$ and HTS$_N$ approaches pave the road to solutions for the multidimensional nodal transport equations.

r/R	Hankel	F_N
0.0	0.201898(−1)	0.201898(−1)
0.1	0.216079(−1)	0.216079(−1)
0.2	0.261624(−1)	0.261624(−1)
0.3	0.348252(−1)	0.348252(−1)
0.4	0.494720(−1)	0.494721(−1)
0.5	0.733358(−1)	0.733358(−1)
0.6	0.1111811	0.1111811
0.7	0.173872	0.173872
0.8	0.274762	0.274762
0.9	0.442665	0.442664
1.0	0.781243	0.781243

Table 2. The function $F(r)$ for $R = 10$ cm and $c = 0.9$.

References

1. M.T. Vilhena and L.B. Barichello, An analytical solution for the multi-group slab geometry discrete ordinates problems, *Transport Theor. Stat.* **224** (1995), 1337–1352.

2. C.F. Segatto and M.T. Vilhena, State-of-art of the LTS$_N$ method, in *Mathematics and Computation, Reactor Physics and Environmental Analysis in Nuclear Applications*, Proc. of M&C'99, J.M. Aragonés, C. Ahnert, and O. Cabellos (eds.), Senda Editorial, Madrid, 1618–1631, 1999.

3. M.T. Vilhena and R.P. Pazos, Convergence in transport theory, *Appl. Numer. Math.* **30** (1999), 79–92.

4. G.J. Mitisis, *Transport Solutions to the Monoenergetic Critical Problems*, PhD Thesis, Report ANL-6787, Argone National Laboratory, Chicago, 1963.

5. R. Piessens, The Hankel transform, in *The Transforms and Applications Handbook*, 2nd ed., A. Poularikas (ed.), CRC Press, 1996.

6. I.N. Sneddon, *The Use of Integral Transforms*, MacGraw-Hill, 1972.

7. Bateman manuscript project, in *Tables of Integral Transforms*, vol. II, McGraw-Hill, 1954.

8. C.E. Siewert and R.J. Thomas, Jr., Neutron transport calculations in cylindrical geometry, *Nuclear Sci. Engrg.* **87** (1984), 107–112.

9. R.P. Pazos, M.T. Vilhena, and E.B. Hauser, Solution and study of two-dimensional nodal neutron transport equation, *Tenth International Conference on Nuclear Engineering ICONE 10*, Arlington, 2002.

42 Integral Equation Methods for Scattering by Periodic Lipschitz Surfaces

Bo Zhang and Guozheng Yan

42.1 Introduction

In this paper we consider the two-dimensional Dirichlet and impedance boundary value problems for the Helmholtz equation, $\Delta u + k^2 u = 0$, in a non-locally perturbed half-plane with a periodic Lipschitz boundary. The Dirichlet problem arises in a study of time-harmonic acoustic scattering of an incident field by a sound-soft, non-smooth (Lipschitz) periodic surface where the total field u_t (the sum of the incident field u^i and the scattered field u) vanishes. The impedance problem, with the boundary condition $\partial u/\partial \nu + i\lambda u = 0$, where $\lambda \in \mathbb{C}$ is a constant, models acoustic or electromagnetic scattering (in both polarization cases) by a one-dimensional Lipschitz periodic boundary of finite surface impedance.

The problem of scattering of waves by a periodic surface has many important applications, e.g., in antenna theory, filter theory, or optics (holography). Much attention has been devoted to the case where the periodic surface is assumed to be smooth (e.g., C^2). For example, uniqueness and existence of solutions to the scattering problems have been established using both the integral equation method and the variational method (see, e.g., [2]–[8] and the references quoted there). However, for the non-smooth (Lipschitz) surface case which is the realistic case in practical applications, the results obtained are unfortunately not applicable. The problem is that the integral operators are now strongly singular, which is fundamentally different from the smooth (Lyapunov) surface case where the integral operators are only weakly singular. In the case when the boundary is a bounded Lipschitz surface, boundary value problems for the Laplace equation, corresponding to zero wavenumber, have been extensively studied using integral equation methods since [10], where the invertibility of integral operators on Lipschitz domains was first proved in L^2 by means of a Rellich-type integral identity, substituting for compactness in the case of Lipschitz domains. Extensions to non-zero wavenumbers are obtained and described, e.g., in [9].

In this paper, we extend such an integral equation approach to the case of non-smooth (Lipschitz) periodic surfaces. Precisely, we will show the

This work was supported by the UK Engineering and Physical Science Research Council under Grant GR/N14415.

existence of a unique solution to the Dirichlet and impedance boundary value problems satisfying the Rayleigh expansion condition at infinity and the boundary conditions almost everywhere on each period part of the boundary. Note that the existence of a unique weak solution to the Dirichlet problem has also been studied in [6] using a variational method.

42.2 The Scattering Problem

Given $f \in C_p^{0,1}(\mathbb{R})$, i.e., f is a periodic Lipschitz function of period 2π : for some positive constant $M > 0$,

$$|f(s) - f(t)| \leq M|s - t|, \quad s, t \in R,$$

define the two-dimensional region D by

$$D := \{x = (x_1, x_2) \in R^2 | x_2 > f(x_1)\}$$

so that the boundary of D is $\partial D = \{(x_1, f(x_1)) | x_1 \in \mathbb{R}\}$.

We consider the problem of scattering of a field u^i incident on the infinite boundary Γ. We assume that k is a real positive constant, i.e., $k > 0$, and restrict our attention to two cases: the case where the total field vanishes on the boundary, so that the scattered field u, a solution of the Helmholtz equation in D, satisfies the Dirichlet boundary condition $u = -u^i$ on ∂D, and the case when the total field satisfies the homogeneous impedance boundary condition, $\partial u_t/\partial \nu + i\lambda u_t = 0$ on ∂D, where, and subsequently, $\nu(x)$ stands for the unit normal vector at $x \in \Gamma$ pointing out of D, $\partial/\partial \nu$ is the rate of change in this direction and $\lambda \in \mathbb{C}$ is a constant with $\text{Re}(\lambda) < 0$.

We consider the case where a plane wave, given by $u^i = \exp(i\alpha x_1 - i\beta x_2)$, is incident on the periodic surface ∂D from the top, where $\alpha = k \sin \theta$, $\beta = k \cos \theta$, and $\theta \in (-\pi/2, \pi/2)$ is the incident angle. Since the incident field u^i is α-quasi-periodic, that is, u^i satisfies

$$u^i(x_1 + 2\pi, x_2) = u^i(x_1, x_2)\exp(i2\alpha\pi),$$

we may assume that the scattered field u is also α-quasi-periodic. Moreover, u is required to satisfy a Rayleigh expansion radiation condition, that is, the scattered field can be expanded as an infinite sum of plane waves

$$u(x) = \sum_{n \in \mathbb{Z}} u_n e^{i(\alpha_n x_1 + \beta_n x_2)}, \quad x_2 > \|f\|_\infty := \max_{0 \leq t \leq 2\pi} |f(t)|, \quad (42.1)$$

with the Rayleigh coefficients $u_n \in \mathbb{C}$. Here $\alpha_n = n + \alpha$ and β_n is defined by

$$\beta_n := \begin{cases} (k^2 - \alpha_n^2)^{1/2}, & \text{if } |\alpha_n| \leq k, \\ i(\alpha_n^2 - k^2)^{1/2}, & \text{if } |\alpha_n| > k. \end{cases} \quad (42.2)$$

Since β_n is real for at most a finite number of indices, then only a finite number of plane waves in the sum (42.3) propagate into the far field, with the remaining evanescent waves decaying exponentially as $x_2 \to \infty$.

The mathematical formulation of the scattering problems will involve the non-tangential maximal functions. Let us define $D_\pi := \{x \in \mathbb{R}^2 | x_2 > f(x_1), 0 < x_1 < 2\pi\}$, $\mathbb{R}^2_\pi := \{x \in \mathbb{R}^2 | 0 < x_1 < 2\pi\}$ and $\Gamma := \{x \in \partial D | 0 < x_1 < 2\pi\}$. Then for a function u defined on $\mathbb{R}^2_\pi \backslash \Gamma$, the non-tangential maximal functions of u, denoted by $u_\pm^*$, are defined for $x \in \Gamma$ by

$$u_\pm^*(x) = \sup_{X \in \Gamma_\pm(x)} |u(X)|,$$

where $\Gamma_+(x)$ and $\Gamma_-(x)$ are two components in D_π and $\mathbb{R}^2 \backslash \Gamma$, respectively, of a truncated cone with vertex $x \in \Gamma$. Boundary values are defined in the non-tangential sense and almost everywhere with respect to the surface measure ds : we say that $u = g$ a.e. on Γ if for a.e. $y \in \Gamma$,

$$\lim_{x \to y,\, x \in \Gamma_+(y)} u(x) = g(y).$$

Similar definitions apply for derivatives of a function.
Define

$$C_q^2(D_\pi) := \{v \in C^2(D_\pi) \big| v(x_1 + 2\pi, x_2) = v(x_1, x_2)\exp(i2\alpha\pi)\},$$

$$L_q^2(\Gamma) := \{v \in L^2(\Gamma) \big| v(x_1 + 2\pi, x_2) = v(x_1, x_2)\exp(i2\alpha\pi)\}.$$

Then the above problems of scattering of an incident plane wave by an infinite, periodic, Lipschitz surface can be formulated as the following boundary value problems for the scattered field u.

Dirichlet Problem (DP): given $g \in L_q^2(\Gamma)$, find $u \in C_q^2(D_\pi)$ such that (i) $\Delta u + k^2 u = 0$ in D_π, (ii) $u = g$ a.e. on Γ, (iii) $\|u_+^*\|_{L^2(\Gamma)} < \infty$, and (iv) u satisfies the radiation condition (42.1).

Impedance Problem (IP): given $g \in L_q^2(\Gamma)$, find $u \in C_q^2(D_\pi)$ such that (i) $\Delta u + k^2 u = 0$ in D_π, (ii) $\partial u/\partial \nu + i\lambda u = g$ a.e. on Γ, (iii) $\|u_+^*\|_{L^2(\Gamma)} + \|(\nabla u)_+^*\|_{L^2(\Gamma)} < \infty$, and (iv) u satisfies the radiation condition (42.1).

42.3 Uniqueness of Solutions

To prove uniqueness of solutions to the Dirichlet and impedance boundary value problems, it is enough to prove that both problems (DP) and (IP) with $g \equiv 0$ have only the trivial solution.

To do this, let $f_j \in C^{1,1}(R)$ be periodic functions with the period of 2π such that

$$\|f_j - f\|_{C_p^0(R)} \to 0, \qquad \sup_j \|f_j\|_{C_p^{0,1}} < \infty, \qquad \text{as} \quad j \to \infty, \qquad (42.3)$$

and $D_j^\pi := \{x \in \mathbb{R}^2 : x_2 > f_j(x_1),\, 0 < x_1 < 2\pi\} \subset D^\pi$ with the boundary $\Gamma_j = \partial D_j^\pi$. Apply Green's theorem to the solution u of the impedance problem (IP) with $g \equiv 0$ in D_j^π and make use of the radiation condition (42.2). Letting $j \to \infty$ and using the dominated convergence theorem together with (IP) (iii) we obtain the following result [11].

Theorem 1. *If* $\operatorname{Re}(\lambda) < 0$, *then the impedance problem* (*IP*) *has at most one solution.*

The above argument does not work for the Dirichlet problem (DP) since we only assume that $\|u^*\|_{L^2(\Gamma)} < \infty$ which is not enough to justify the application of the dominated convergence theorem. However, it follows from [8] that the Dirichlet problem (DP) has a unique solution u if Γ and D_π are replaced by Γ_j and D_j^π, respectively, and

$$u(x) = \int_{\Gamma_j} \frac{\partial G_q(x,y)}{\partial \nu(y)} \phi(y) ds(y), \quad x \in D_j^\pi, \qquad (42.4)$$

where $\phi = \left(\frac{1}{2}I - K_j\right)^{-1} u_j$, $u_j = u|_{\Gamma_j}$,

$$K_j \phi(x) = \int_{\Gamma_j} \frac{\partial G_q(x,y)}{\partial \nu(y)} \phi(y) ds(y), \quad x \in \Gamma_j,$$

$$G_q(x,y) = \sum_{n=-\infty}^{\infty} G_i(x,(y_1 + 2\pi n, y_2)) \exp(i\alpha n 2\pi), \quad x,y \in \overline{U}, \ x \neq y,$$

and G_i is the impedance Green's function of the Helmholtz equation in the upper half-plane. It can be shown by using results in [1] and [8] that $\|\phi\|_2 \leq C\|u_j\|_2$ for some constant C independent of j. This and (42.4) imply that

$$|u(x)| \leq \frac{C}{d(x, \Gamma_{j_0})} \|u_j\|_{L^2(\Gamma_j)},$$

which tends to zero as $j \to \infty$ from the Dirichlet boundary condition. We thus have the following result (see [11] for details).

Theorem 2. *The Dirichlet problem* (*DP*) *has at most one solution.*

42.4 Existence of Solution

In this section we show existence of solution to the problems (DP) and (IP), using integral equation methods. To this end, define the free space quasi-periodic Green's function by

$$G(x,y) = \frac{i}{2\pi} \sum_{n \in \mathbb{Z}} \frac{1}{\beta_n} \exp[i\alpha_n(x_1 - y_1) + i\beta_n|x_2 - y_2|],$$

where $\beta_n \neq 0$ or $|n + \alpha| \neq k$ for all $n \in \mathbb{Z}$. Then G has the same singularity as the fundamental solution $\Phi(x,y) = \frac{i}{2} H_0^{(1)}(k|x - y|)$ (see [7]).

For $\phi \in L^2(\Gamma)$, let

$$S\phi(x) = \int_\Gamma G(x,y)\phi(y)ds(y), \quad x \in \mathbb{R}_\pi^2,$$

$$D\phi(x) = \int_\Gamma \frac{\partial G(x,y)}{\partial \nu(y)}\phi(y)ds(y), \quad x \in \mathbb{R}_\pi^2 \backslash \Gamma.$$

Then $S\phi$ and $D\phi$ satisfy the Helmholtz equation in $\mathbb{R}_\pi^2 \backslash \Gamma$. We seek a representation of the solution u in the form of combined single- and double-layer potential

$$u(x) = D\phi(x) + i\eta S\phi(x), \quad x \in D_\pi \tag{42.5}$$

for the Dirichlet problem (DP) or in the form of a single-layer potential

$$u(x) = S\phi(x), \quad x \in D_\pi \tag{42.6}$$

for the impedance problem (IP), where $\phi \in L^2(\Gamma)$. Using the properties of the layer potentials it can be shown [11] that (42.5) satisfies the Dirichlet problem (DP) provided $\phi \in L^2(\Gamma)$ satisfies the boundary integral equation

$$\left[\left(\tfrac{1}{2}I - K\right) + i\eta S\right]\phi = -g, \tag{42.7}$$

and (42.6) satisfies the impedance problem (IP) provided $\phi \in L^2(\Gamma)$ satisfies the boundary integral equation

$$\left[\left(\tfrac{1}{2}I + K^*\right) + i\lambda S\right]\phi = -g, \tag{42.8}$$

where K and its adjoint K^* are bounded in $L^2(\Gamma)$ and are defined by

$$K\phi(x) = p.v. \int_\Gamma \frac{\partial G(x,y)}{\partial \nu(y)}\phi(y)ds(y), \quad x \in \Gamma,$$

$$K^*\phi(x) = p.v. \int_\Gamma \frac{\partial G(x,y)}{\partial \nu(x)}\phi(y)ds(y), \quad x \in \Gamma.$$

The following results are proved in [11] on the unique solvability of the integral equations (42.7) and (42.8).

Theorem 3. *If* $\operatorname{Re}(\eta) > 0$ *and* $\operatorname{Re}(\lambda) < 0$, *then the operators* $\tfrac{1}{2}I - K + i\eta S$ *and* $\tfrac{1}{2}I + K^* + i\lambda S$ *are invertible on* $L^2(\Gamma)$. *Thus, the integral equations (42.7) and (42.8) have exactly one solution for every* $g \in L^2(\Gamma)$.

Using Theorems 1, 2, and 3, we deduce existence and uniqueness results for the problems (DP) and (IP) (see [11] for details).

Theorem 4. *The Dirichlet problem (DP) has exactly one solution. If* $\operatorname{Re}(\lambda) < 0$, *then the impedance problem (IP) also has exactly one solution.*

References

1. T. Arens, S.N. Chandler-Wilde, and K.O. Haseloh, Solvability and spectral properties of integral equations on the real line. II. L^p spaces and applications, *J. Integral Equations Appl.* **15** (2003), 1–35.

2. G. Bao, D.C. Dobson, and J.A. Cox, Mathematical studies in rigorous grating theory, *J. Opt. Soc. Amer.* **A12** (1995), 1029–1042.

3. S.N. Chandler-Wilde, C.R. Ross, and B. Zhang, Scattering by infinite one-dimensional rough surfaces, *Proc. Roy. Soc. London A* **455** (1999), 3767–3787.

4. X. Chen and A. Friedman, Maxwell's equations in a periodic structure, *Trans. Amer. Math. Soc.* **323** (1991), 465–507.

5. D. Dobson and A. Friedman, The time-harmonic Maxwell equations in a doubly periodic structure, *J. Math. Anal. Appl.* **166** (1992), 507–528.

6. J. Elschner and M. Yamamoto, An inverse problem in periodic diffractive optics: reconstruction of Lipschitz grating profiles, preprint no. 718, Weierstrass Institute für Angewandte Analysis und Stochastik, Berlin, 2002.

7. A. Kirsch, Diffraction by periodic structures, in *Inverse Problems in Mathematical Physics,* L. Paivarinta and E. Somersalo (eds.), Springer-Verlag, 1993, 87–102.

8. C.R. Ross, *Direct and Inverse Scattering by Rough Surfaces,* PhD thesis, Brunel University, 1996.

9. R.H. Torres and G.V. Welland, The Helmholtz equation and transmission problems with Lipschitz interfaces, *Indiana Univ. Math. J.* **42** (1993), 1457–1485.

10. G. Verchota, Layer potentials and regularity for the Dirichlet problem for Laplace's equations in Lipschitz domains, *J. Functional Anal.* **59** (1984), 572–611.

11. B. Zhang, G. Yan, and C. Miao, Diffraction by Lipschitz periodic structures (in preparation).

Index